Assessment and Remediation of Petroleum Contaminated Sites

G. Mattney Cole

CRC Press
Taylor & Francis Group
Boca Raton London New York

CRC Press is an imprint of the
Taylor & Francis Group, an **informa** business

CRC Press
Taylor & Francis Group
6000 Broken Sound Parkway NW, Suite 300
Boca Raton, FL 33487-2742

© 1994 by Taylor & Francis Group, LLC
CRC Press is an imprint of Taylor & Francis Group, an Informa business

First issued in paperback 2019

No claim to original U.S. Government works

ISBN 13: 978-0-367-44954-4 (pbk)
ISBN 13: 978-0-87371-824-0 (hbk)

**Visit the Taylor & Francis Web site at
http://www.taylorandfrancis.com**

**and the CRC Press Web site at
http://www.crcpress.com**

Library of Congress Card Number 93-39378

Library of Congress Cataloging–in–Publication Data

Cole, G. Mattney
      Assessment and Remediation of Petroleum Contaminated Sites / G.
   Mattney Cole
            p.   cm.

      Includes bibliographical references, glossary, and index.
      1. Oil Pollution of soils.   2. Soil remediation   I. Title
      TD879.P4C65   1994
      628.5'5- -dc20   93-39378
      ISBN 0-87371-824-0

For Jane and Michael and Brian

Table of Contents

Chapter 1   Introduction 1

Chapter 2   Environmental Legislation and Regulations 13

Chapter 3   Petroleum Hydrocarbons 37

Chapter 4   Soils and Subsurface Characteristics 75

Chapter 5   Environmental Assessments 91

Chapter 6   Site Assessments 111

Chapter 7   Environmental Sampling and Laboratory Analysis 123

Chapter 8   Data Integration and Technology Selection:
            The Corrective Action Plan 155

Chapter 9   In Situ Remediation Technologies 197

Chapter 10 Non–In Situ Soil Treatment Technologies 237

References 257

Appendix A   Petroleum Products 271

Appendix B   Federal Regulations 287

Appendix C   The Unified Soil Classification System 299

Appendix D   Documents for Environmental Sampling 303

Appendix E   CERCLA Case Law 307

Glossary 313

Index 331

Extended Table of Contents

1 **Introduction** 1
   The Problem 1
   LUST and Remediation 2
   Scope and Purpose 3
   Overview of the Book 4

2 **Environmental Legislation and Regulations** 13
   The Legislative Environment 13
   Publications 15
   National Environmental Protection Act of 1970 15
   Solid Waste Disposal Act of 1965 17
   Resource Conservation and Recovery Act 18
   Definitions in RCRA 18
   UST Regulations Under RCRA 20
   The Petroleum Exclusion 22
   Comprehensive Environmental Response, Compensation, and
   Liability Act 22
   Superfund Amendments and Reauthorization Act of 1986 24
   Clean Water Act of 1987 26
   Clean Air Act of 1990 27
   Safe Drinking Water Act of 1986 28
   Oil Pollution Act of 1990 28
   Toxic Substances Control Act of 1976 28
   Occupational Safety and Health Act of 1970 29
   Other Statutes 30
   Federal Regulations for Owners/Operators of UST Systems 32
   Summary of Federal Regulations for Reporting, Investigation, and
   Cleanup 33
   Leak Prevention vs Leak Detection 34
   Offsite Impacts 34
   Initial Abatement Measures 34
   Initial Site Characterization 35
   Soil and Groundwater Cleanup 35
   Corrective Action Plan (CAP) 35
   Public Right To Know 36

3 Petroleum Hydrocarbons..37
The Good News and the Bad News.......................................37
What Is Petroleum? 38
Environmental Effects 40
The Simplest Hydrocarbons 42
Isomeric Hydrocarbons 44
Unsaturated Hydrocarbons..47
Cyclic Hydrocarbons 48
Aromatic Hydrocarbons 50
Alkylbenzenes...50
Polycyclic Aromatic Hydrocarbons 53
Refining Petroleum 53
Gasolines...57
Oxygenated Fuels 58
Diesel Fuels 60
Physical Properties of Hydrocarbons 61
Toxicity and Health Effects 67
Other Additives 73
Middle Distillates 73

4 Soils and Subsurface Characteristics 75
Soil Characteristics 76
Hydrocarbon Phases..80
Soil Permeability 84
Contaminant Migration 88
Soil Classification System 90

5 Environmental Assessments...91
The Legal Environment 91
Overview of CERCLA 92
Definitions in CERCLA...93
The Secured Creditor Exemption 95
The CERCLA Concept of Liability.......................................96
The Innocent Purchaser Defense 97
Prior Case Law 97
The "Black Hole" of Lender Liability 98
Conflicting Case Law 99
Legal Outcomes 99
Financial Outcomes 100
Proposed Legislation...101
The EPA Rule on Lender Liability 101

Overview of Environmental Assessments 102
Property Risk Analysis 103
Phase I Environmental Assessments 105
Search Radius 107
Phase I Site Inspection 107
Phase I Records Review 109
Phase II and Phase III 110

6 Site Assessments 111
Initial Site Characterization 111
Summary of Initial Site Characterization 115
Detailed Site Characterization 117
Sampling and Sampling Protocols 120
Planning 121

7 Environmental Sampling and Laboratory Analysis 123
Sampling Procedures 125
Planning Sampling Protocols 126
Operating Procedures 127
Field Sampling 130
Chain-of-Custody 132
Decontamination and Cross-Contamination 133
Sample Storage And Holding Times 135
Problems with Sampling 138
Laboratory Methods 139
Data Reporting 142
Methods Appropriate to Petroleum Contaminated Soils 144
TCLP 149
Data Validity and Authenticity 150
Measurement Integrity 152

8 Data Integration and Technology Selection:
The Corrective Action Plan 155
Overview 155
Emergency Response 156
Regulatory Agencies 157
Project Planning and Priorities 158
Risk Assessment 160
Regulatory Guidelines 162
Data Acquisition 166

Technology Selection 167
Performance Monitoring 168
Evaluation Parameters for In Situ Technologies 171
Evaluation Fact Sheets 172
Volatilization Fact Sheet 172
In Situ Bioremediation Fact Sheet 174
In Situ Passive Bioremediation Fact Sheet 176
In Situ Leaching and Chemical Extraction Fact Sheet 179
In Situ Vitrification Fact Sheet 180
Isolation and Containment Fact Sheet 182
Solidification and Stabilization Fact Sheet 184
Evaluation Parameters for Non–In Situ Technologies 186
Excavation and Landfilling Fact Sheet 186
Landfarming Fact Sheet 188
Low Temperature Thermal Stripping Fact Sheet 190
High Temperature Thermal Treatment Fact Sheet 192
Asphalt Incorporation Fact Sheet 194

9 In Situ Remediation Technologies 197
In Situ vs Non–In Situ 197
Volatilization 198
Bioremediation 206
In Situ Passive Bioremediation 212
Soil Leaching 214
Chemical Extraction 215
In Situ Vitrification 216
Linear Interception 218
Isolation and Containment 219
Solidification and Stabilization 221
Groundwater Extraction and Treatment 222
Anaerobic Biodegradation 227
Other Technologies 228
Well Design 228

10 Non-In Situ Soil Treatment Technologies 237
In Situ vs Non–in Situ 237
Excavation and Landfilling 238
Landfarming 241
Land Treatment for Contaminated Groundwater 244
Thermal Treatment Techniques 246
Low Temperature Thermal Stripping 248

High Temperature Thermal Treatment 250
Asphalt Incorporation 252

References 257
General References 257
Chapter 1 Introduction 259
Chapter 2 Environmental Legislation 259
Chapter 3 Petroleum Hydrocarbons 260
Chapter 4 Soils and Subsurface Characteristics 260
Chapter 5 Environmental Assessments 261
Chapter 6 Site Assessments...262
Chapter 7 Environmental Sampling and Laboratory Analysis 263
Chapter 8 Data Integration and Technology Selection:
        The Corrective Action Plan 265
Chapter 9 In Situ Remediation Technologies..267
Chapter 10 Non–In Situ Soil Treatment Technologies..............................269

Appendix A Petroleum Products 271

Appendix B Summary of Federal Regulations 287
Petroleum Storage Sites 287
Subpart E — Release Reporting, Investigation, and Confirmation 287
Subpart F — Release Response and Corrective Action for UST
Systems Containing Petroleum or Hazardous Substances 290
Subpart G — Out–of–Service UST Systems and Closure......................295

Appendix C The Unified Soil Classification System 299

Appendix D Documents for Environmental Sampling 303

Appendix E CERCLA Case Law... 307
Fleet Factors Decision 307
Lender Liability 308
Petroleum Byproducts and The Petroleum Exclusion 308
Parent/Subsidiary Liability 310
Mixed Waste 310
Passive Migration–Definition of Disposal..311

Glossary of Terms 313

Index 331

List of Figures

2 Environmental Legislation and Regulations 13
    II–1   The Resource Conservation and Recovery Act Chain of
            Legislation 19

3 Petroleum Hydrocarbons..37
    III–1   Combustion of Hydrocarbons 39
    III–2   The First Three Saturated Hydrocarbons 41
    III–3   Normal Heptane 42
    III–4   Alternative Representations of Hydrocarbons 44
    III–5   The Butane Family of Alkanes 44
    III–6   The Pentane Family of Alkanes 45
    III–7   Two Isomers of the Octane Family of Alkanes 46
    III–8   The First Three Unsaturated Hydrocarbons...............47
    III–9   The First Three Acetylenic Hydrocarbons 48
    III–10   Cycloalkanes 49
    III–11   The Chemical Structure of Benzene...........................50
    III–12   The Chemical Structure of Toluene 51
    III–13   The Chemical Structure of Ethylbenzene and Alkylbenzenes 52
    III–14   The Xylenes..52
    III–15   Characteristic Polycyclic Aromatic Hydrocarbons (PAHs) 54
    III–16   Petroleum Refining...55
    III–17   Oxygenated Fuel Additives 58
    III–18   The Chemical Structure of Phenol.............................59
    III–19   Volatility of Selected Alkanes 64
    III–20   Boiling Point Distribution of Petroleum Products 65
    III–21   Kinematic Viscosity 66
    III–22   Solubilities of Selected Hydrocarbons 68

4 Soils and Subsurface Characteristics 75
    IV-1   Generalized Soil Column 77

IV-2   Generalized Soil Column in Microview 78
IV-3   Distribution of Hydrocarbon Phases in Soils 79
IV-4   Microview of Petroleum Contaminated Soils 81
IV-5   Seasonally Saturated Soils 82
IV-6   Importance of Volatilization as a Migration Pathway 83
IV-7   Distribution of Contaminant Phases for Typical Soils 86
IV-8   Potential Routes of Migration of Hydrocarbon
        Contaminants 89

6 Site Assessments

6   Site Assessments 111
    VI-1   Subsurface Profiles 116
    VI-2   Decision Tree for Contaminant Determination 120

7 Environmental Sampling and Laboratory Analysis

7   Environmental Sampling and Laboratory Analysis 123
    VII-1   Sampling Wheel 124
    VII-2   Obtaining Usable Data 127
    VII-3   Data Flow in Sampling and Analyses 128
    VII-4   Sample Contamination Sources 133
    VII-5   Blanks and Controls 134
    VII-6   Precision and Accuracy 140
    VII-7   Liquid Chromatography Separations 141
    VII-8   Quality Control/Data Validation Techniques 151

8 Data Integration and Technology Selection: The Corrective Action Plan

8   Data Integration and Technology Selection:
    The Corrective Action Plan 155
    VIII-1   Site Assessment and Evaluation...161
    VIII-2   Site Assessment, Planning, and Evaluation. 163
    VIII-3   The Planning Phase 164
    VIII-4   Initial Site Assessment 165
    VIII-5   On/Off Site Decision Tree 168
    VIII-6   Performance Monitoring and Evaluation Flow Chart 169
    VIII-7   Technology Decision Tree. 170
    VIII-8   Evaluation Parameters for Soil Venting 173
    VIII-9   Evaluation Parameters for In Situ Bioremediation.........175
    VIII-10   Evaluation Parameters for Passive Remediation 177
    VIII-11   Evaluation Parameters for Soil Washing and
             Chemical Extraction 179
    VIII-12   Evaluation Parameters for In Situ Vitrification 181
    VIII-13   Evaluation Parameters for Isolation and Containment ...183
    VIII-14   Evaluation Parameters for Solidification and Stabilization........185

VIII–15   Evaluation Parameters for Excavation 187
VIII–16   Evaluation Parameters for Land Farming 189
VIII–17   Evaluation Parameters for Low Temperature
           Thermal Stripping 191
VIII–18   Evaluation Parameters for High Temperature
           Thermal Treatment 193
VIII–19   Evaluation Parameters for Asphalt Incorporation 195

9   In Situ Remediation Technologies 197
    IX–1   Soil Vapor Extraction System 201
    IX–2   Subsurface Phase Equilibria 203
    IX–3   Soil Vapor Recovery Manifolds 204
    IX–4   Extraction Wells and Nutrient System for In Situ
            Bioremediation 208
    IX–5   Solubility of Hydrocarbons in Water 223
    IX–6   Groundwater Extraction and Injection System 224
    IX–7   pϵ vs pH Diagram for Natural Degradation Pathways 227
    IX–8   Groundwater Sampling Well 231
    IX–9   Product Recovery and Groundwater Extraction Well 233
    IX–10   Product Recovery System 234

Appendix C   The Unified Soil Classification System 299
    C–1   Generalized Soil Column..300
    C–2   Unified Soil Classification System 301
    C–3   Classification Chart for Fine–Grained Soils 302

Appendix D   Documents for Environmental Sampling 303
    D–1   Chain–of–Custody Record 304
    D–2   Well Log Record 305
    D–3   Request For Analytical Services...306

List of Tables

3   Petroleum Hydrocarbons...37
    3–1   The Normal Alkanes 43

3–2   Components of Gasoline 56
3–3   Volatility, Flashpoint, and Flammability Data for Representative
      Petroleum Products 62
3–4   Solubility and Viscosity Data for Representative Petroleum
      Products 63
3–5   Health Effects of Hydrocarbons 72

4   Soils and Subsurface Characteristics 75
    4–1   Distribution of Subsurface Contaminants.......................84
    4–2   Phase Migration Routes 87
    4–3   Classification of Soil Types by Coefficient of Permeability 88

5   Environmental Assessments...91
    5–1   Examples of High Environmental Risk 104
    5–2   Visual and Physical Site Reconnaissance Checklist...................108
    5–3   Resources Available for Historical and Records Review 109

6   Site Assessments 111
    6–1   Initial Site Characterization Report 113
    6–2   Detailed Site Assessment Checklist 117
    6–3   A Summary of Project Planning...................................122

7   Environmental Sampling and Laboratory Analysis............................123
    7–1   Field Logbook Checklist 129
    7–2   Chain–of–Custody Record Checklist.............................132
    7–3   Laboratory Methods 136
    7–4   Conversion Multipliers for Common Fractions 143
    7–5   Laboratory Methods for Petroleum Contaminated Soils 144
    7–6   Analytical Methods for Quantifying Petroleum
          Hydrocarbons..146
    7–7   Analytical Methods for Petroleum Contaminated
          Soils and Groundwater 147
    7–8   TCLP Hazardous Levels for Inorganic Elements...................148
    7–9   TCLP Hazardous Levels for Organic Compounds...................148
    7–10 Summary of Measurement Integrity Factors 152

8 Data Integration and Technology Selection:
   The Corrective Action Plan 155
   8–1 Contaminant Phases and Mobility 167

9 In Situ Remediation Technologies 197
   9–1 Considerations and Recommendations: Volatilization 205
   9–2 Considerations and Recommendations: In Situ
        Bioremediation 211
   9–3 Considerations and Recommendations: Passive
        Bioremediation 213
   9–4 Considerations and Recommendations: Leaching and
        Chemical Extraction 215
   9–5 Considerations and Recommendations: In Situ Vitrification ...217
   9–6 Considerations and Recommendations: Linear Interception 219
   9–7 Considerations and Recommendations: Isolation
        and Containment 221
   9–8 Considerations and Recommendations: Solidification
        and Stabilization 222
   9–9 Considerations and Recommendations: Groundwater
        Extraction and Treatment 226
   9–10 Considerations and Recommendations: Well Documentation
        Checklist 235

10 Non–In Situ Soil Treatment Technologies 237
   10–1 Consideration and Recommendations: Excavation and
         Landfilling 240
   10–2 Considerations and Recommendations: Landfarming 243
   10–3 Considerations and Recommendations: Land Treatment
         of Contaminated Groundwater 245
   10-4 Considerations and Recommendations: Low Temperature
         Thermal Stripping 249
   10–5 Considerations and Recommendations: High Temperature
         Thermal Treatment 251
   10–6 Considerations and Recommendations: Asphalt
         Incorporation 253

Appendix A Petroleum Products 271
   A–1 Components of Arabian Crude Petroleum 272
   A–2 Components of Gasoline 273
   A–3 Chromatograms of Representative Petroleum Products 278

G. Mattney (Matt) Cole is an Adjunct Associate Professor at the Colorado School of Mines in Golden, Colorado. Currently, Dr. Cole is Director of Environmental Programs for the Special Projects and Continuing Education (SPACE) Office at CSM. He organizes and presents workshops and short courses on environmental topics and technologies for professionals in the environmental industry and related fields. From the original short course on Underground Storage Tank Management, the environmental program has grown to include some 15 short courses and workshops on USTs, field sampling, groundwater chemistry, Phase I assessments, data validity, special problems in remediation, and of course, assessment and remediation of petroleum contaminated sites. In addition Dr. Cole gives numerous presentations to state regulatory agencies throughout the country.

Dr. Cole received a Ph.D. in Inorganic Chemistry from Florida State University in 1972. After a Postdoctoral Fellowship at the University of Wyoming, he joined the faculty at the University of Georgia where he did research in corrosion on synthetic fuel systems. In 1982 he joined the Department of Chemistry and Geochemistry at the Colorado School of Mines. He is the author of more than 50 scientific papers and abstracts, including **Underground Storage Tanks: Installation and Management,** published by Lewis Publishers in 1992. He can be reached through the Department of Chemistry and Geochemistry, Colorado School of Mines, Golden, Colorado 80401.

Dr. Cole is an avid outdoors person who enjoys backpacking, hiking, backcountry and downhill skiing, and climbing mountains in Utah and Colorado. He lives in Lakewood, Colorado with his wife, two teenage sons, a Golden Retriever and a crazy Corgi.

Acknowledgments

It goes without saying that no book of this magnitude can be put together without a great deal of help. I am indebted to many who have been willing to share their knowledge and experience.

Chapters 2 and 5 were reviewed by Mike Glade, Esq., an associate with the Denver law firm of Inman, Flynn, and Biesterfeld, P.C., and an able, experienced environmental attorney. The contribution of material in Appendix E is gratefully acknowledged. His clarifications of liability case law and the intricacies of CERCLA liability have been extremely helpful..

I am indebted to Pat McGuckin of ENTRAC of Denver for many suggestions and clarifications of Phase I environmental assessments, particularly document review.

Chapter 7 on field sampling and laboratory methods was reviewed by Dave Osborne, Laboratory Supervisor for Waste Tech, Inc., a national hazardous waste disposal company. I am indebted to Dave for several figures in Chapter 7 and for clarifying explanations of laboratory procedures.

Dennis Hotovec, lead technician with the UST Section of the Colorado Department of Health, has been very helpful in explaining and clarifying the intent of state regulations and the fine details of compliance.

Scot Donato, Senior Project Manager of Industrial Compliance, Inc., and Joby Adams, of CGRS, Inc. have been very helpful in explaining the intricacies of wells, well design, remedial technologies, and the real world of budgetary constraints. Both have always been willing to share their knowledge.

I am also indebted to Gary Baughman, Director of Special Projects and Continuing Education at the Colorado School of Mines for support and encourage-ment to pursue new goals and topics. A special thanks goes to all the participants in the Site Assessment and Remediation short courses. The hundreds of questions and interesting "off–the–cuff" discussions have inevitably found their way into this book. Thanks to one and all.

This book was prepared in "camera ready" format. I am indebted to Robert A. "Skip" DeWall, Jr., Elise Hoffman, and Vivian Collier of Lewis Publishers who have been patient and helpful throughout the evolution of the manuscript. They have done a wonderful job of proofing the manuscript. Any errors that remain are mine.

A very special thanks to my family for patience and understanding. This book has been a difficult project because of the open-ended nature of some of the topics. Almost every chapter could be expanded into its own volume. Through it all my wife, Jane, and two boys, Michael and Brian, have always been helpful and encouraging.

1 Introduction

The Problem

Until quite recently underground petroleum storage tanks were peacefully out of sight and out of mind. Underground petroleum storage tank systems, USTs, have been buried since the early decades of the twentieth century by directive of the National Fire Protection Association and the Uniform Fire Code for reasons of safety. Gasoline is a Class A flammable liquid and aboveground storage was considered an unacceptable hazard.

Not until the mid–1970s did anyone seriously consider the fact that all these buried USTs might be leaking gasoline into the environment. The fact that approximately 85% of all USTs were steel with little or no corrosion protection went unnoticed. By the mid–eighties, enough incidents had been reported that Congress, through amendments to the Resource Conservation and Recovery Act, and the Comprehensive Emergency Response, Compensation and Liability Act, empowered the Environmental Protection Agency (EPA) to set regulations to ensure proper operation of UST systems.

The EPA issued final regulations on September 8, 1988. The document entitled, *Technical Standards for Operation of Underground Storage Tank Systems*, was published in the *Federal Register* on September 23, 1988. The regulations consist of a set of requirements for the owners and operators of UST systems. The principal requirements include notification, corrosion protection, leak detection, and spill and overfill prevention. The leak detection regulations were to be phased in over a period of 10 years. USTs are required to be replaced or upgraded to meet these standards in increments, beginning with oldest tanks first. Tanks installed prior to 1965, or those whose age is unknown, must have been in compliance by 1989. Those installed between 1966 and 1969 must have been in compliance by 1990. Those installed between 1970 and 1974 must be in compliance by 1991; between 1975 and 1979 by 1992; and between 1980 and 1988 by 1993 (Cole, 1992).

Most of the compliance deadlines have already passed. By December 1993, *all* tanks and piping must have leak detection in place or have a design that is exempt. The requirements for spill and overfill prevention and for corrosion protection come due in 1998. The most recent existing tanks, those installed from 1980 to 1988, must replaced or upgraded to be in compliance by 1998. New UST systems, those installed after 1988, must be in compliance at the time of installation.

The regulations specify that compliance will be enforced at the state level. The federal regulations are the minimum standard for the country, but states may be more stringent. Compliance is delegated to an implementing agency at the state level. Most states generally follow the federal regulations, but a few are more stringent.

Leaking Underground Storage Tanks and Remediation

December 1998 is approaching rapidly. As a result, thousands of buried USTs have been dug up and removed or replaced over the past few years. Estimates vary, but something in the neighborhood of one–third to one–half of all UST systems that have been exposed are associated with moderate to severe contamination. Virtually all existing USTs, those installed prior to 1988, have contamination at least in the backfill due to overfills.

The sheer number of petroleum contaminated sites is staggering. The list of sites on the National Priorities List, the so–called Superfund list, runs to a few thousand sites; the list of registered USTs in the U.S. runs into the millions. The location of petroleum contaminated sites also presents problems. Frequently the contamination arises from the corner gas station near commercial and residential areas. The contaminant plume migrates offsite to pose problems both for remediation and nearby residents.

In a recent report to the House Appropriations Committee, the Office of Underground Storage Tanks (OUST) of the U.S. EPA noted that there are about 1.6 million USTs and 37,000 hazardous substances tanks as of 1992. The EPA estimates that about 20% of the 1.6 million USTs are leaking, and approximately 1,000 confirmed new releases are reported each week. States and responsible parties (RPs) are initiating cleanups at a rate of about 36,000 sites per year and completing about 16,000 sites annually. (Guide, 1993)

Remediation costs vary depending on the complexity of the site. On average, cleaning up contaminated soil costs between $10,000 and $125,000, but can run higher. Groundwater cleanups average between $100,000 and $1 million. According to EPA estimates, costs of remediation are expected to rise as states and RPs address more complicated and expensive sites.

Add in all the abandoned USTs, aboveground petroleum storage tank sites, railroad fueling operations, airports, refineries, and production facilities and it should not be surprising that remediation and restoration of petroleum contaminated sites has become a multibillion dollar a year industry in the United States.

Four factors are important in the overall remediation process. The primary goal of remediation of petroleum contaminated sites is the preservation of public health and safety. Equally important is restoration of the environment in compliance with regulatory guidelines. Then there is the necessity of carrying out the remediation in a cost–effective manner to keep the owner/operator of the site in business and out of bankruptcy court. Finally, the remediation must be carried out to protect the owner/operator from future liability.

As the number of petroleum contaminated sites has risen, the complexity and variety of sites have also increased. At the same time, the importance of legally defensible data has increased until this has become the most important consideration in many remediation decisions. A recent report by the Government Accounting Office notes that environmental sampling and data acquisition are notably flawed and inconsistent.

Remediation of petroleum contaminated sites is not as complicated as remediation of a Superfund site. A petroleum site is not as sensitive and not as contested, and not as expensive. However, a petroleum contaminated site requires careful planning, attention to detail, and a very careful eye on the budget. Even though the site may qualify for reimbursement, regulatory agencies will not reimburse frivolous or unnecessary expenses.

All of these factors combine to place new responsibilities and demands on the environmental professional supervising a remediation project. At one time, the supervisor of a remediation project at a petroleum contaminated site needed to be skilled only in excavating. Now, the individual must combine the skills and talents of engineer, lawyer, scientist, and negotiator. He or she must also be aware of the fine line between the needs of the client and the requirements of regulatory agencies.

Scope and Purpose

The purpose of this book is to present the broad scope of the remedial process, from initial site assessment to closure, in an integrated, understandable, and coherent format. The technical aspects of remediation are but one portion of a complex problem. It is shortsighted to begin remediation without adequate planning and forethought.

First, the eventual remediation is more cost effective if the site is approached in a coherent manner rather than piecemeal.

Second, relations with the regulatory agency or agencies are smoother and less expensive if the remediation plan is in the context of a thorough site assessment.

Finally, the potential for future liability exposure is reduced significantly if the consultant has proceeded with "all due diligence."

A number of very good specialty books on various technical aspects of remediation of petroleum contaminated soils or groundwater exist — including several from Lewis Publishers. However, none approach remediation from the integrated point of view found in this book and none are concerned with liability avoidance. The central feature of the book and of Chapter 8 is the Corrective Action Plan and closure with as little future liability exposure as possible.

The flow of the book is from regulations and legislation through the various assessment levels to remediation and closure in three broad sections. The beginning is a set of basic information on regulatory requirements, petroleum, and soils and groundwater. These chapters establish an essential foundation before proceeding into the applied material. The middle section describes environmental assessments, site assessment, sampling, and evaluation. This section is essentially data gathering, acquiring information. The results of evaluation and interpretation are only as good as the data on which they are based. However, it should be stressed that data are only numbers and not particularly useful without interpretation. Information, which is useful, is data that have been interpreted and evaluated.

The final section is integration and application of the information acquired. Data integration leads to technology selection and application of the appropriate remediation method or methods. Remediation is accompanied by an ongoing process of evaluation and monitoring leading eventually to closure.

Overview of the Book

Remediation of petroleum contaminated sites is a subject almost without limit. Sites can be as simple as a corner service station with little contamination to a refinery contaminated with hundreds of compounds over every square inch of ground. The field has reached the point where it is no longer a matter of digging out to the property line and backfilling the excavation. Addressing a site requires an understanding of regulations, sampling, corrective action plans, reimbursement funds, budgeting, and the needs of owners and operators. The intent of this book is to provide environmental consultants, managers, and owners and operators with the tools needed to follow a remediation through to completion.

Remediation must necessarily begin with the laws and regulations governing all remedial efforts. Popular support for the environment stimulated the United States to become the world leader in efforts to clean up polluted air and waterways.[‡] The

---

[‡]   In fairness it should be noted that the EC had UST regulations in place considerably before the U.S. In the process of developing Technical Standards, many of the European regulations, including suction pumps and double wall construction, were given careful consideration by U.S. EPA engineers and policy makers.

results have generally been positive, in spite of the fact that efforts to clean up hazardous waste — Superfund or NPL — sites are mired in litigation. Enforcement of provisions of the Clean Air and Clean Water Acts, for example, have reduced air and water pollution from very large point sources, such as foundaries, petrochemical plants, and refineries. Large intractable problems remain: agricultural runoff and non–point sources of water pollutants, air quality in cities, sulfur dioxide, SO_2, emissions and acid rain, and large amounts of petroleum wastes.

Completely new legislation is unlikely at this time; however, it is equally unlikely that legislation already in place will be repealed or allowed to expire. Although political support for reenactment of RCRA and the Clean Air Act waned through the recession of 1990–92, prospects for eventual passage of these bills are good. At the time of writing the Comprehensive Emergency Response, Compensation and Liability Act, CERCLA, is also being considered for reauthorization. Given the hostility in some sectors to its liability provisions, passage of a reauthorization bill will happen, but it will not be easy.

Chapter 2 presents a summary of federal environmental statutes that affect petroleum contaminated sites and remedial activities either directly through regulation and permitting, or indirectly through liability considerations. The statutes can be divided roughly into the following classifications:

- Environmental policy
- Hazardous materials
- Water quality
- Air quality
- Protection of national resources

In the above list, hazardous materials legislation can be divided approximately into two branches:

- Technical aspects that are included in RCRA;
  and
- Legal considerations included in CERCLA.

All the legislation discussed is rarely applicable to a single site, but taken together it defines the climate in which the UST owner/operators and the environmental companies must function. Federal and state regulations must be factored into the assessment/remediation equation at a very early stage.

Remediation of a petroleum contaminated site is a mixture of "good news/bad news." The good news is that petroleum compounds are relatively straightforward to treat for the following reasons:

- They form a homogeneous class of compounds;

- They are less dense than water;

and

- They have very low solubility in water.

Since contaminants float on water rather than sinking and dissolve in groundwater only slowly, a single technology is often adequate to treat a site. This contrasts with the hazardous wastes sites for which multiple technologies is the norm.

Most petroleum hydrocarbons are not highly toxic with respect to the environment or to human health; heavier hydrocarbons are not harmful to groundwater, but may constitute a fire hazard. Those hydrocarbons that are toxic have low solubility compared to other common solvents such as methyl ethyl ketone or acetone. Under RCRA, products derived from crude petroleum are *regulated substances,* but not hazardous substances.[‡] All of these features combine to make petroleum contaminated sites generally easier and cheaper to remediate than hazardous waste or mixed waste sites.

The bad news is that even under normal conditions petroleum contamination can be quite hazardous to work around, particularly gasoline. This is due to the more volatile products constituting a fire or explosive hazard. In at least one instance tunneling equipment set off explosions in highly contaminated soils. Fortunately the explosions were small.

Most petroleum compounds are chronic toxic hazards to humans, especially the aromatic compounds routinely found in the most common contaminant, gasoline. The hazards are particularly prevalent for workers routinely exposed to gasoline vapors since gasoline contains substantial percentages of benzene. Workers in this category include those doing tank removals, environmental sampling and remediation, and inspectors.

The good news is that many petroleum products are readily consumed by soil bacteria with the result that a petroleum contaminated site may, in effect, remediate itself; the bad news is that typically a contaminated area is sufficiently deep that the soil bacterial population is too low to be effective.

This mixture of "good news/bad news" is a result of the particular nature of petroleum products — the chemistry is rather simple, but the formulation is often complex. Understanding the chemical nature of the contaminating agents is an important step in assessing the possibilities for remediation. Chapter 3 is an introduction to the physical and chemical properties of petroleum hydrocarbons and some gasoline additives that can complicate the remedial process.

The chapter is not a course in Organic Chemistry; rather, it is an explanation of the special properties of hydrocarbons that make remediation of petroleum contaminated sites different from remediation of sites contaminated with hazardous

---

[‡]   Some states are more stringent and do consider gasoline contaminated soil to be hazardous.

chemicals. The most common and abundant petroleum product available is gasoline. Not surprisingly, therefore, the most common petroleum contamination is due to leaking gasoline storage tanks. The fact that gasoline is a regulated substance and not stored as a hazardous substance nor are gasoline contaminated soils hazardous wastes is of crucial importance in cost calculations. However, legal definitions should not obscure the hazardous nature of benzene and its aromatic relatives. This chapter contains a review of the relevant toxicology of aromatic petroleum hydrocarbons, which supplements the comments on health and safety features found in Chapters 8 and 9.

The two subsurface areas that are affected by contaminants are groundwater and soil. Thus, one of the more important physical parameters required throughout the planning phase is the nature and type of the soil or soils in which the contamination is located. An open porous soil will offer more options for remediation, whereas in a tight clay, excavation may be the only available option. Conversely, an open, porous soil will allow the contamination to migrate more rapidly, while a contaminated clay retards migration. Whatever the soil type, it is essential to know what that type is in order to predict the phases and distribution of contaminants.

Chapter 4 describes the various levels and types of soils that must be considered including soil types, classification, and subsurface characteristics from the point of view of remedial choices and planning. Soil porosity and adsorption have significant impact on volatility and retention of petroleum products. Soil moisture and groundwater levels are also important in deciding on a specific remediation technology. Fluctuations in groundwater levels can disperse a plume over a much larger area than might otherwise be expected and cause hydrocarbons to be "smeared" onto soil particles. Because there are infinitely varied combinations of subsurface conditions, it is impossible to discuss all possibilities. However, we can look at some of the more common and more difficult structures.

Throughout the discussion in this chapter, it is assumed that groundwater is a factor to be considered. That is, groundwater lies reasonably close to the surface — an arbitrary boundary is permanent groundwater at less than 40 feet. If groundwater is more than 100 feet deep, it is probably not worth worrying about unless the aquifer is a source of drinking water. One major oil company estimates that 60 to 70% of all their USTs are set in groundwater, regardless of the region of the country.

Also, there is a distinct slant toward gasoline as the contaminant. Gasoline accounts for approximately 90% of all remedial sites and it is the most volatile, most soluble, and most mobile of all contaminants. In most circumstances any product heavier than the kerosene derivatives, diesel or jet fuel, will not migrate rapidly enough to create a problem, at least in the short term. Diesel and jet fuel are intermediate in volatility and mobility.

The demise of so many Savings and Loan companies in the S&L scandal of the late eighties was brought about less by the nefarious dealings of a few executives — although that made the best stories on the evening news — than by the unforeseen consequences of S&Ls foreclosing on property that turned out to be contaminated. A typical example was a commercial property worth $500,000 that required $1,000,000 to clean up. The result was a loss of $500,000 in real dollars, not paper. It should come as no surprise that the largest owner of contaminated property in the U.S. is the Resolution Trust Corporation, RTC, the quasi–governmental organization set up to liquidate properties owned by bankrupt S&Ls.

The other complicating factor in the liability equation was the exposure of lending institutions to CERCLA liability through the Fleet Factors decision that held that lenders having the capacity to influence hazardous materials operations incurred CERCLA liability. The risk of foreclosure on property made worthless by contamination and the risk of exposure to CERCLA liability combined to create the rush to Phase I environmental assessments now required on all commercial property transactions.

From time to time critics suggest that RTC simply sell off all the property and save taxpayers the cost of operating RTC and managing the properties. That is easier said than done. Before a property can be sold, RTC must either clean up the contamination or make cleanup part of the sales agreement. That causes enormous complexity. For example, just one property adjacent to an interstate highway interchange is worth $8 million, but has an estimated $30 million in cleanup liability.

The first half of Chapter 5 reviews the present legal climate from the point of view of environmental liability. Existing case law, the facts leading to the Fleet Factors decision, and recent (at the time of writing) EPA final regulations designed to clarify the secured creditor exemption are discussed along with the implications for managers of petroleum contaminated sites. Liability issues are so important that no site manager can afford to ignore them.

The national standards for Phase I environmental assessments[‡] developed by the E50.01 and E50.02 subcommittees of ASTM, formerly the American Society for Testing and Materials, are summarized in the second half of the chapter, but space does not permit a full discussion of all the features of this interesting document. These standards — actually guidelines — are expected to become *de facto* national standards for Phase I assessments and the innocent landowner defense. At the time of writing, a full discussion of the ASTM standard in book form is due out in late 1993 or early 1994 from McGraw–Hill, Inc. The author is Pat McGuckin of ENTRAC of Denver, Inc. Mr. McGuckin was active in formulating the standard and has given many presentations explaining the requirements and pitfalls associated with using the standards. His comments and advice have been very helpful in the preparation of this book.

[‡]   Published June 1993.

The purpose here is not to examine Phase I assessments since that is a separate standalone topic. Phase I assessments are relevant in the context of remediation for two reasons. First, the information contained in a Phase I report overlaps considerably the information required in an initial site assessment. Since it is likely that a Phase I has already been done on the property, the project supervisor needs to know what information is included and how it was obtained. Second, the context of a Phase I is used to establish a framework around which to build the data required by regulatory agencies in an initial site assessment. Historical site use, surrounding land use and population, permitted wells, utility corridors, etc., are all required data for initial risk assessment and management.

Chapter 6 presents an overview of site characterizations often referred to as Phase II assessments or second level site characterizations. Regulatory compliance requirements for corrective action plans are heavily weighted toward groundwater, and especially toward drinking water wells. It is worth the time and effort to adequately characterize a site with respect to possible third party contamination, drinking water wells, surface waters and wetlands, and other points of potential impact.

Phase II site assessments depend on intrusive sampling. This is where the data that will ultimately dictate not only remediation decisions, but liability questions as well, are generated. Chapter 7 is a discussion of sampling protocols, laboratory methods, and data quality. Sampling procedures must lead to data that is not only useful, but relevant and defensible as well. Likewise, laboratory methods must be useful, relevant, and defensible. Having gasoline samples tested for total petroleum hydrocarbons is not relevant if the state regulatory authority needs benzene levels in a corrective action plan. Likewise, testing diesel contaminated soils for BTEX levels is wasteful and not cost–effective.

Chapter 8 is the central feature of the book not only in its physical location, but also because lack of adequate planning is the primary cause of problems in remedial projects. This chapter is both an overview of project planning and a summary of data integration and completion of the Corrective Action Plan (CAP). Tables are presented for evaluation of the variables that should be considered in arriving at a sensible, efficient remediation process.

The owner or operator's needs and goals must enter into consideration. Liability avoidance is paramount. All of these sometimes conflicting areas must be integrated into the requirements of the regulatory agencies to arrive at a workable, cost–effective plan.

This chapter sets up a logical chain of procedures for integrating data into the remediation scheme. The process is somewhat formalized, however, it does reduce the probability of running into unexpected, expensive, and possibly hazardous, unknowns. The chapter sets up sampling protocols, technology screening, and evaluation parameters.

There are extensive tables of evaluation parameters in Chapter 8 to guide a decision on a particular technology. These tables are stand-alone, but may also be combined with tables of considerations and recommendations found in Chapters 9 and 10.

The final two chapters, 9 and 10, are reviews of the state–of–the–art of the technical features of remediation methods. The two chapters are divided into in situ methods (Chapter 9) and non–in situ (also called ex situ) methods (Chapter 10). This treatment follows conventional engineering terminology, although new methods are blurring the distinction. Each of these is, in turn, divided into soil treatment and groundwater treatment.

In situ technologies approach remediation at the site of contamination without prior excavation. These technologies include volatilization, active and passive bioremediation, soil leaching, chemical (detergent) extraction, in situ vitrification, and groundwater treatment. Isolation and containment and linear interception are included despite the fact that these are not remedial technologies.

Non–in situ technologies are those that approach remediation after a substantial fraction of the contaminated soil has been excavated. In situ normally refers to soil treatment since groundwater treatment is usually carried out onsite. These technologies include bioremediation (landfarming), land treatment (for treating contaminated groundwater), asphalt incorporation, high temperature and low temperature thermal treatments (thermal stripping or incineration), surface (chemical) extraction, and stabilization.

In much of the literature the terms LNAPLs and DNAPLs are used. These acronyms stand for Light Non–Aqueous Phase Liquids and Dense Non–Aqueous Phase Liquids, respectively. These terms are used rarely in this book since it deals with specific contaminants, namely petroleum hydrocarbons. LNAPLs is an appropriate term to use in connection with mixed waste or generic sites, but not when the contaminant is known.

This book is not an academic exercise. It is meant to aid the environmental professional through the many levels of data acquisition and evaluation to a successful conclusion and closure. Therefore, most references are to secondary literature, such as the excellent annual reviews edited by Paul Kostecki and Edward Calabrese.

Interestingly, technologies have not changed drastically since publication of *Remedial Technologies for Leaking Underground Storage Tanks* in 1988 (biotechnologies excepted). However, there is an increasing use of combined technologies; for example, bioventing combining bioremediation and soil venting, or air sparging combining groundwater treatment and soil venting. The combined technologies are discussed in context.

Each technology is factored into its technical basis, health and safety features, feasibility, and engineering requirements. Each section is concluded with a table of "Considerations," which summarizes each technology according to the parameters that are important. These tables provide an effective way to compare the technologies and match them to local site conditions. For that reason the tables are repeated in Chapter 8 for convenient reference.

Much of the material in the middle chapters, especially Chapters 6, 7, and 8, is drawn from publications and regulations of the Colorado Department of Public Health and Environment (CDPHE) and the publication, *Leaking Underground Fuel Tank — Field Manual* (LUFT), published by the California State Water Board, Sacramento, CA. 1989.

. Colorado is (happily) neither as large nor as populous a state as is our neighbor to the west, California. However, in terms of regulations, Colorado is a "generic" state; that is, the compliance guidelines for petroleum contaminated sites generally follow federal requirements. State regulations may be more stringent than federal requirements, but not less. States that are more stringent, such as California, New Jersey, and Florida, usually have special problems that warrant special regulations. Colorado's regulations are typical of the majority of states.

2 Environmental Legislation and Regulations

The Legislative Environment

The last four decades of the twentieth century certainly deserve to be designated the beginning of the Environmental Era by future historians. The United States and much of the developed world have become increasingly aware that the environment is not limitless and cannot continue to serve as the repository for the refuse of affluent industrial societies. The awareness includes not only politicians and policy makers, but also extends to a grassroots level.

In particular, publicity surrounding heavily polluted lakes and rivers and deteriorating air quality in the United States, Canada, and the nations of the European Economic Community prompted the introduction, passage, and enactment of far–reaching and expensive environmental legislation. All of these elements combine to create a legislative climate in which owners and operators, regulators, and consultants must find a way to coexist — to clean up the environment and stay in business at the same time.

Public sentiment in the U.S. during the decades of the Sixties and Seventies was so strongly in favor of environmental protection that most legislation passed relatively easily and became law. Because the environment enjoys widespread popular support, taxpayers have been willing to pay the cost of cleanup and remediation.

The United States has been a world leader in efforts to reduce air and water pollution. Efforts to control and reduce new contamination have met with considerable success although efforts to address existing contamination have been less successful. For example, throughout the 1980s the official response of the U.S. to the international problem of acid rain resulting from coal–burning power plant emissions was to recommend yet another study. Efforts to clean up hazardous waste (NPL) sites are so mired in litigation and so complex that the majority of public and private money spent so far has gone to lawyers, with very little going to actual remediation.

If the decades of the '60s and '70s saw passage of the initial legislation, the decade of the '80s was a period of education and compliance. The early '90s is certain to be a period of enforcement. As knowledge and awareness of applicable regulations permeates society, regulatory agencies are increasingly turning to enforcement. Enforcement and compliance have not come as swiftly as some would like, and, of course, much too swiftly for others. So far public support for a cleaner, safer environment has remained solid, in spite of the cost. Although political support for reenactment of the Resource Conservation and Recovery Act, better known as RCRA, and the Clean Air Act waned through the '90–'92 recession, prospects for passage of these bills look better as the economy turns around.

Some of the statutes such as the National Environmental Protection Act (NEPA) are mainly statements of national policy and have not cost the taxpayer much in terms of real dollars, although some would argue that the Environmental Impact Statement requirement has severely hindered development and has been very expensive in terms of lost opportunities. Some, such as the Occupational Safety and Health Act (OSHA) are very expensive in terms of both administration and compliance. Others, such as the Comprehensive Environmental Response, Compensation, and Liability Act, better known as CERCLA, and RCRA have proven costly for everyone; regulators, manufacturers, and consumers alike.

In a society with 250 million people and a trillion dollar plus economy it is simply not possible to have instant compliance with any law or set of regulations, regardless of the intent of Congress, the Executive branch, and the regulated community. Regulations must first be adopted, a process that requires input from all members of the regulated community and compromise by the regulating agency. Next, the regulations must be published and accepted by the regulated community. Finally, the regulations must become a way of life throughout all sectors of that community.

Under the best of conditions a law requiring new regulations affecting large segments of society takes years from enactment to more or less full compliance. The UST regulations are a good example. The requirements for the Administrator of the EPA to set regulations for the ownership and operation of underground storage tank systems were included in the 1984 Hazardous and Solid Waste Amendments to the Resource Conservation and Recovery Act.

Formulating regulations and obtaining public comment took three years (1984 – 1987). Another two to four years (1987 – 1990) went by before the impact of the regulations was pervasive throughout the country. By the end of 1991 an estimated 75% of the UST operations covered by the regulations were in compliance or in the process of coming into compliance. In urban areas the percentage compliance is probably higher, while the compliance in largely rural areas is somewhat lower.

On the other hand, it is possible that years can pass with the intent of regulatory laws still unfulfilled. This scenario can arise because the law is poorly written, because the regulated industry stonewalls, because the scientific or engineering basis for regulation is inadequate, or a combination of all of the above. Difficulties in implementing and enforcing the Toxic Substances Control Act is an example of the latter. Fortunately, the UST regulations are fairly well understood by the petroleum industry, which has cooperated with the U.S. EPA and state regulatory agencies.

This chapter presents a summary of federal environmental statutes that affect petroleum contaminated sites and remediation activities either through direct legislation or indirectly through other effects such as liability or other regulation. The statutes divide roughly into the following classifications:

- Environmental policy

- Hazardous materials

- Water quality

- Air quality

- Protection of national resources

At the end of the chapter is a lengthy, but by no means complete, list of peripheral legislation which may or may not bear directly on remedial operations, but nevertheless, taken together, form the climate in which the UST owner/operators and the environmental companies must function.

Publications

The *Federal Register*, FR, is a daily official publication of the federal government. Regulatory agencies use this mechanism to disseminate information relevant to the regulated community. Rules and regulations are formulated through a hierarchy of proposed rules, interim rules, interim final rules, and final rules. An integral part of the rule–making process is solicitation of comments from interested parties. Once all commentary has been received and reviewed, the new rules or regulations are codified in the Code of Federal Regulations (CFR).

Each federal agency is assigned a title (or chapter) in CFR for certain categories of regulation: Title 29 is assigned to the Occupational Safety and Health Administration; Title 32 is for defense; Title 40 is assigned to the EPA for protection of the environment; Title 50 is wildlife. Thus, "40 CFR" refers to regulations generated by the EPA for protection of the environment. UST regulations are found in 40 CFR §280 and 281.

National Environmental Protection Act of 1970

This act has no direct bearing on UST operations, but may indirectly affect remediation operations through its policy statement, or conceivably through a requirement for an environmental impact statement prior to cleanup of an extensively contaminated site. The possibility may be remote for petroleum contaminated sites, but stranger legal things have happened.

NEPA was one of the first comprehensive environmental laws to mandate national goals and guidelines for cleaning up the environment. The preamble declares that

"...it is the continuing policy of the Federal Government, in cooperation with State and local governments, to use all practicable means and measures, including financial and technical assistance, ... to create and maintain conditions under which man and nature can exist in productive harmony, and fulfill the social, economic, and other requirements of present and future generations of Americans.."

NEPA established the Council on Environmental Quality (CEQ), a committee appointed by the president within the Executive Department. The CEQ advises the President on environmental policy and, by executive Order 11514, 1976, can issue guidelines regarding the interpretation and implementation of NEPA and environmental impact statements. In February 1993, President Clinton abolished the CEQ and replaced it with a cabinet level White House Office on Environmental Policy (WTN, 1993). The new office is intended to put sharper focus on environmental issues and make it easier to integrate environmental policy with both economic and foreign policies. At the same time, the Environmental Protection Agency is being proposed as a new cabinet department (UST Bulletin, May 1993).

NEPA also requires environmental assessments embodied in the Environmental Impact Statement (EIS) as part of the government's decision–making process. Through NEPA...

The Congress authorizes and directs that, to the fullest extent possible:...

(2) all agencies in the federal government shall —

(c) include in every recommendation of report on proposals for legislation and other major federal actions significantly affecting the quality of the human environment, a detailed statement by the responsible official on –

   (i)   the environmental impact of the proposed action,

   (ii) any adverse environmental effects which cannot be avoided should the proposal be implemented,

   (iii) alternatives to the proposed action,

   (iv) the relationship between local short–term uses of man's environment and the maintenance and enhancement of long–term productivity, and

   (v)   any irreversible and irretrievable commitment of resources which would be involved in the proposed action should it be implemented [NEPA §102(2)(c)].

The effect of this statute has been in three areas:
- It has forced federal agencies to reorder their priorities from straight development/exploitation to a broader view of economic equivalence;

- It has mandated a more equitable and democratic determination of land and resource use;

- It has made it the policy of the federal government to use all practicable means to administer federal programs in the most environmentally sound manner possible.

NEPA requires that there be some degree of cooperation and uniformity in the decision–making process across agency boundaries. Under NEPA all federal agencies must exercise some degree of uniformity in planning and approving changes in federal lands.

It is always a temptation to interpret the term "environment" to include only the natural environment—the "bugs 'n bunnies" part—rivers, forests, wilderness, beaches, etc. However, Congress clearly had broader concerns (West, 1991) since NEPA speaks of the need to assure all Americans "safe, healthful, productive, and esthetically and culturally pleasing surroundings." In Metropolitan Edison Co. v. People Against Nuclear Energy, the Supreme Court held that an EIS was required prior to reopening one of the reactors at Three Mile Island because the psychological stress caused by reopening the plant would have a significant health effect on the residents of the surrounding communities.

Given the right court interpretation, NEPA's requirement for an EIS could conceivably extend to just about anything! Building an apartment complex or shopping mall could conceivably require an EIS. A remedial project that would affect wetlands or have an adverse psychological effect on the neighbors might require an EIS. Very few actions of the federal government can be construed to have no effect on the quality of life somewhere.

Solid Waste Disposal Act of 1965

The Solid Waste Disposal Act was the original waste disposal act passed in 1965 to regulate solid and hazardous waste disposal at landfills, and to address the ongoing problem of resource recycling, especially the problem of recycling used motor oil. SWDA is Public Law 82 as amended. SWDA was amended in 1970, 1973, 1976, 1984, and 1986. Each amendment expanded the law and made it more inclusive.

This act provided most of the legislative regulatory authority for the Environmental Protection Agency when the EPA was created by Executive Order in 1970 by President Nixon.

SWDA established regulations for:
- Solid waste management and disposal;
- Hazardous waste management and disposal;
- Standards and definitions for hazardous waste generators;
- Standards and definitions for operators of hazardous waste storage facilities;
- Resource recovery (used motor oil).

Resource Conservation and Recovery Act

RCRA was originally passed in 1976 in response to public concerns over improper disposal of hazardous wastes. The EPA estimated that prior to regulation under RCRA, 90% of all hazardous waste was disposed of in an improper manner. As late as 1983 an EPA survey indicated that one in seven generators disposed of their hazardous wastes illegally (West, 1991).

The hazardous waste provisions are found in Subtitle C, as amended in 1984 by the Hazardous and Solid Waste Amendments (HASWA) and in 1986 by the Superfund Amendments and Reauthorization Act (SARA). The EPA is given broad authority to set regulations for management of hazardous substances, and for the generation, transportation, and disposal of hazardous wastes.

Generators of hazardous wastes are required to set up a "cradle-to-grave" management system that tracks hazardous materials from the "front door to the back door." Hazardous wastes must be packaged appropriately and manifested prior to shipment. Transporters are also regulated by Department of Transportation regulations.

Subtitle D of RCRA establishes a program for controlling solid wastes and sanitary (municipal) landfills. The 1984 amendments of HASWA shifted the focus of concern from disposal to treatment alternatives such that… "land disposal… should be the least favored method for managing hazardous wastes" [RCRA §1002(b)(7)]

SWDA, RCRA, SARA, and HASWA are all pieces of the same federal legislation (**Figure II–1**). The regulations which implement environmental legislation are published as Title 40 Code of Federal Regulations (40 CFR). As of this writing, the reauthorization of RCRA is stalled in Congress awaiting the resolution of controversial recycling requirements, and for some improvement in the economy which has been in recession through 1991–92.

Definitions in RCRA

RCRA contains a number of definitions that are essentially the same as those found in CERCLA. The definitions summarized here are only those required for the ensuing discussion. The list is by no means complete.

The first significant term to be defined is a "facility." A "facility" means [RCRA §101, (A)]:

"(A) any building, structure, installation, equipment, pipe or pipeline (including any pipe into a sewer or publicly owned treatment works), well, pit, pond, lagoon, impoundment, ditch, landfill, storage container, motor vehicle, rolling stock, or aircraft, or (B) any site or area where a hazardous substance has been deposited, stored, disposed of, or placed, or otherwise come to be located; but does not include any consumer product in consumer use or any vessel."

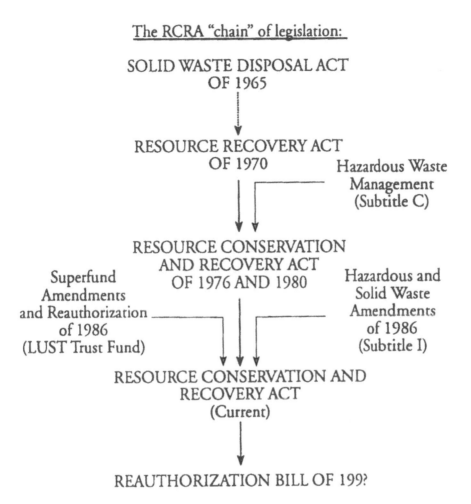

The RCRA "chain" of legislation:

SOLID WASTE DISPOSAL ACT
OF 1965

RESOURCE RECOVERY ACT
OF 1970

Hazardous Waste
Management
(Subtitle C)

RESOURCE CONSERVATION
AND RECOVERY ACT
OF 1976 AND 1980

Superfund
Amendments
and Reauthorization
of 1986
(LUST Trust Fund)

Hazardous and
Solid Waste
Amendments
of 1986
(Subtitle I)

RESOURCE CONSERVATION AND
RECOVERY ACT
(Current)

REAUTHORIZATION BILL OF 199?

Figure II–1. The Resource Conservation and Recovery Act Chain of Legislation. Numerous pieces of legislation have contributed to the current version of RCRA, including two sets of amendments incorporated into RCRA through SARA and HASWA. The dotted line indicates an indirect relationship. SWDA is still a separate law. Regulations are established by the Administrator of the U.S. EPA, not by RCRA itself.

RCRA defines "hazardous substances" as any substances having characteristics defined in RCRA and the Solid Waste Disposal Act. The definition of hazardous substances in RCRA contains a notable exclusion by excepting petroleum and petroleum products from the definition of hazardous substances, the so–called "petroleum exclusion [CERCLA 42 U.S.C.A. §101(14)].

"The term {hazardous substance} does not include petroleum, including crude oil or any fraction thereof which is not otherwise specifically listed or designated as a hazardous substance under {other definitions in this act}, and the term does not include natural gas, natural gas liquids, liquefied natural gas, or synthetic gas usable for fuel (or mixtures of natural gas and such synthetic gas)."

RCRA and CERCLA definitions are consistent. The same exclusion is found in the definition of hazardous substances in RCRA. The exclusion was deemed necessary due to the enormous volume of stored petroleum products in the U.S. A substantial number of owners and operators of petroleum storage systems would have been unable to financially comply with regulations requiring gasoline to be stored as a hazardous substance. Strictly speaking, owners and operators of petroleum storage facilities or petroleum contaminated sites should not need to be concerned about CERCLA because of the petroleum exemption. However, as explained below, reality has proven somewhat different.

Another definition in RCRA is "owner or operator." The term, as defined in RCRA and also in CERCLA, is sufficiently imprecise to have caused great difficulty. "Owner or operator" means [RCRA §101 (2) (A)]

"(i) in the case of a vessel, any person owning, operating, or chartering by demise, such vessel, (ii) in the case of an onshore facility or offshore facility, any person owning or operating such facility, and (iii) in the case of any facility, title or control of which was conveyed due to bankruptcy, foreclosure, tax delinquency, abandonment, or similar means to a unit of State or local government, any person who owned, operated, or otherwise controlled activities at such facility immediately beforehand."

In other words, an owner or operator owns or operates a facility. But this apparently simple definition has proven difficult to apply. This subject is discussed in more detail in Chapter 5.

RCRA provides for administrative, civil, and criminal penalties for violation of regulatory requirements. Actions or non–actions that lead or could lead to "serious bodily harm" have been prosecuted in federal court. To date, criminal prosecutions have generally been in connection with RCRA's hazardous waste provisions rather than Subtitle I. One feature of the enforcement activities forecast for the '90s is increased criminal prosecution by state regulatory agencies and the U.S. EPA for willful non–compliance, including violations under Subtitle I.

UST Regulations Under RCRA

Congress provided for regulation of underground storage tanks containing petroleum and hazardous liquids for the first time through inclusion of Subtitle I in HASWA. The UST amendments were introduced as a direct result of exposés on the television programs, *60 Minutes* and *Nightline*. The exposés featured one story of gasoline contamination of municipal drinking water wells in a New York suburb in 1978. A second incident concerned a gasoline leak that resulted in contamination of basements and sewers in a subdivision north of Denver, Colorado during 1979–1980. Some of the affected homes were purchased at more than twice their appraised value. The source of contamination in both cases was leaking underground storage tanks at neighborhood service stations.

During the summer of 1983 the Administrator of the EPA testified before the Senate Environmental and Public Works Committee's subcommittee on groundwater contamination to the effect that leaking USTs were a major unaddressed source of groundwater pollution in all areas of the country. At the time, the EPA had little or no data on the number of leaking tanks or even the number of USTs in the country. (Guide, 1990)

In January 1984 the UST amendments were introduced in the Senate by David Durenberger (R-Montana) and in the House by Don Ritter (R-Pennsylvania). The UST amendments were incorporated into HASWA as Subtitle I. A compromise bill designated H.R. 2867 passed both the House and Senate on October 3, 1983 and was signed into law on November 8, 1984, shortly after President Reagan's reelection. (Findley and Farber, 1992)

Under this authority the EPA administrator created first the Office of Leaking Underground Storage Tanks (LUST). After government watchdogs decided that it was not appropriate for the federal government to be involved in LUST, the name was changed to OUST, the Office of Underground Storage Tanks, which is a division within the Office of Solid Waste and Emergency Response of the EPA.

As noted in **Figure II–1,** Subtitle I became part of RCRA through the provisions of HASWA. Subtitle I defines an underground storage tank system (UST) as a system of tanks, plus piping and equipment that routinely contains product, and that is 10% or more underground. There are nine exclusions from the definition of an UST system:

- Farm or residential tanks of 1,100 gallons or less capacity storing fuel for non-commercial purposes;

- Heating oil tanks where the heating oil is used on the premises;

- Septic tanks;

- Pipeline facilities;

- Surface impoundments, pits, ponds, and lagoons;

- Stormwater or waste water collection systems;

- Flow–through process tanks;

- Liquid traps or lines directly related to oil or gas production operations;

  (Note: This does **not** include oil or grease traps in service stations.)

- Storage tanks in an underground area, such as a basement or cellar.

EPA also defines four exemptions from the requirements of *Technical Standards*:

- USTs containing hazardous wastes or mixtures of hazardous wastes and regulated substances;
- Machinery containing regulated substances for operational purposes, including hydraulic lift oil storage tanks;
- Waste water treatment tanks (regulated under the Clean Water Act);
- Tanks of less than 110 gallons capacity, and other small quantity tanks.

Since state regulations can be more stringent than federal, many states disregard the above exclusions and regulate some or all of the above, particularly heating oil tanks, farm tanks, and hydraulic oil tanks. Finally, RCRA requires all owners/operators of USTs to notify the EPA of the UST's existence. The EPA estimates that in 1993 approximately 70–75% of all UST systems are registered.

The Petroleum Exclusion

RCRA defines "hazardous substances" as any substances having characteristics defined in RCRA and the Solid Waste Disposal Act. The definition of hazardous substances in RCRA contains a notable exclusion by excepting petroleum and petroleum products from the definition of hazardous substances, the so–called "petroleum exclusion" [RCRA §101 (7)(A)].

"The term {hazardous substance} does not include petroleum, including crude oil or any fraction thereof which is not otherwise specifically listed or designated as a hazardous substance under {other definitions in this act}, and the term does not include natural gas, natural gas liquids, liquefied natural gas, or synthetic gas usable for fuel (or mixtures of natural gas and such synthetic gas)."

In some states this exclusion does not apply to contaminated soils containing benzene. It also does not apply to some petroleum wastes (see *Cose vs. Getty Oil*, Appendix E). The U.S. EPA has proposed a final rule to permanently exclude gasoline contaminated soils from being designated as a hazardous waste. However, states can set regulations more stringent than federal regulations. This topic will be covered in more detail in Chapter 7.

Although petroleum products are covered under Subtitle I of RCRA, petroleum based products that contain a listed or characteristic hazardous material (Subtitle C), and are not used as a fuel or other typical petroleum product, must be treated as a hazardous material, not as a petroleum product. That is, products made from chemicals derived from petroleum, but manufactured into non–petroleum products are not entitled to protection under the petroleum exclusion.

For example, antifreeze contains ethylene glycol, a listed hazardous material derived from petroleum. Used antifreeze is often collected along with used motor oil, particularly at used oil collection facilities. Used oil contaminated with ethylene glycol must be treated as a hazardous waste regulated under Subtitle C. Other examples of industrial or commercial products derived from petroleum that are hazardous include vinyl chloride, all chlorofluorocarbons, CFCs, PCBs, acetylene, most pesticides, and so on — the list is long!

Comprehensive Environmental Response, Compensation, and Liability Act

This act, commonly called "CERCLA" or "The Superfund Law," was passed in an effort to cleanup, establish responsibility for, and compensate victims of highly contaminated sites. Superfund sites are those 10,000 or so sites that are listed on the National Priorities List (NPL). These sites are so extensively contaminated with hazardous substances as to be outside normal remediation efforts. In other words, the cost of remediation is astronomical even by federal government budgetary standards. NPL sites are also complicated by the fact that responsibility for cleanup is vigorously disputed among all potentially responsible parties; that is, among those who may have contributed to the contamination, those who owned all or part of the site, and/or those who managed or operated all or part of the site.

CERCLA set new definitions of who is liable for cleanup of a site found to be contaminated. Potentially responsible parties, "PRPs," who may be liable for cleanup of a site contaminated by hazardous substances include:

- The owner and/or operator of the vessel or facility;
- The owner of the hazardous storage facility;
- The generator of the hazardous substances;
- The person who transports the hazardous substance.

The CERCLA definitions of liability fundamentally altered the way in which courts viewed liability, at least in environmental impairment litigation. Prior to CERCLA, proving liability usually involved proving negligence. After CERCLA, responsibility in environmental liability cases often involved simply showing ownership. One commentator has stated that the impact of CERCLA has been such that there are only two types of environmental liability cases, CERCLA and everything else. And everything else constitutes only about 10% of the total (Dunn, 1990).

Like RCRA, CERCLA specifically exempts petroleum products from the definition of hazardous substances. Petroleum products are defined in CERCLA as "including crude oil or any fraction thereof which is not otherwise listed or designated as a hazardous substance..., and natural gas, natural gas liquids, liquefied natural gas, or synthetic natural gas." Under this exemption petroleum products are by law not hazardous materials. However, petroleum contaminated sites are a different matter. As of this writing, the EPA has issued a Final Rule to exempt petroleum contaminated soils from CERCLA liability — and Subtitle C disposal — but the exemption is not yet permanent (see Appendix E). Therefore, at some point in the future, petroleum contaminated sites could fall under the CERCLA definition of liability. Alternatively, states can designate benzene contaminated soils as hazardous on their own.

CERCLA liability for contaminated sites contains three components. Under CERCLA liability is

- strict;
- joint and several ;
- retroactive.

Strict liability means "no excuses accepted." In effect, in the view of the implementing agencies and the courts, the contaminated site must be cleaned up regardless of who or what originally caused the contamination.

Joint and several means that all PRPs may be held responsible for cleanup costs. One of the reasons for the prolonged response time in addressing NPL sites is the necessity for courts to apportion remediation costs among PRPs. The court will try to determine what percentage of the contamination was contributed by each PRP and apportion remediation costs appropriately. It does not always work out quite that way. Generally it is felt that this results in the "deepest pockets" paying a disproportionate share of the remediation costs.

Retroactive means that PRPs may be held responsible for contamination that existed before CERCLA was enacted. In fact, CERCLA liability cannot be truly retroactive since *ex post facto* laws are prohibited by Section 9, Article 3 of the U.S. Constitution. However, the courts have consistently ruled that fines and penalties are levied for contamination that exists *now*, at the present time, regardless of when or how the contamination came to be. That interpretation has lead to nearly endless complaints about the CERCLA concept of liability.

CERCLA allows only three defenses against liability:

- an act of God;
- an act of war;
- an act or omission of a third party who had no relation to the owner [of the site], and the owner took all due care to prevent such an act or omission.

Defenses against CERCLA liability are discussed in detail in Chapter 5. For the moment it is sufficient to note that, in spite of the "petroleum exclusion," owners of petroleum contaminated sites have been caught up in the wasteland of CERCLA liability and must be prepared to deal with the reality.

Superfund Amendments and Reauthorization Act of 1986

CERCLA was reauthorized and extended by the Superfund Amendments and Reauthorization Act of 1986 (SARA).

A major portion of SARA deals with emergency response to spills, accidents, and other sudden releases of hazardous and regulated substances, including petroleum products. The emergency response section is implemented through the following agencies:

- Local Emergency Response Centers (LERCs);

- State Emergency Response Centers (SERCs);

- Federal Emergency Response Centers (FERCs).

In the event an emergency response is required, one or more of the above agencies must be contacted. The regulatory features of the Clean Air and Clean Water Acts have been described as "regulation through permits." If that is true, SARA can be considered as "regulation through paperwork."

SARA contains a number of definitions that must be considered in the course of routine activities by businesses that use or store hazardous substances. Definitions include the following:
- threshold planning quantities (TPQs);

- extremely hazardous substances (EHSs);

and

- reportable quantities (RQs).

"SARA Title III," that portion of SARA which is also known as the Emergency Planning and Community Right to Know Act, contains a "community right to know" provision that requires disclosure of chemical information to states, local fire departments, and communities, and directly affects owners/operators and managers of UST, AST, and chemical storage facilities. The requirements apply to extremely hazardous substances and hazardous substances present onsite in more than reportable quantities. Compliance is complex, burdensome, and expensive.

Congress was concerned that Subtitle I of RCRA would not address "past sins;" that is, petroleum contamination that had existed prior to passage of HASWA. In addition, there was evidence that owners and operators of UST's would not have the financial resources to clean up, upgrade, or replace "this historic source of groundwater contamination."

Therefore, as part of SARA, Congress established a five-year Leaking Underground Storage Tank (LUST) Trust Fund to pay for corrective action when EPA or a state deemed it necessary, and authorized recovery of corrective action expenses from any solvent responsible party. The LUST Trust Fund was funded by a 0.1 cent per gallon tax on motor fuels (UST Bulletin, 1990). The Fund is available to state implementing agencies for reimbursement for expenses incurred in assessment and cleanup activities.

The fund was allowed to expire August 31, 1990, but then was reauthorized in October 1990. The 0.1 cent per gallon tax was reimposed and will extend through December 31, 1995. Currently, the federal LUST fund is not accessible to private UST owners/operators for reimbursement of cleanup costs.

Clean Water Act of 1987

A series of acts have all been called "Clean Water Acts" beginning with the original Water Pollution Control Act of 1948, which was amended twice, once in 1956 and again in 1965. These acts established a municipal grant program to build municipal sewage treatment plants. The acts establish federal enforcement authority in the area of pollution discharge, but also provided for State Implementation Programs (SIPs) to set and enforce water quality standards. SIPs can be equal to or more stringent than the federal standards.

The Federal Water Pollution Control Act of 1972 established a system of funding for municipal publicly-owned water and sewage treatment plants, usually called publicly–owned treatment works (POTWs).

It also set up a system of enforcement through issuance of discharge permits administered through the EPA. A pervasive feature of the 1972 Act defined a National Pollutant Discharge Elimination System (NPDES) to control the amount of toxic material discharged by point sources. Since virtually all discharges into waterways require a permit, the system provides an effective means of controlling the content and amount of discharge. EPA regulations under this authority affect virtually every facility in the U.S. that discharges directly into surface waters. NPDES regulates all "point source" operations that discharge into surface waters That is, it is illegal for a person or operation to discharge a pollutant from a point source into navigable waters without a permit.

The Federal Water Pollution Control Act of 1972 established procedures to prevent or contain the discharge of oil onto "navigable waters or adjoining shorelines." Recent court interpretations have included wetlands as "navigable waters." Under current court interpretation, a wetland is anything "deep enough to float a toothpick." Therefore this feature applies to aboveground storage tanks (ASTs) which currently lack the consistent set of regulations such as those that apply to USTs.

The Clean Water Act of 1977 made a clear distinction between conventional pollutants, which include the so-called BOD (Biological Oxygen Demand) pollutants — microorganisms and organic waste — and toxic pollutants, which include hazardous material and wastes, sometimes called COD (Chemical Oxygen Demand) pollutants.

The Water Quality Act of 1987, usually called the Clean Water Act of 1987, was a major rewrite of the 1977 law that continued funding municipal water treatment plants, established a comprehensive program for toxic pollutant control, and set new requirements for non-point source discharges.

At the time of writing, renewal legislation is currently in Congress as S1081 introduced by Senators Max Baucus (D-Mont.) and John Chafee (R.-R.I.). The renewal bill is stalled for the same reason as the RCRA renewal; namely, a slow economy and for political reasons. When environmental legislation is presented in Congress, proponents often seek wide powers and new regulatory authority. Opponents seek to restrict regulations and limit the power conferred by the legislation. Wider regulatory authority is always perceived as being more expensive. Passage then consists of finding some middle

ground that all parties can accept. The new Clean Water Act is expected to pass —
eventually.

The CWA affects UST and AST owners/operators by requiring Spill Prevention,
Control, and Countermeasures, (SPCC) plans. That is, in the event of a petroleum
product spill, plans must be in place to deal with the spill — particularly to prevent
spread of the spill on an interim basis until emergency response personnel arrive on the
scene.

Clean Air Act of 1990

This act is the latest in a series of acts designed to protect the quality of the
atmosphere, primarily in urban areas. The original legislation was the Air Pollution
Control Act passed in 1955. The first Clean Air Act was passed in 1963 and renewed in
1970, 1977, and in 1990.

The CAA is designed to control and abate air pollution from mobile and stationary
sources. The 1970 law required the EPA administrator to establish national ambient air
quality standards (NAAQSs) for air pollutants. States in turn were required to develop
state implementation plans (SIPs) which could be equal to or more stringent than federal
standards. The CAA specified a "reasonable time" in which to develop and implement
SIPs.

The philosophy and mechanism of CAA is similar to that found in CWA; that is,
regulation by permit. Industries are allowed emissions of air pollutants within certain
limits. The new CAA of 1990 has already generated its share of controversy because it
allows industry to trade emission rights or exemptions, called "allowances," rather than
banning the emissions outright.

The CAA 1990 also requires stringent reductions in hydrocarbon emissions from
mobile sources. The term "mobile sources" means automobiles, trucks, and buses. By
1998, tailpipe hydrocarbon emissions will have to be cut by 30% over previous
standards, and by 2003 emissions must be cut by a further 50%.

It is too soon to tell, but the CAA 1990 is expected to be very expensive since it will
impact many small and very small quantity generators through Stage II vapor recovery
requirements similar to those already in effect in California.[‡] A great many "Ma and Pa"
operations, which are just recovering from replacing or upgrading their UST systems and
complying with leak detection regulations, are now faced with expensive add-ons to
limit hydrocarbon vapor emissions at the point of individual fueling operations. Most
service stations and other marketing operations in moderate and severe ozone noncom-
pliance areas will be affected sooner or later.

[‡] At the time of writing, the EPA has mandated add-on canisters on new
automobiles which represents a victory for petroleum marketers and bad
news for automakers. In effect, the decision defers Stage II vapor recovery
except in severe non-attainment areas, and unless required by state implementation plans.

Safe Drinking Water Act of 1986

This act requires the following from the U.S. Environmental Protection Agency:

- To set health–based standards for contaminants in drinking water;

- To require water supply operators to come as close as possible to meeting the standards by using the "best available technology" (BAT) that is economically and technically "feasible;"

  and

- To set maximum levels for contamination (MCLs) in water delivered to users of public water systems.

Primary enforcement is delegated to the states, provided that state drinking water regulations are no less stringent than federal regulations. To facilitate standard methods of analyses of drinking water, the U.S. EPA has published regulations and protocols entitled, *Drinking Water Methods*, found in 40 CFR §141 and 142.

Oil Pollution Act of 1990

Passed in response to the 1989 *Exxon Valdez* spill in Prince William Sound, Alaska, the Oil Pollution Act (OPA) includes regulations that primarily affect ASTs. OPA's approach to liability is derived from CERCLA and is expected to impact business transactions. Under the petroleum exclusion, CERCLA liability is presumed not to apply to petroleum spills; however, court interpretation has varied.

OPA establishes a "worst case scenario" planning requirement more stringent than SPCC, expands the maximum recovery costs for a spill, and further establishes almost any wetlands as "navigable waters."

Toxic Substances Control Act of 1976

The Federal Insecticide, Fungicide and Rodenticide Act (FIFRA) originally passed in 1972 and amended in 1988, left massive numbers of chemical products unregulated. Consequently, legislation in the form of the Toxic Substances Control Act (TSCA) was enacted in 1976 to control the production and use of most industrial and agricultural chemicals. (Note that storage and disposal of *hazardous* chemicals are covered under RCRA). The Office of Toxic Substances (OTS) of the EPA was set up to administer TSCA and other related laws.

The law regulates the production, use, and disposal of pesticides and other chemical products and requires manufacturers to accumulate data on the effects of new chemicals on the environment and on human health.

Occupational Safety and Health Act of 1970

OSHA regulations apply only to the workplace and to workers in private industry; that is, nongovernmental workers. On March 6, 1989, OSHA adopted final worker protection rules to minimize worker exposure to toxic wastes, reduce likelihood of accidental spills of hazardous substances, and reduce severity of spills. On June 23, 1989 EPA announced identical standards for all workers not covered by OSHA regulations.

Title III of SARA and subsections known as the Emergency Planning and Community Right–to–Know Act of 1986 (EPCRTKA) require OSHA or EPA to set standards for workers engaged in operations involving hazardous chemicals, hazardous waste, and emergency response. Collectively these are known as the hazardous waste operation and emergency response, or "HAZWOPER," regulations.

OSHA hazard communication standards require the following:

- •. Material Safety Data Sheets, MSDSs;

    These data sheets, which list potentially hazardous properties of a chemical or product, must be available on site where the chemical or product is stored or used. Typical data include flashpoint, boiling point, corrosivity, reactivity, toxicity, antidote, fire extinguisher class, etc.

- • OSHA Chemical Hazards Lists;

    A list of chemical hazards must be available at the site where the chemical(s) are stored or in use.

- • Worker training;

    Workers must be informed of potentially hazardous conditions and trained to deal with emergencies arising from those conditions.

Contractors and environmental companies must ensure that workers have adequate training, which usually includes the OSHA 40–hour course, although this is not particularly well suited for UST removals and petroleum contaminated site remediations. Nevertheless, workers must be trained in emergency response and protected from exposure to listed chemical hazards, including benzene and other aromatic petroleum hydrocarbons.

Despite the petroleum exclusion, these regulations directly affect workers at petroleum site remediation operations since worker health and safety are primary concerns, and by Department of Labor interpretation of OSHA regulations.

Petroleum operations affected include:
- • UST removals;
- • Site investigations;
- • Cleanup operations;
- • Corrective actions;and
- • Emergency spill response operations.

Leak detection, leak prevention, tank cleaning, and closure are covered by OSHA regulations 29 CFR 1910.120 if any of the following conditions apply:

1. A government body is requiring removal because of the potential threat to the environment or to the public.

2. The activities are necessary to complete a corrective action.

3. A government body has recognized the site to be uncontrolled hazardous waste.

4. There is a need for emergency response procedures.

Other Statutes

Many other laws have been passed that relate to some portion of the environment, either urban or otherwise. Most of these laws have little or nothing to do with petroleum or petroleum contaminated sites; however, all taken together serve to create a climate of preservation that pervades American society. These laws may be grouped according to their relevance to specific objectives. For each set of objectives, the acts are listed chronologically.

Protection of water quality

- **The Federal Water Pollution Control Act Amendments of 1972** — which replaced the Federal Water Pollution Control Act of 1965, addressed industrial water pollution by providing for area wide waste treatment plants (POTWs) and management plans.

- **The Coastal Zone Management Act of 1972** — provides federal funds to assist states to develop land use plans in the coastal areas.

- **The Safe Drinking Water Act of 1974** — increased control of the quality of drinking water and controlled underground injections of pollutants.

Protection of air quality

- **The Air Pollution Control Act of 1963** — authorized air pollution control research and assistance to state and local governments.

- **The Air Quality Act of 1967** — intensified efforts to abate air pollution by establishing air quality standards.

- **The Noise Control Act of 1972** — requires that Americans must be free of noise that jeopardizes their health and welfare.

Pesticide regulation

- **The Federal Environmental Pesticide Control Act of 1972;**

- **The Federal Pesticide Act of 1978;**

- **The Federal Insecticide, Fungicide and Rodenticide Act of 1988** — originally passed in 1972 and amended in 1988. All these laws regulate the production, distribution, sale, and use of pesticides by requiring permits and testing.

Protection of wildlife

- **The Endangered Species Act of 1973** — requires that all federal agencies must take active steps to conserve endangered and threatened species, including both flora and fauna; and prohibits the harm, harassment, trade, or capture of endangered or threatened species. Acts to protect individual species or groups include the following:
  - The Migratory Bird Conservation Act of 1929;
  - The Bald Eagle Protection Act of 1940;
  - The Golden Eagle Protection Act of 1962;
  - The Anadromous Fish Conservation Act of 1965;
  - The Wild Horse and Burro Protection Act of 1971;
  - The Marine Mammal Protection Act of 1972.

Protection of habitat

- **The Fish and Wildlife Coordination Act of 1958** — requires that any federal agency proposing a project which encroaches on a water resource must develop plans for the protection and enhancement of the wildlife associated with that resource.
- **The Wilderness Act of 1964** — recognizes the value of preserving certain areas that should be left "unimpaired for future use and enjoyment as wilderness."
- **The Water Bank Act of 1970** — established a system by which the Secretary of Interior can take easements on wetlands to preserve and improve them as habitat for wildlife.

Preservation of recreational areas

- **The National Wildlife Refuge System Administration Act of 1966** — established a central administration control of lands acquired for habitat and habitat restoration.
- **The Wild and Scenic Rivers Act of 1968** — authorizes the Departments of the Interior and Agriculture to study certain rivers and to acquire land along them for protection from development.

- The National Trails System Act of 1968 — provides for the development and protection of hiking trials.

- The Land and Water Conservation Fund Act of 1965 — provides funds to be matched by the state to plan, acquire, and develop land and water areas for recreational use.

Protection of historical and cultural sites

- The Historic Sites Act of 1935 — enacted a national policy of preservation for historic properties, and established the National Registry of Natural Landmarks.

- The National Historic Preservation Act of 1966 — expanded the national inventory known as the National Register of Historic Places and created a grant to help preserve nonfederal properties.

Land use control

- The Federal Aid Highway Act of 1968 — requires the preservation of parklands and refuge areas in highway planning.

- The Federal Land Policy and Management Act of 1976 — provides guidelines for public land use planning, protection of the "quality of scientific, scenic, historical, ecological, environmental, air and atmosphere, water resource, and archeological values," preservation and protection of certain public lands in their natural condition, and outdoor recreation and human occupancy.

Federal Regulations for Owners/Operators of UST Systems

The authority to establish regulations for the operation of underground storage tank facilities is provided by Subtitle I of the 1984 Hazardous and Solid Waste Amendments to the Resource Conservation and Recovery Act and is published in 40 CFR §280–281. Under the U.S. EPA's final UST regulations, owners and operators of UST systems must investigate, report, abate, and remedy releases of regulated substances into the environment, and must take quick action to identify and reduce any immediate health and safety threats posed by the release (FR, 1988; and FR, 1990).

The principal steps that must be taken by an owner or operator in the event of a release are as follows:

- Rapid notification of the appropriate agencies that a release has occurred;

- Investigation to mitigate fire, explosion, and vapor hazards;

- Preventing further release of the regulated substance from the leaking UST system;

    and

- Recovering free product from the environment.

The regulations require an initial site investigation that includes estimating the amount of product released and gathering information regarding surrounding populations, routes of migration, and the location of drinking water wells. Initial site characterizations are discussed in Chapters 5, 6, and 8. Corrective action plans are discussed in Chapter 7 and 8.

This section provides a summary of the most important portions of §280 relating to reporting, investigation, and cleanup. These regulations are referred to in subsequent chapters. The complete text of the final regulations for response, investigation, and cleanup are given in Appendix B. The regulations summarized below and listed in Appendix B are federal regulations issued by the U.S. EPA and are the minimum standards for the country. Implementing agencies in individual states can set regulations that are more stringent, but not less stringent. Indeed, several states have UST remediation regulations that are considerably more stringent than federal regulations. Most states have adopted the federal standards more or less intact.

Summary of Federal Regulations for Reporting, Investigation, and Cleanup

EPA's final regulations (40 CFR §280) define mandatory and discretionary requirements for UST owners/ operators.

Federal regulations mandate:

- Response;

- Investigation;

- Cleanup.

Federal regulations require under certain conditions:

- Site assessment;

- Remediation.

In the event of a release, the regulations require:

- Rapid notification that a release has occurred;

- Investigation to mitigate fire, vapor, and explosive hazards;

- Removing free product from the environment.

It is usually necessary to:

- Establish the nature and quantity of the release;

- Remove as much of the regulated substance as possible from the leaking UST;

- Gather information about the location of wells, sewer lines, and population in areas surrounding the release site.

Leak Prevention vs Leak Detection

The primary thrust of the EPA regulations is to prevent leaks from occurring. Release detection is an essential backup.

Release detection falls into one of five categories:

- Inventory control plus tank tightness tests;
- Automatic tank gauging systems;
- External monitoring;
  - groundwater monitoring;
  - vapor monitoring;
- Internal monitoring:
  - interstitial monitoring
- Other (Currently this is statistical inventory reconciliation).

If a release is indicated, owners/operators generally have 24 hours to report the release to the implementing agency.

Release confirmation must be by actual sampling.

- There are no depth requirements, sampling must be in the areas most likely occupied by the contaminant
- If drinking water is threatened, samples must be taken at the property line.

Offsite Impacts

EPA requires owners/operators to report offsite conditions or third party reports that indicate a release has occurred. This includes complaints of gasoline odors, sewer alarms, stained or discolored soils, distressed vegetation, seeps, or any other indication of released petroleum products.

If a release has occurred, EPA may then require the owner/operator to finance cleanup.

Initial Abatement Measures

The owner/operator must take steps to abate and correct a release. The owner/operator must report the initial abatement steps to the implementing agency within 20 days. Note: several states have regulations more restrictive than this.

Initial Site Characterization

Owners and operators must provide information regarding the site itself and surrounding property.

The information must cover the following:

• Nature of the product;

• Extent of release;

• Water quality in nearby wells;

• Groundwater quality if contamination is detected;

If readily available, the following should be included:

• Surrounding populations;

• Soil conditions;

• Land use; and

• Climate.

Soil and Groundwater Cleanup

The extent of soil and groundwater contamination must be investigated if:

• Release confirmation or previous corrective action measures indicate that drinking water wells have been contaminated;

• Free product is found to need recovery to be in compliance with the regulations;

• The initial response measures indicate that contaminated soils may have been in contact with groundwater;

• The implementing agency requests an investigation based on the potential effects of contaminated soil or groundwater on nearby surface water and groundwater resources.

Corrective Action Plan (CAP)

The implementing agency will request a CAP.

A corrective action plan must consist of a minimum of the following information:

• Measures to be used to remediate the site;

• Measures to be used to recover any free product;

• Physical and chemical characteristics of the regulated substance, including toxicity, persistence, and potential for migration;

- Hydrogeological characteristics of the area;
- Proximity to, and current use of, nearby surface water and groundwater, especially drinking water wells;
- Potential for exposure; and
- Anything else.

Public Right To Know

The implementing agency must notify the public of any confirmed release that requires a corrective action plan. All site information becomes public information, and must be made available on request.

3 Petroleum Hydrocarbons

The Good News and the Bad News

Remediation of a site contaminated with petroleum products is a mixture of "good news/bad news." The good news is that petroleum compounds are relatively straightforward to treat. They form a chemically homogeneous class of compounds, have a limited solubility in water, are lighter than water, and are not magnified through the biological food chain. In contrast, hazardous waste contaminated sites are frequently mixtures of dissimilar compounds that may be quite soluble in water, resistant to bacterial degradation, and/or very toxic. Most petroleum hydrocarbons are not particularly harmful with respect to groundwater or to human health. It is only the lighter hydrocarbons that create human health hazards. Long–term exposure to benzene in drinking water is certainly hazardous to human health; however, by comparison, exposure to pesticides, mercury, and other common contaminants is acutely hazardous, and these substances are more soluble. The bad news is that gasoline is nearly ubiquitous in the environment.

By law, products derived from crude petroleum are *regulated*, but not hazardous. Under RCRA, petroleum products are excluded from the requirements of Subtitle C. Similarly, waste from petroleum contaminated sites, such as contaminated soil, is a regulated waste, not a hazardous waste Some exceptions to this are discussed later. All of these features combine to make petroleum contaminated sites generally easier to remediate and much less expensive than sites contaminated with hazardous wastes.

The bad news is that under some conditions petroleum contamination can be quite hazardous to work around since the more volatile products constitute a fire and explosive hazard. In at least one instance tunneling equipment set off explosions in highly contaminated soils. Fortunately, the explosions were small. Some petroleum compounds also may present toxic hazards, especially compounds routinely found in gasolines, the most common contaminant. The hazards are particularly relevant for workers routinely exposed to gasoline vapors since gasoline contains high percentages of benzene.

The good news is that many petroleum products are readily consumed by soil bacteria with the result that a petroleum contaminated site could, in effect, remediate itself; the bad news is that typically a contaminated area is sufficiently deep in the soil that the bacterial population is too low to be effective. The good news is that petroleum products are less dense than water and float on the surface. That is, petroleum products are Light Non–Aqueous Phase Liquids, or LNAPLs. Halogenated hydrocarbons, for example, are Dense Non–Aqueous Phase Liquids, or DNAPLs. LNAPLs are relatively easy to find and recover in the subsurface environment. DNAPLs sink to bedrock and are neither easy to find nor easy to recover.

This mixture of "good news/bad news" is a result of the particular nature of petroleum products — the chemistry is rather simple, but the product formulation is often complex. Understanding the chemical nature of the contaminating agents is an important step in assessing the possibilities for remediation. Which compounds are expected to be volatile? Which are expected to be toxic? Therefore, it is appropriate at this point to look briefly at the chemistry of petroleum hydrocarbons, and at the composition of some typical petroleum products.

What Is Petroleum?

Petroleum products are mixtures[‡] of hydrocarbons; that is, a class of chemical compounds that contain only the elements *hydro*gen and *carbon;* hence, *hydrocarbon*. Petroleum hydrocarbons are obtained from naturally–occurring reservoirs of crude petroleum, which is the organic debris of ancient plant and animal origin. The detritus was gradually buried in sediments over geologic time and subsequently subjected to heat and pressure as the sediments themselves were buried.

Biological compounds present in the original deposits were slowly reduced under anaerobic conditions to relatively pure hydrocarbons. Residues of the ancient biological compounds are used today as markers to identify particular crudes. Since the original organisms contained elements other than carbon and hydrogen, virtually all petroleum deposits contain some sulfur, nitrogen, and small trace amounts of metals.

The reduction process is energy intensive. The necessary energy and reaction conditions are supplied by the high temperatures deep within the earth in a process that occurs over geological time. Petroleum formation requires time periods on the order of millions of years and is still occurring. Algal beds, for example, are thought to be "young" petroleum.

All this energy is stored in the formation of C–H bonds; the energy, on the order of 5,200 Joules per kilogram (11,500 Joules per lb.), is released when C–H bonds are

---

‡   Strickly speaking, petroleum and petroleum products are solutions, not mixtures. All petroleum hydrocarbons are soluble (or miscible) in other petroleum hydrocarbons. Thus, gasoline, diesel fuels, and motor oils, for instance, are liquid solutions. Mixtures are dispersions or emulsions made by mixing imiscible compounds, as for example, water, detergents, and oils.

broken and new H—O and C=O bonds are made. During the combustion process O_2 molecules are split into atomic oxygen and C—H bonds are broken as hydrocarbons react with oxygen (**Figure III–1**).

Fossil fuels are therefore a source of energy derived partly from the sun since photosynthesis is required to manufacture the biological compounds and partly from the heat of the earth itself. When hydrocarbons are burned, the stored energy is made available to power machines.

Although the detailed composition of crude petroleum deposits depends upon the origin and location of the petroleum, there are considerable similarities among various sources. A typical composition for Arabian light crude is shown in Appendix A.

Petroleum products have a vast array of uses. In approximate order of importance the uses are: fuels for vehicles and industry, heating oils, lubricants, raw materials in manufacturing petrochemicals and pharmaceuticals, and solvents. By a wide margin, most of the products derived from petroleum find use as fossil fuels to run vehicles, produce electricity, and to heat homes and businesses. About 65% of the petroleum used as fuel is consumed as gasoline in automobiles. Thus, petroleum products are ubiquitous in the modern environment which leads to contamination problems both for the environment and in sampling activities (see Chapter 7).

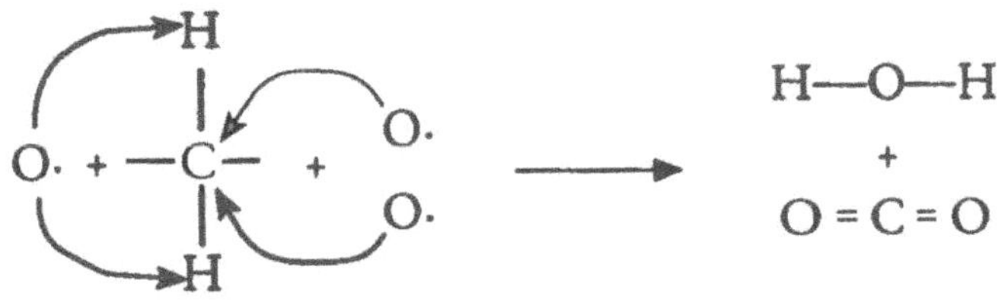

Figure III–1. Combustion of Hydrocarbons. A schematic view of the process of combustion of petroleum hydrocarbons. Atomic oxygen atoms attack the hydrogens and carbons of the fuel. The result is production of water and carbon dioxide. A quantitative version of the same process is shown on the next page. Since the reaction is kinetically controlled, too much of any reactant results in no reaction or incomplete reaction; that is, no combustion occurs when there is too much or too little oxygen. Too little oxygen is a "too rich" mixture, too much oxygen is a "too lean" mixture.

The vast majority of fossil fuels used in this country is in the form of gasoline, which is stored underground in a currently estimated 1.5 million storage tanks. Almost all of the tanks installed prior to 1988 were unprotected steel underground storage tanks that have leaked or have the potential for leaking gasoline into the environment.[‡]

[‡] These data come from a survey conducted by the U.S. EPA during the mid–'80s to determine the causes of leaking underground fuel tanks, to try to estimate the extent of the problem. Much of the information is published in Thompson's UST Guide. (See General References.)

There may be an equal number of unused or abandoned USTs that have not been drained and still have the potential to leak. Unused tanks are those left in the ground when a service station is converted to other uses; abandoned tanks are those that have been paved over and forgotten. Many of these orphan tanks are forgotten or in unknown locations, and only become apparent when contamination is discovered through excavation or an environmental assessment.

Environmental Effects

Fuels and other petroleum products are complex solutions of up to several hundred different chemical compounds; more than 100 different compounds may be found in a typical gasoline fuel blend (see, for example, Appendix A, p. 271). All petroleum–derived fuels have the similar feature of being predominantly hydrocarbons, plus small percentages of additives such as detergents, alcohols, and amines. Thus, combustion is mainly oxidizing carbon and hydrogen to carbon dioxide and water. Use of petroleum fuels has two potentially adverse effects on the environment.

First, there is production of CO_2 and other atmospheric pollutants. Petroleum products must be burned to release the stored chemical energy. As described in a previous section, hydrocarbons react chemically with molecular oxygen, O_2, to produce carbon dioxide, CO_2, and water, H_2O. The actual process is a series of several steps involving free radicals and formation of atomic oxygen; however, the combustion can be approximated. Using the complete combustion of octane as an example, the overall process can be written as follows:

$$2\ C_8H_{18} \quad + \quad 34\ O_2 \quad \longrightarrow \quad 16\ CO_2 + \quad 36\ H_2O \quad + \quad Heat$$

(Octane)   (Oxygen)   (Carbon   (Water)   (Energy)
                              dioxide)

Carbon dioxide is the principal culprit in models of the greenhouse effect. In the simplest possible model, increasing amounts of atmospheric carbon dioxide store heat in the atmosphere leading to an eventual increase in average global temperature. In more refined models using more accurate variables, there are various possible outcomes of increased CO_2 production depending on which model is being quoted.

However, from a fuel manufacturer's point of view CO_2 is an ideal emission product; it is non–toxic and nonpolluting. At least CO_2 is not considered a pollutant in either a toxic or legal sense. Carbon dioxide will not support life, but it is not an acute metabolic toxin as is carbon monoxide.

In the real world, air is combined with gasoline in internal combustion engines, and air is 80% nitrogen, N_2. Therefore, the above reaction is not exactly accurate. A better approximation is to write the reaction for the combustion of octane as follows:

$$C_8H_{18} + \boxed{O_2 + N_2} \longrightarrow CO_2 + H_2O + \text{Heat}$$

C$_8$H$_{18}$ + | O$_2$ + N$_2$ | → CO$_2$ + H$_2$O + Heat
(Octane)   (Oxygen) (Nitrogen)   (Carbon   (Water) (Energy)
                                    dioxide)

+ Sulfur   ∟— (Air)
(in petroleum)   + | SO$_x$ + NO$_x$ + CO |
                                      (Sulfur   (Nitrogen (Carbon
                                      oxides)   oxides)   monoxide)

                                  ∟— (Air pollution)

All fossil fuels contain some sulfur left over from the original plant material. Sulfur and nitrogen combine with O_2 at high temperatures to form oxides. The compounds, SO_x and NO_x are responsible for the polluted air in urban areas and for acid rain.

The second potentially adverse effect is the threat to public health when petroleum products are spilled or leaked into the environment. The environment is threatened with an array of chemical pollutants; petroleum products, and gasoline in particular, is in such widespread use that it is more abundant than any other contaminant.

Public health is threatened by gasoline contamination in several ways. The most immediate threat is through the accumulation of vapors which pose a fire or explosive hazard.

```
      H                  H  H              H  H  H
      |                  |  |              |  |  |
   H—C—H             H—C—C—H          H—C—C—C—H
      |                  |  |              |  |  |
      H                  H  H              H  H  H
   Methane             Ethane            Propane
    CH4                 C2H6              C3H8
```

Figure III–2. The First Three Saturated Hydrocarbons. These hydrocarbons are all gases at room temperature. They occur in nature both dissolved in crude petroleum deposits and in natural gas reservoirs. At one time natural gas was considered a nuisance and burned off at the well head. Now it is recognized as a clean–burning fuel in its own right. Most of the energy in fossil fuels is stored in the C–H bonds. Breaking the C—H bonds releases energy and makes the energy available as work.

The second route is by contamination of groundwater, surface waters, and soils. In this case, the principal risk is the result of exposure to aromatic compounds. Indeed the federal UST regulations themselves resulted in part because of contaminated drinking water wells (see Chapter 2).

An important step in assessing the effects of petroleum products that have been released into the environment is to evaluate the nature of the particular mixture and eventually select an optimum remediation technology for that mixture. Some hydrocarbons tend to volatilize easily, some are adsorbed onto soil particles, some do not migrate readily and can be addressed in a less aggressive manner.

Therefore, before plunging headlong into remediation technologies it is necessary to try to understand exactly *what* it is that is being cleaned up. The consequences of a gasoline release are very different from a diesel release. Regulatory agencies tend to get unhappy if the contaminant is simply treated as a "black box."

At this point it is time to look at some basic ideas from organic chemistry — just enough to become conversant with the compounds that make up petroleum products and that are important in remediation activities.

The Simplest Hydrocarbons

All hydrocarbons are composed only of the elements hydrogen and carbon in a vast array of possible combinations. Carbon forms a very large number of compounds because it forms chemical bonds with itself, building chains of carbons ranging from a single carbon through hundreds and even thousands of carbons. The arrangement of carbon atoms may be an approximately straight chain or a treelike branched arrangement (see Figure III–3 and Figure III–4).

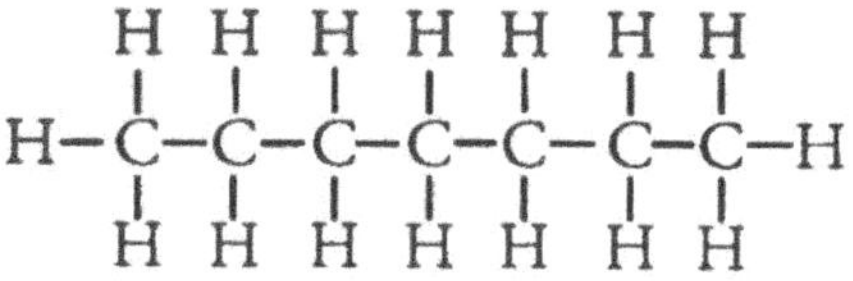

$$
\begin{array}{ccccccc}
& H & H & H & H & H & H & H \\
& | & | & | & | & | & | & | \\
H-&C&-C&-C&-C&-C&-C&-C&-H \\
& | & | & | & | & | & | & | \\
& H & H & H & H & H & H & H \\
\end{array}
$$

n–Heptane
C_7H_{16}

Flat projection

Figure III–3. Normal Heptane. Hydrocarbon molecules are visualized by thinking of carbon atoms being connected together, one after another in a "straight" chain of carbon atoms; each carbon atom is then surrounded by hydrogen atoms. Carbon atoms are actually arranged in a staggered formation with approximately 104° between bonds (see Figure III–4). Regardless, it is still customary to refer to these compounds as "straight–chain" hydrocarbons.

There is a rule, learned in a basic chemistry class, that carbon always forms four bonds. The immense number of carbon compounds comes from the variety of ways that carbon can form the four bonds. If carbon is constrained to form bonds only with itself and hydrogen, then hydrocarbons are formed. The simplest possible hydrocarbons are methane, ethane and propane, containing one–, two–, and three–carbons, respectively. These hydrocarbons are illustrated in **Figure III–2.**

Table 3–1. The Normal Alkanes. The name, formula and simplest "straight chain" structure of the first 10 alkanes (paraffin hydrocarbons) are listed. The first four have "trivial" names; that is, names with no special meaning. The rest use Latin prefixes: penta – five, hexa – six, and so forth. The general formula for alkanes is C_nH_{2n+2}.

Name	Formula	Structure
Methane	CH_4	CH_4
Ethane	C_2H_6	CH_3CH_3
Propane	C_3H_8	$CH_3CH_2CH_3$
Butane	C_4H_{10}	$CH_3CH_2CH_2CH_3$
Pentane	C_5H_{12}	$CH_3CH_2CH_2CH_2CH_3$
Hexane	C_6H_{14}	$CH_3CH_2CH_2CH_2CH_2CH_3$
Heptane	C_7H_{16}	$CH_3CH_2CH_2CH_2CH_2CH_2CH_3$
Octane	C_8H_{18}	$CH_3CH_2CH_2CH_2CH_2CH_2CH_2CH_3$
Nonane	C_9H_{20}	$CH_3CH_2CH_2CH_2CH_2CH_2CH_2CH_2CH_3$
Decane	$C_{10}H_{22}$	$CH_3CH_2CH_2CH_2CH_2CH_2CH_2CH_2CH_2CH_3$

By simply extending the chain of carbons one carbon at a time, adding a CH_2 unit each time, an entire family of analogous compounds can be assembled. The first ten straight–chain hydrocarbons, which are members of the alkane family, are listed in **Table 3–1**.

Families of straight–chain alkanes are built by linking carbon atoms one after another in a straight line (**Figure III–3**). The chain is more–or–less straight since the actual angle between adjacent carbons is approximately 104° (**Figure III–3**) regardless of the number of carbons in the chain.

Because of the arrangement of atoms in normal straight–chain alkanes, the carbon atoms are relatively exposed to the outside world. This means that the carbons are shielded from the outside world only by a single layer of hydrogens. However, from butane (C_4) on, for a given number of carbons and hydrogens, there are multiple possible structures.‡ Since there is no particular reason to add the fourth carbon on the end of the chain, it can just as well be added in the middle. The result is two possible structures (**Figure III–5**). Both exist in nature and have quite different physical and chemical properties.

---

‡   For higher molecular weight hydrocarbons, the number of possible isomers grows extremely large. Most are of academic interest only. In this section, the isomers discussed are those that are either very simple or those that are important to gasoline formulation. Branched isomers and cyclic alkanes and alkenes are created during the refining process, specifically during cracking and reforming.

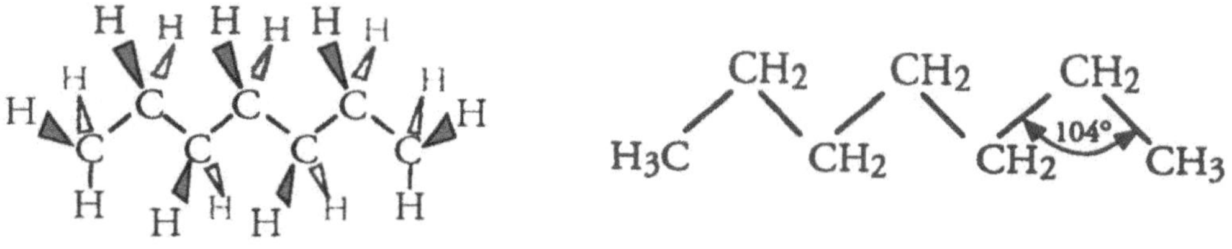

3–Dimensional view

"Staggered" view

n–Heptane
C_7H_{16}

Figure III–4. Alternative Representations of Hydrocarbons. Three dimensional and "staggered" representations of n–heptane. The three dimensional representation shows the spatial distribution of hydrogens around each carbon. A solid wedge means the hydrogen is "coming out of the page;" an open wedge means the hydrogen is "behind the page." The significance of the various structural forms of hydrocarbons shown here and in Figure III–3 is in the rate of burning in internal combustion engines and the effects of adsorption onto the surface of soil particles. Straight–chain hydrocarbons tend to be less volatile and less soluble than the corresponding branched isomers.

normal–Butane
C_4H_{10}

iso–Butane
C_4H_{10}

Figure III–5. The Butane Family of Alkanes. The two isomers of butane are shown. Normal–butane is a gas at room temperature; isobutane is a liquid. The volatility of butane fuels can be adjusted by varying the ratios of the two isomers. The butane in butane lighters is maintained as a liquid with a slight positive pressure.

Isomeric Hydrocarbons

When two or more compounds have the same empirical formula— the same number of carbons and hydrogens — the different compounds are collectively called *(structural) isomers*. Isomers have the same empirical formula but different structural arrangements. Isomers are important in gasoline formulation.

The simplest hydrocarbon for which isomers are possible is butane, which has the empirical formula, C_4H_{10}. There are two possible ways to arrange four carbons to form the two isomers of butane. One isomer has a "straight" chain of arrangement of carbons and one has a "branched" chain arrangement of carbons (**Figure III–5**). That is, in extending the chain of carbons from propane to butane by adding one more CH_3– group, the CH_3– can be added in the middle of the chain or at the end to form a continuous straight chain. Carbon chains are not exactly straight (see **Figure III–3**); straight in this context simply means that the carbons are attached one after another with no branches. The straight chain isomer is always called the *normal* isomer and prefixed with *n-*; one other branched isomer is called the *iso* isomer and prefixed with an *i-*.

Clearly, as the number of carbons increase the number of possible isomers increases rapidly (in a geometric proportion). The isomers of pentane, which are among the more important volatile components of gasoline, are illustrated in **Figure III–6**. Branched isomers are important in formulating gasolines since branched isomers burn at a different rate from the corresponding straight–chain isomer.

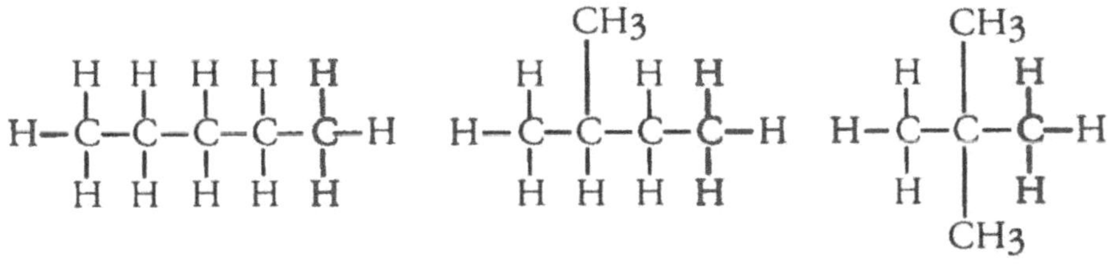

normal–Pentane C_5H_{12} iso–Pentane C_5H_{12} "neo–Pentane" 2,2–Dimethylpropane C_5H_{12}

Figure III–6. The Pentane Family of Alkanes. There are three possible isomers of C_5H_{12}, five isomers of C_6H_{14}, etc. Pentanes are among the more abundant volatile components of gasolines. The branched isomers are produced through catalytic cracking and reforming.

Combustion is the combination of atomic oxygen atoms with the hydrogens and carbons in hydrocarbon molecules (see **Figure III–1**). The easier it is for an oxygen atom to combine with hydrogens and carbons during the combustion process, the faster the *rate* of combustion. Since carbon atoms in hydrocarbons are always protected by hydrogen atoms, then the fewer hydrogens between oxygen and carbon in the chain, the faster the rate of combustion. In other words, straight–chain hydrocarbons burn easily and quickly because the carbons are more exposed. Similarly branched hydrocarbons burn more slowly since the carbons are buried deeper inside layers of hydrogens. Burn rate is a major factor in gasoline formulation, and is measured by *octane number*.

$$CH_3-CH_2-CH_2-CH_2-CH_2-CH_2-CH_2-CH_3$$

n–Octane
C_8H_{18}

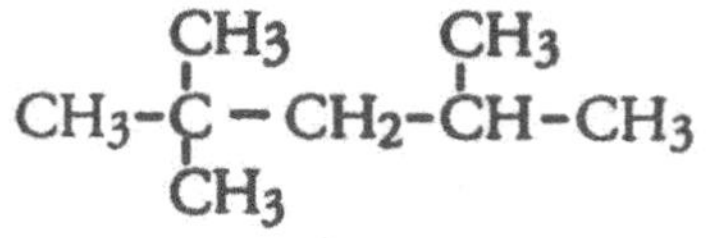

$$
\begin{array}{ccc}
CH_3 & & CH_3 \\
| & & | \\
CH_3-C- & CH_2-CH- & CH_3 \\
| & & \\
CH_3 & &
\end{array}
$$

"iso–Octane"
2,2,4–Trimethylpentane
C_8H_{18}

Figure III–7. Two Isomers of the Octane Family of Hydrocarbons. Two important isomers of octane, C_8H_{18}, are normal–octane and the so–called "iso"– octane. Isooctane is one of two benchmarks for the octane scale of fuel formulations. Pure isooctane has an octane number (RON) of 100; the RON of n–octane is close to zero. To burn pure isooctane in a car would cost about $150 per gallon.

The branched isomers are desirable components in automotive fuels since they burn more slowly in engines than the corresponding straight–chain isomers and, hence, give a fuel a higher octane rating. The prototype branched hydrocarbon is isooctane which has an octane number of 100 in its pure state (**Figure III–7**). An apparently simple solution to the problem of having to add aromatic compounds to gasoline to increase the octane rating would seem to be merely formulate gasoline from isooctane or a mixture of isooctane and the corresponding "isoheptane" or "isononane". Unfortunately, even if the refineries could produce such a formulation in sufficient quantities, gasoline would cost in excess of $150 a gallon! At that rate the great American love affair with the automobile would sour in a hurry.

The entire family of straight– and/or branched–chain hydrocarbons that are saturated — having no carbon–carbon double bonds — goes by two synonymous names:

- *Alkane* is the correct collective term for these hydrocarbons and includes straight, branched, and cycloalkanes (see **Figure III–10**).

- *Paraffin* hydrocarbons is the older, probably more common name.

This family of compounds comprises about 40 to 70% of gasoline products and at least 90 to 100% of all heavier petroleum products. Alkanes are not very soluble in water and, at least for the lighter members of the family, are readily degraded by indigenous soil organisms.

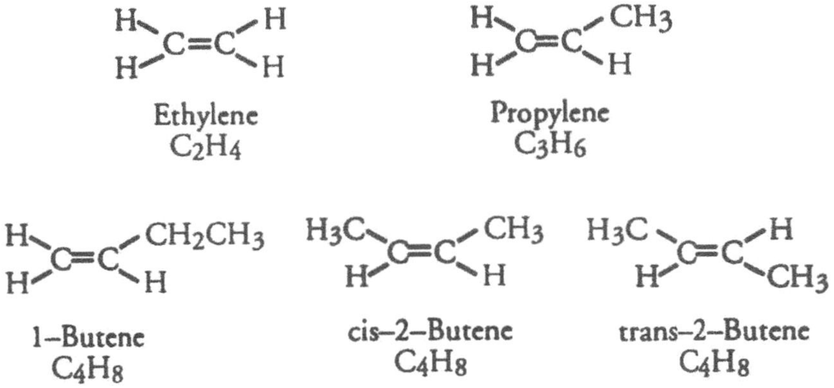

$$H_2C=CH_2$$

Ethylene
C_2H_4

$$H_2C=CH(CH_3)$$

Propylene
C_3H_6

1-Butene
C_4H_8

cis-2-Butene
C_4H_8

trans-2-Butene
C_4H_8

Figure III–8. The First Three Unsaturated Hydrocarbons. The compounds are produced in refinery operation during hydrocracking. The three shown are important as feedstocks for the polymer industry. Unsaturated compounds containing six to ten carbons make up 1–3% of gasoline blends.

Unsaturated Hydrocarbons

Recall the rule that carbon always forms four bonds in stable hydrocarbon compounds. The rule is valid, but there are several ways to form the bonds. One way is for each carbon atom to form one bond to four other elements, including itself. That arrangement leads to the alkane family described above. One way is for carbon to form multiple bonds with other carbons, which leads to another family of hydrocarbons, the *alkenes* or, as they are sometimes called, the *olefins* (**Figure III–8**).

Collectively, unsaturated compounds containing double bonds are known as *olefinic* hydrocarbons or *alkenes*. Alkenes are created through cracking and reforming operations and are relatively abundant in gasolines where they make up from 1 to 3% by weight. As a family, alkenes tend to be more soluble and more toxic than their alkane counterparts.

Unsaturation in this context means exactly the same thing as in ads for margarines. Unsaturated fats or oils contain double bonds in the carbon chain and as a result cannot be converted by the body into cholesterol. The hydrocarbon residues from unsaturated fats and oils can be converted into saturated triglycerides as storage fat, but not into the more harmful cholesterol.

From a remediation standpoint, olefins are relatively more soluble in water than the corresponding alkane hydrocarbons. There is no routine specific laboratory test for alkenes. An IR scan for carbon–carbon double bonds will give a total contribution.

Alkenes contribute collectively to the results of TPH, total petroleum hydrocarbons, analyses.

$$H-C\equiv C-H$$

Acetylene
C_2H_2

$$H-C\equiv C-CH_3 \qquad H-C\equiv C-CH_2CH_3 \qquad H_3C-C\equiv C-CH_3$$

Propyne   1–Butyne   2–Butyne
C_3H_4 C_4H_6 C_4H_6

Figure III–9. The First Three Acetylenic Hydrocarbons. These compounds, collectively called alkynes or acetylenic hydrocarbons, have considerable economic importance as welding gases and in polymer production, but are not significant in remediation of petroleum contaminated sites. All of these compounds have low flashpoints and are explosive. None of these compounds occur in nature.

A second type of unsaturation arises when there is a triple bond between carbons (**Figure III–9**). The prototype for this family is acetylene, widely used as a welding gas and in polymer manufacture. The family of hydrocarbons containing one or more triple bonds are called *acetylenic hydrocarbons* or *alkynes*. Alkynes are generally very reactive and do not occur in nature. This family of hydrocarbons has great economic importance throughout the petrochemical industry, but is not significant in gasoline or in remedial activities.

Cyclic Hydrocarbons

Yet another way to arrange carbons is to bring the end of the chain around to form a circle of carbons. The result is a family of *cyclic* hydrocarbons. The members of the hydrocarbon family are called *cycloalkanes* and *cycloalkenes*, and have octane numbers similar to the branched hydrocarbons (**Figure III–10**).

Cyclopropane and cyclobutane are known but are not very stable. Cyclopentane and cyclohexane occur naturally and are quite stable. The only cyclic alkenes found in petroleum products in significant amounts contain five or six carbons in the ring and are thus cyclopent*enes* or cyclohex*enes*.

In addition to cyclopentane and cyclohexane themselves, a variety of alkyl cycloalkanes are found in gasolines. That is, the parent cycloalkane is substituted with methyl, CH_3-, ethyl, CH_3CH_2-, n–propyl, $CH_3CH_2CH_2-$, and isopropyl, $(CH_3)_2CH-$, groups. Cyclohexene may also be substituted with alkyl groups. Gasoline contains significant amounts of cycloalkanes and cycloalkenes. Cyclopentane occurs in a range from 0.19 to 0.58 percent by weight.

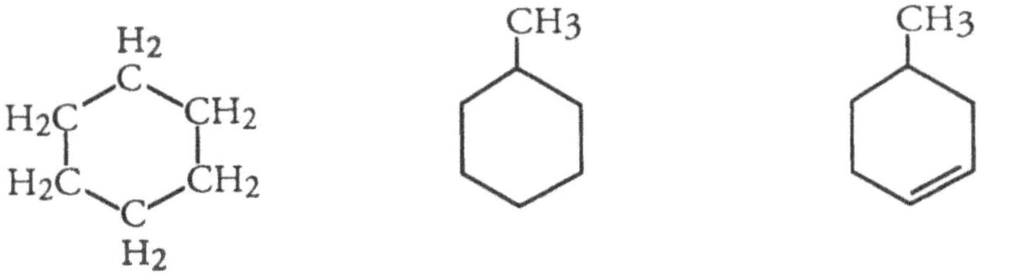

Cyclohexane
C_6H_{12}

Methyl cyclohexane
C_7H_{14}

Methyl cyclohex*ene*
C_7H_{12}

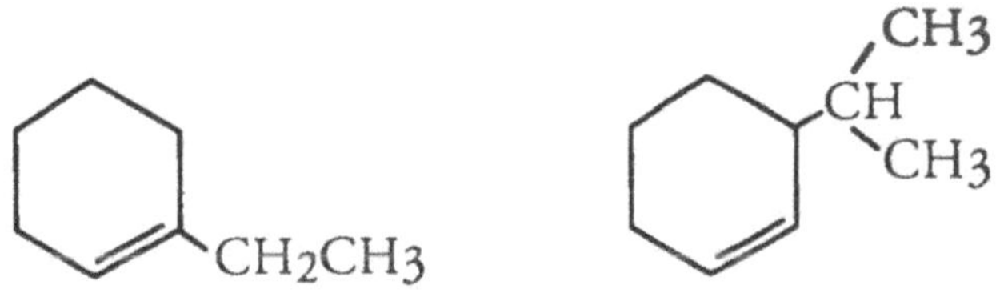

2–Ethyl cyclohexene
C_8H_{14}

3–Isopropyl cyclohexene
C_9H_{16}

Figure III–10. Cycloalkanes. Cyclohexane and methyl cyclohexane are two alkanes commonly found in gasolines. Cyclohexenes are produced by the cracking process. Cycloalkanes and cycloalkenes have burn rates similar to the branched isomers or aromatic compounds. Methyl cyclohexane and methyl cyclohexene are shown in stylized representations. In the figure the hexagon is a representation of the cyclohexane ring.

Collectively the entire family of non–aromatic hydrocarbons is called *aliphatic hydrocarbons*. This family includes normal, branched, and cyclic alkanes, alkenes, alkynes, or all of the paraffins, cycloparaffins, olefins, cycloolefins, and acetylenic hydrocarbons.

In older literature authorities often did not include cyclic alkanes, alkenes, and alkynes under the term aliphatic hydrocarbons. That is, hydrocarbons were divided into three groups: aliphatic, cyclic, and aromatic. Modern terminology reduces this usage to just two groups, aliphatic and aromatic. These two groups are all that are necessary for regulatory or remedial purposes.

Aromatic Hydrocarbons

Naturally-occurring petroleum contains still another class of unsaturated hydrocarbons called *aromatic* hydrocarbons. The name derives from the fact that many members of this family have distinctive, not unpleasant, odors. Unfortunately, most are toxic to some extent. The prototype for this family is benzene, C_6H_6, (**Figure III–11**).

Benzene is a small, flat molecule with three conjugated (alternating) double bonds. The small flat shape of benzene is partly responsible for many of its toxic characteristics. Benzene is chemically and physically distinct from the cycloalkane analogs such as cyclohexadiene, and forms a separate class of compounds. The aromatic ring is very stable and inert to many chemical reagents. It is more resistant to oxidation than cyclohexane or cyclohexene.

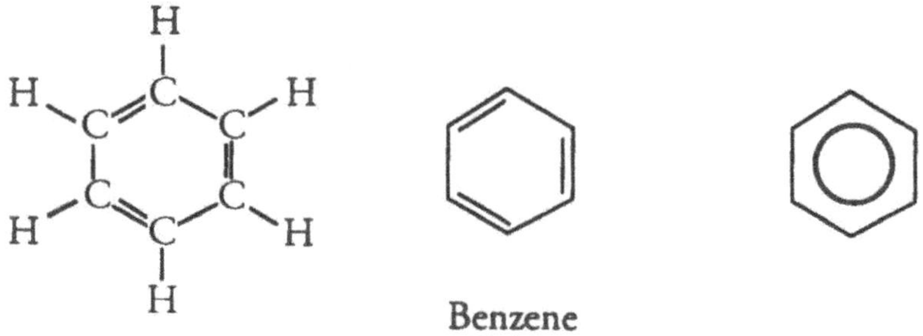

Benzene
C_6H_6

Chemical structure

Shorthand representation

Chemist's stylized representation

Figure III–11. The Chemical Structure of Benzene. Benzene is a cyclic molecule with three alternating double bonds, called conjugated double bonds. It is a flat, compact molecule, both structurally and in its electron configuration. The representation in the middle is a shorthand, schematic notation which summarizes the structure of benzene. The representation on the right is used by chemists to denote the benzene ring. Benzene is generally unreactive and very difficult to oxidize by normal chemical reagents. Indigenous soil bacteria will oxidize benzene and other aromatic compounds. Benzene is relatively soluble for a hydrocarbon and adsorbed fairly well onto soil particles. The rate of combustion makes benzene a desirable component in gasolines, but its toxicity limits use. Workers exposed to benzene in gasolines or in contaminated soils should be monitored for adverse health effects.

Alkylbenzenes

An extensive family of aromatic hydrocarbons can be derived from benzene by figuratively replacing hydrogens with other groups of atoms called substituents. The benzene ring can be substituted with alkyl groups or with other elements such as chlorine. Many of these chlorinated compounds have great commercial and economic importance. Most are hazardous to a degree; some less so than benzene itself, some more so. The molecular fragment C_6H_5— is called phenyl, two phenyls connected end–to–end is "biphenyl." A well–known class of contaminants is the PCBs. The acronym is derived from polychlorinatedbiphenyl, which simply means chlorine atoms substituted around the two rings of a biphenyl molecule. In gasoline formulation, an important class of compounds is the alkylbenzenes used to increase the octane number of fuel blends.

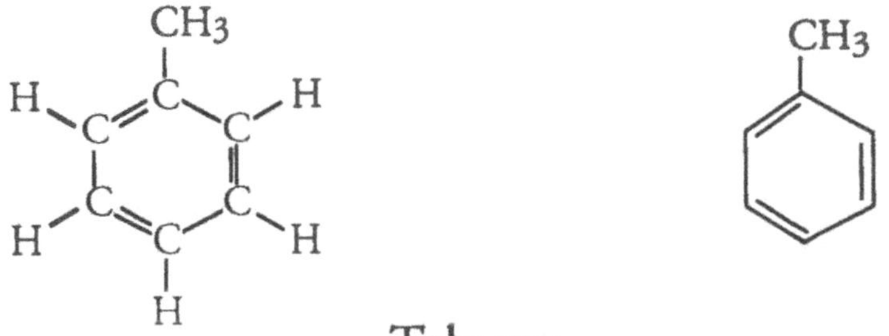

Toluene
C_7H_8

Figure III–12. The Chemical Structure of Toluene. Toluene has many applications in addition to gasoline. It is widely used in cleaning solutions since it is an excellent solvent for oils and greases. Other uses include solvents for plastics, adhesives and polymers, and in gasoline additives. As an alkylbenzene, toluene is more reactive, less soluble, and much less toxic than benzene. Nevertheless, workers routinely exposed to toluene in gasoline or through remedial activities should be monitored for adverse health effects.

Replacing a hydrogen on the benzene ring with a CH_3— unit gives a compound which could be called phenylmethane or methylbenzene, but is actually called toluene[‡]. Toluene occurs naturally in crude petroleum and is found in a variety of commercial petroleum products (**Figure III–12**).

Toluene is considerably less toxic and less soluble in water than benzene itself and is frequently used in commercial applications in place of benzene. The extra CH_3— group is more readily oxidized by liver enzymes to benzoate, which is picked up by the blood stream, carried to the kidneys, and excreted through the urine. The time–weighted average value for exposure to toluene vapor is 100 ppm, which should be compared to the corresponding value of 10 ppm for exposure to benzene vapor.

---

‡   There is only one "l" in *toluene*. For some reason, many people insist on pronouncing the word as *tolulene*.

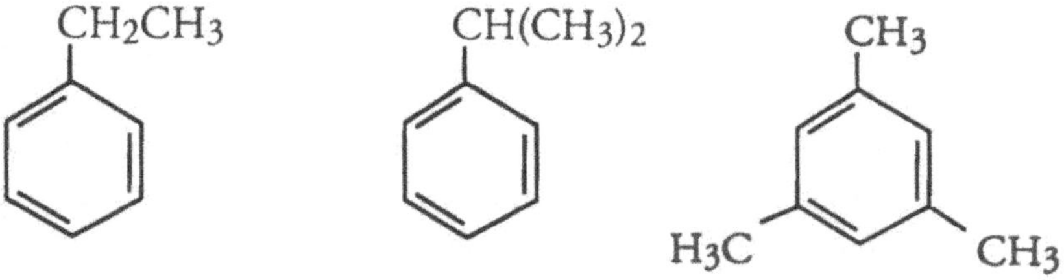

CH₂CH₃

CH(CH₃)₂

CH₃

H₃C   CH₃

"Cumene"
Ethylbenzene   Isopropylbenzene   "Mesitylene"
C₈H₁₀   C₉H₁₂   Trimethylbenzene
                                    C₈H₁₀

Figure III-13. The Chemical Structure of Ethylbenzene and Alkylbenzenes. Ethylbenzene and other alkylbenzenes such as cumene are added to unleaded gasolines to improve octane number. Trimethylbenzene has several isomeric structures; mesitylene is the most common. Like toluene, alkylbenzenes are wide spread in commercial and industrial applications. Plastic solvents, degreasers, and carriers agents are common uses.

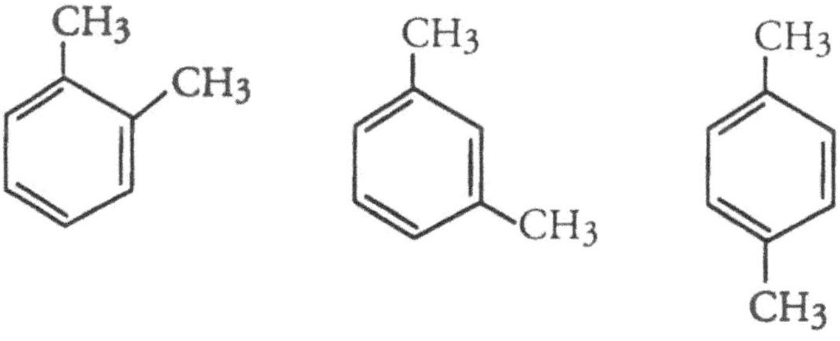

CH₃
  CH₃

CH₃

CH₃

CH₃

CH₃

CH₃

ortho–Xylene   meta–Xylene   para–Xylene
o–Xylene   m–Xylene   p–Xylene

The isomers of xylene
C₈H₁₀

Figure III–14. The Xylenes. The three isomers of xylene are indistinguishable from one another in field activities and in their health effects. Xylenes are used in commercial and industrial manufacturing processes. Laboratory analyses for volatile aromatic hydrocarbons or "BTEX" analyses give average values for all xylene isomers. The xylenes have a narcotic effect at high concentrations, but are less toxic than benzene.

Adding a CH₃CH₂— unit, an ethyl fragment, to benzene gives ethylbenzene (**Figure III–13**). Like toluene, ethylbenzene is less toxic than benzene, and more readily excreted. The time–weighted average value for exposure to ethylbenzene vapor is 125 ppm. Cumene, mesitylene, and other similar alkyl aromatic compounds are found in gasolines, especially in premium grades (see Appendix A). The maximum solubility of mesitylene in water is about 2 ppb (2 μg/Kg H_2O).

Adding two methyl groups, CH₃—, to the benzene ring produces several compounds collectively called xylenes. Since two CH₃— groups are being added, there are three distinct possibilities (**Figure III–14**). Xylenes are relatively abundant in gasoline fuels and in a wide variety of solvents, adhesives, and other consumer products. The xylenes are considerably less toxic than benzene. The three isomers are called *ortho*, *meta*, and *para*, and are distinguishable chemically under laboratory conditions. Therefore, in lab analysis reports "xylenes" is always plural even though there is little reason to distinguish among the three isomers in the field.

Polycyclic Aromatic Hydrocarbons

The final class of aromatic compounds contains multiple, fused benzene rings, sometimes called "chickenwire" compounds, but properly known as polycyclic aromatic or polyaromatic hydrocarbons, PAHs, or, in older literature, polynuclear aromatic hydrocarbons, PNAs. These compounds are naturally–occurring in many products and in combustion residues. (**Figure III–15**)

Diesel fuels contain PAHs as does the exhaust from gasoline and diesel engines. Some PAHs are known to be potent human carcinogens. In other applications PAHs are sometimes called "coal tars" and are present in cigarette and other tobacco smoke. Nowadays charcoal briquettes are no longer made from wood; rather, they are a petroleum product. Consequently, coal tars are present in the smoke from charcoal fires. They are, in fact, the reason why charcoal-broiled steaks are said to cause cancer!

Since PAHs are not degraded very well by microbial action and are strongly adsorbed onto soil particles, particularly onto clays, they can be very difficult to remediate. However, they are not particularly mobile, either through soils or groundwater. Care must be taken if humans are to be exposed to high levels of PAHs, either by way of sampling or remedial activities.

Refining Petroleum

Crude petroleum, which may consist of hundreds of individual compounds must be refined to separate the constituents into useful fractions. The fractions themselves are often further distilled to produce the desired commercial product. A refinery is basically a distillation tower in which the crude petroleum is introduced at the bottom of the tower and heated to a vapor. As the vapor rises through the tower, it contacts relatively cooler condensing caps. The vapors condense back to a liquid beginning with the highest boiling compounds first (**Figure III–16**). The highest boiling fractions are heavy, high molecular weight hydrocarbons suitable for lubricants and heating oils (see **Figures III– 21, 22, and 23**).

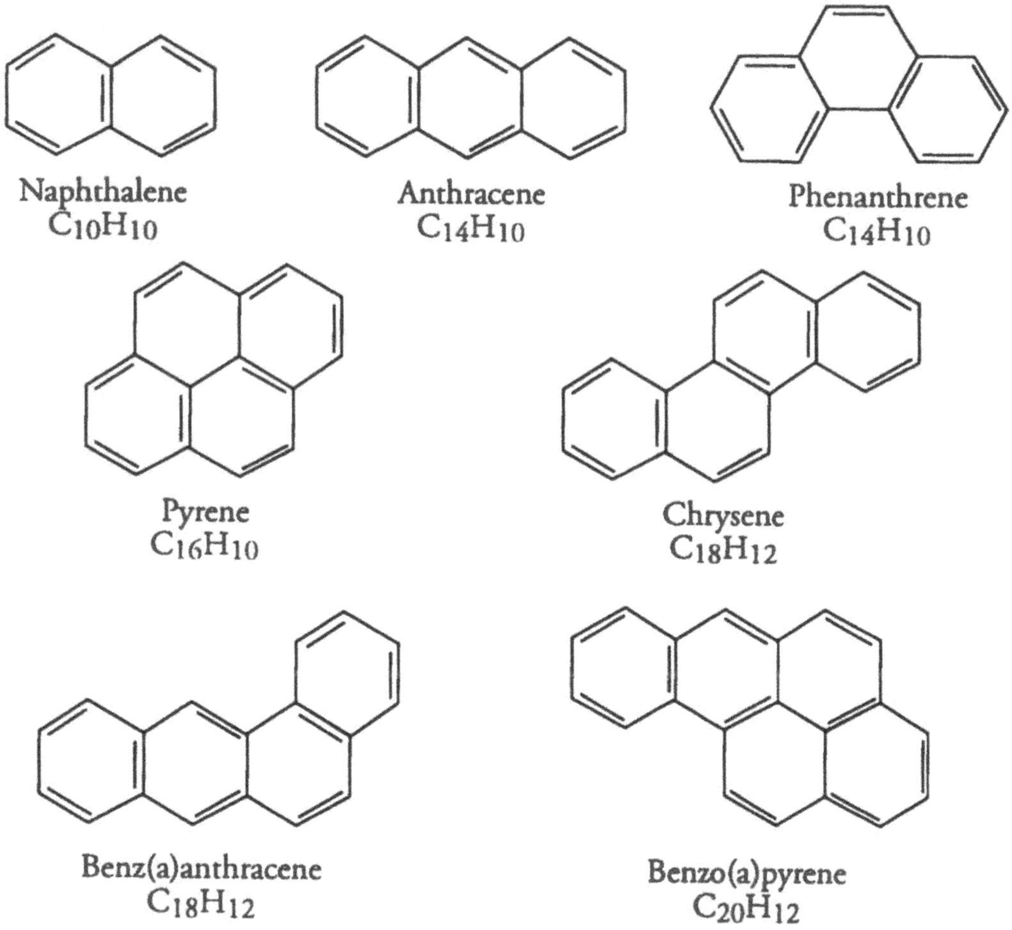

Naphthalene
$C_{10}H_{10}$

Anthracene
$C_{14}H_{10}$

Phenanthrene
$C_{14}H_{10}$

Pyrene
$C_{16}H_{10}$

Chrysene
$C_{18}H_{12}$

Benz(a)anthracene
$C_{18}H_{12}$

Benzo(a)pyrene
$C_{20}H_{12}$

Figure III–15. Characteristic Polycyclic Aromatic Hydrocarbons (PAHs). These compounds are formed from multiple benzene rings sharing one or more sides. A synonym is "chicken wire" compounds. Older literature refers to these compounds as polynuclear aromatic compounds, "PNAs." Another term that is often used is polyaromatic hydrocarbons, also "PAHs." Whatever the name the acronym "PAH" is common. They are found in diesel fuels and in fumes from partially burned middle distillates. The compounds shown are classed as probable carcinogens in humans and animals.

Vapors condensing in the middle range or "middle distillates" are characteristic of kerosene hydrocarbons. These hydrocarbons are used for heavier fuels, diesel and jet fuels, and a variety of commercial products (see Appendix A). The next lighter hydrocarbons occur in the C_4 to C_9 range and are used in gasoline formulation. Finally, the lightest hydrocarbons are gases at room temperature, and are collected and used in heating gas (LP) mixtures and in the petrochemical industry.

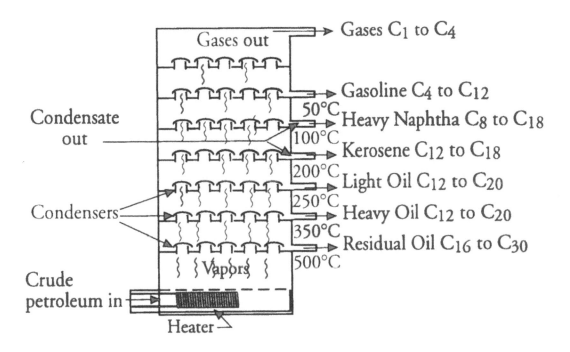

Figure III–16. Petroleum Refining. A schematic diagram of a petroleum refinery distillation tower. Crude petroleum is introduced at the bottom, passed through heaters, and vaporized. Vapors migrate upward through bubble cap condensers set into perforated plates. The figure shows a few plates; in practice there are hundreds of plates in a refinery tower. The highest boiling vapors condense first and are collected as a condensed liquid phase. Liquids are removed from the distillation tower as they form. Lower and lower boiling compounds are collected successively up the tower. Compounds that are vapors at room temperature exit through the top. The typical fractions shown are recovered from distillation of an Arabian light crude. Fraction names can vary depending on the precise method of distillation. The gasoline fraction is also known as the light and heavy naphtha fractions and the kerosene fraction is also called "middle distillates". The heaviest molecular weight compounds that do not vaporize are asphalts or paraffins, depending on the source of the crude.

Table 3–2. Components of Gasoline. The more abundant components of gasoline include both regulated and non–regulated hydrocarbons under Subtitle C of RCRA. That is, some of the constituents of gasoline are listed hazardous substances under Subtitle C. Since the compounds occur in gasoline they are considered regulated substances. Therefore, "hazardous" is referenced in this table for health and safety purposes in remedial activities. For regulatory purposes petroleum hydrocarbons fall under the petroleum exclusion. Composition percentages are % by weight and represent average values for typical brands; actual values will vary according to manufacture, season, and region of the country. For remediation purposes there is relatively little difference in formulations among major brands of gasoline, except for possible identification purposes. (See also Appendix A).

Hydrocarbon	% by weight		Hazardous Substance[a]	CAS Number[b]
	Unleaded	Premium Unleaded		
n-, i-Pentane	9	13	No	
n-Hexane	11	6	No	
2-Pentane	7	3	No	
2-Methylpentane	5	4	No	
n-Octane	1	1	No	
i-Octane	2	4	No	
Benzene	4	5	Yes	71432
Toluene	4	7	Yes	108883
Ethylbenzene	2	2	Yes	100414
Xylenes	6	6	Yes	1330207[c]
Cumene	1	2	Yes	98828
MTBE, methanol, or ethanol	11 – 15	11 – 15	Yes[d]	67561[d]

[a] "Yes" means the hydrocarbon is a listed hazardous substance under Subtitle C of RCRA.

[b] Chemical Abstract Service number; useful for looking up further information.

[c] For mixed isomers of xylene.

[d] For methanol; MTBE and ethanol are not hazardous substances.

As early as 1912 it was recognized that since the economically important gasoline fraction comprises only about 10–20% of a typical crude, refinery output would fall far short of the demands of a modern society addicted to automobiles. In order to supply market demands for gasoline 45–70% — the amount varies by season — of a refinery's output must go to gasoline.

Therefore, a substantial portion of the heavy naphtha, kerosene, and fuel oil fractions are "hydrocracked"; that is, catalytically converted from higher molecular weight compounds to lower molecular weight compounds. In the process a relatively high proportion of branched, olefinic, and cyclic isomers is produced. Hydrogen and alkenes are by-products of the cracking process. These by-products are used to manufacture fertilizer, polymers, and commercial plastics.

Gasolines

The energy to run vehicles is derived from "fossil fuels". Burning hydrocarbons releases chemical energy stored in these compounds by microorganisms when plant material was reduced under anaerobic conditions.

Most gasoline blends are complex solutions (fuels are solutions, not mixtures), containing 50 to 150 components, formulated for burn rate (octane number), volatility (for constant performance in hot and cold weather), and emission control (oxygenated fuels). Table 3–2 gives typical percentages for aliphatic and aromatic compounds for regular, unleaded, and premium unleaded fuels.

Gasoline fuels are optimized to *burn* at a certain rate, not explode[‡]. Octane number is a measure of burn *rate*, not power as is commonly assumed. Maximum power output is achieved by optimizing the rate at which a fuel burns inside the cylinders of an engine. The *Octane Scale* is defined such that pure n-heptane has an octane number of zero, and iso-octane has an octane number of 100.

In regular gasoline blends lead compounds, such as tetramethyllead and tetraethyllead, are used to increase the octane number. Other hazardous compounds, including ethylene dichloride, EDC, and ethylene dibromide, EDB, are added as lead scavengers to prevent buildup of lead oxide deposits. In the combustion chamber, EDC combines with lead to produce lead chloride, $PbCl_2$, a volatile compound that is carried from the engine with the flow of exhaust gases.

From a fuel manufacturer's standpoint aromatic compounds are desirable in gasolines since they increase the octane rating of gasoline blends. After the EPA ban on lead additives in 1973, aromatic compounds, including benzene and alkylbenzenes have

---

[‡]   All hydrocarbon fuels require initiation, something to ignite the fuel. For gasoline the initiation is provided by a spark; hence the fuel burns. Diesel fuels explode under the heat of compression which gives diesel engines the characteristic "knock". Gasoline-burning engines should not knock; if they do something is wrong.

generally been used to increase the octane number. In present formulations up to 50% of a premium gasoline blend can be aromatic compounds.

Unfortunately, because aromatic compounds are also acutely hazardous to human health, most of the compounds described above are included under RCRA, Subtitle C regulations as hazardous substances. These compounds include benzene, toluene, ethylbenzene, xylenes, and cumenes. There is a clear conflict between EPA's toxicity characteristic (TC) rule and the petroleum exemption under RCRA and CERCLA. Gasoline is clearly hazardous by reason of both the flammability and toxicity. The EPA has attempted to resolve the conflict by regulation, permanently exempting gasoline and gasoline contaminated debris from hazardous regulations.

Oxygenated Fuels

The Clean Air Act of 1990 sets standards for air quality with which all areas must comply. Among the standards CAA sets specific maximum levels of carbon monoxide, CO, which is an acute human health hazard. CO is produced as a result of all fossil fuel combustion, but is particularly prevalent in automobile exhausts. Simply due to the sheer number of vehicles, automobile exhaust is the major contributor of CO in the atmosphere in urban areas.

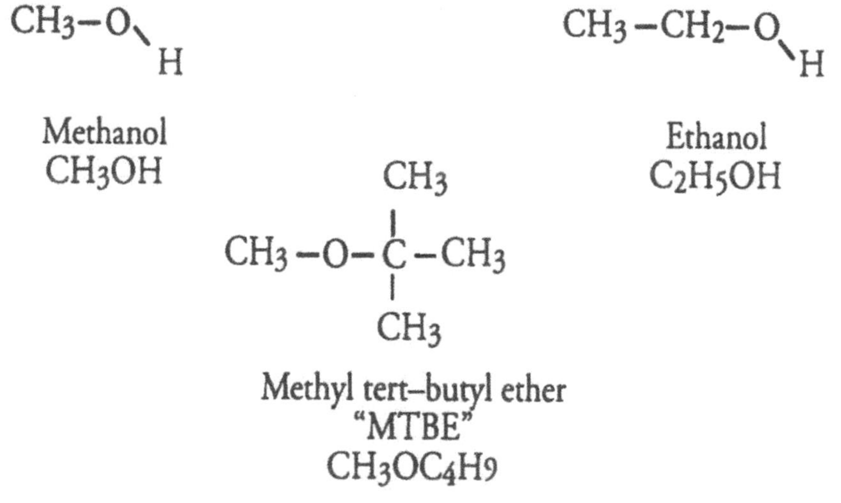

$$CH_3-O \diagdown_H \qquad\qquad CH_3-CH_2-O \diagdown_H$$

Methanol $\qquad\qquad\qquad\qquad$ Ethanol
CH_3OH $\qquad\qquad\qquad\qquad$ C_2H_5OH

$$CH_3-O-\overset{\displaystyle CH_3}{\underset{\displaystyle CH_3}{\overset{|}{\underset{|}{C}}}}-CH_3$$

Methyl tert–butyl ether
"MTBE"
$CH_3OC_4H_9$

Figure III–17. Oxygenated Fuel Additives. Methanol is also called methyl alcohol or wood alcohol since it was once made by distillation of wood chips. Ethanol is also called ethyl alcohol or grain alcohol since it is prepared exclusively from fermentation of grain, usually corn. MTBE is made from petroleum compounds.

CO production arises because of a compromise. Energy is extracted from hydrocarbons by breaking C-H bonds and converting the fuel into carbon dioxide, CO_2, and water, H_2O. If the process were 100% efficient, the only tailpipe emissions would be CO_2

and H_2O. At very high engine operating temperatures this condition comes close to being realized. Unfortunately, engines do not breathe pure oxygen, but rather air which is a mixture of O_2 and nitrogen, N_2, in a 20%:80% ratio, respectively. High engine operating temperatures maximize the production of nitrogen oxides, NO_xs. Therefore, the normal operating temperatures of automobile engines is a compromise between minimizing NO_x production and maximizing CO_2 production (which minimizes CO production).

The result is production of a relatively large amount (up to 11% of exhaust gases by volume) of carbon monoxide, CO. The problem is partially addressed by the use of add–ons such as catalytic converters and air injection systems. But an increasing number of urban areas are requiring mandatory use of "oxygenated fuels," especially during the winter months when cold air inversions are more common. The Denver metro area is currently on 13% oxygenated fuels about seven months out of the year.

The idea is straightforward: the process of combining a fuel with O_2 can be made more efficient by, in effect, "pre–burning" the fuel. Oxygenated fuel additives are already partially oxygenated and, therefore, produce less CO, at the expense of slightly less power (fewer BTUs) per gram of fuel. The three compounds in common use are methanol, ethanol, and methyl tertiary-butyl ether (MTBE). Methanol is obtained from petroleum (from methane or by reacting CO and water); ethanol is obtained from corn by fermentation; and MTBE is made synthetically from petroleum feedstocks. (**Figure III–17.**)

The choice of which of the compounds commonly used in oxygenated fuels to use is dependent on region, cost of production, and availability of raw materials. Methanol and MTBE are made from petroleum feedstocks, ethanol is fermented from corn. Fortunately, methanol is seldom used.

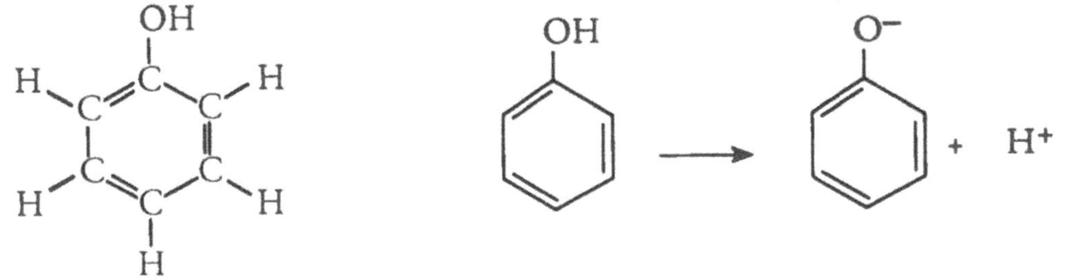

Phenol
$C_6H_5\,OH$

Ionization of phenol
(Corrosivity and toxicity)

Figure III–18. The Chemical Structure of Phenol. Phenol comprises about 0.3% of a typical gasoline blend. It is relatively soluble and quite toxic to humans and aquatic life. The hydrogen associated with the OH group is acidic; that is, it is removable to form H^+ ions which will attack metals, soldered joints and some polymeric resins.

Of the three, methanol is by far the most toxic and hazardous to handle. Methanol is rapidly oxidized by liver enzymes to formaldehyde, CHO, which damages liver and brain cells. The time-weighted average (skin) exposure for methanol for a 40 hour work week is only 200 ppm, compared with 1,200 ppm for ethanol. Methanol is also acidic and corrosive. It will attack ordinary soldered fittings and must be stored in USTs or other containers specifically designed to hold methanol.

The final compound, phenol, is neither additive nor a pure hydrocarbon but does occur naturally in most gasoline blends. Phenol is an aromatic alcohol derived from benzene by replacing a hydrogen with –OH (**Figure III–18**). The –OH hydrogen is acidic; that is, it ionizes to form H^+ ions in solution. This acid characteristic means that phenol is corrosive to metals. It should also be treated as an acute systemic poison. A related set of compounds is collectively known as creosote, derived from distillation of coal tar, and widely used as a wood preservative, especially for railroad ties and telephone poles. Creosote is sometimes found mixed with petroleum wastes. Likewise, the common wood preservative, pentachlorophenol, PCP, is very acidic and is a listed hazardous substance.

Diesel Fuels

As a fuel for internal combustion engines, diesel fuels rank second only to gasoline with a total demand that is about 25% that of gasoline. Diesel fuels are a heavier cut from gasoline ranging from C_6 through about C_{22}. Most diesel hydrocarbons are in the C_{10} to C_{18} range. Diesel fuels actually explode inside cylinders; the activation energy required for ignition comes from heat of compression.

Current diesel fuel blends are a compromise between ease of starting (high volatility) and good fuel economy (low volatility). The three standard grades of diesel are listed below:

No. 1:   Includes volatile fuel oils from kerosene and intermediate distillates. This grade is recommended for high speed engines whose operating conditions involve frequent and wide variations in engine load and speed. Fuels of this grade are also required for use in very cold temperatures.

No. 2:   Includes distillate oils of lower volatility than No. 1 grade. This grade is suitable for high speed engines not subject to wide variations in load and speed; that is, stationary, constant speed engines.

No. 4:   Includes more viscous distillates and blends with residual fuel oils. While more economical than the other grades, this grade is suitable only for engines operating at low to medium speeds under sustained loads and constant speeds.

Physical Properties of Hydrocarbons

Petroleum products have basically similar chemical properties; none of the normally occurring compounds is very reactive or corrosive. That is, petroleum hydrocarbons do not react or react with difficulty with typical oxidizing and reducing agents. The more volatile compounds will react with oxygen, but the reactions have high activation energies, and require some form of initiation. This means that workers must be aware of the potential for fire or explosion, but need not worry about corrosive or reactive hazards.

Physical properties are also similar, although there is a considerable range in magnitude of the properties. From the standpoint of remediation the more important physical properties are volatility, solubility in water, specific gravity, and kinematic viscosity.

Volatility. Volatility is the tendency of a molecule to leave the surface of a liquid. A highly volatile compound is one that vaporizes easily. Highly volatile means the substance has a low boiling point. Some refined petroleum products, especially gasolines, are quite volatile, vaporize readily, and have flashpoints at room temperature or below; that is, they are Class A flammable liquids according to NFPA. At the other end of the spectrum are heavy viscous products such as lubricating oils and fuel oils that vaporize only with difficulty and have flashpoints in excess of 300°F. Flashpoint is proportional to vapor pressure which is in turn proportional to molecular weight (**Table 3–3 and Figure III–20**). A dramatic demonstration of volatility, or lack thereof, is to try to get motor oil to burn; it will, but with great difficulty.

Products having a high vapor pressure, such as gasoline, have the potential to create serious health and environmental problems in subsurface structures such as basements, sewers, and underground parking structures. The vapor concentration required for fire or explosion is generally in the range of a few percent (parts per hundred) relative to air. For example, the lower explosive limit for benzene is 13 mL of benzene per liter of air (1.3% by volume in air).

Even relatively small concentrations of gasoline vapors, possibly undetectable by smell, in underground structures over a long period of time can pose human health problems due to exposure to benzene. Similarly, for volatile hydrocarbons the concentration for acute inhalation toxicity is in the ppm range. For example, the workplace 8 hour averaged threshold exposure limit for benzene is 10 ppm (mg/L) in air (NIOSH).

Boiling point distribution for several representative groups of petroleum products is shown in **Figure III–20**. The significance of the distribution is that only products with high vapor pressures will have a significant vapor phase in the open spaces of the vadose zone. In **Figure III–19** gasoline is clearly in a class by itself and is much more volatile than the kerosene derived products. It is not surprising that under most conditions volatilization (also called vapor extraction) is the method of choice for treatment of soils contaminated with gasoline, but not for soils contaminated with diesel or any of the higher molecular weight hydrocarbon products.

Table 3–3. Volatility, Flashpoint, and Flammability Data for Representative Petroleum Products. The data listed here and illustrated in Figure III–19 indicate the wide variation in volatility among hydrocarbon products. Products less volatile than diesel are essentially nonvolatile for remediation purposes. Products with flashpoints at room temperature or less are potentially explosive (Class A liquids, NFPA = 3). Hydrocarbons burn in air over limited ranges. LFL is the lower flammability limit, UFL is the upper flammability limit. Soils saturated with gasoline or JP–4 must by treated as a potential fire or explosive hazard as well as a potential health hazard.

Product	Volatility (at 70°F in psia)	Flashpoint		Flammability Limits % by vol.	
		in °C	in °F	LFL [5]	UFL
Gasoline [1]	4 – 8	-30 – -43	-36 – -45	1.4	7.6
Benzene	1.6	-11	12	1.3	7.9
Toluene	1.9	4	40	1.2	7.1
Ethylbenzene	2.2	18	68	—	—
Xylenes [2]	2	27	81	1.1	7.0
n-Hexane	1.5	-40	-40	1.2	7.1
JP–4 Jet Fuel	1.6	-10 – +35	-22 – +95	— [3]	— [3]
Diesel	0.009	40 – 65	100 – 130	1.3 [4]	6.0 [4]
Kerosene	0.011	40 – 75	100 – 160	1.4	6.0
Light Fuel Oil #1 and #2	<10-3	40 – 100	100 – 200	— [4]	— [4]
Heavy Fuel Oil #4, #5, and #6	<10-3	65 – 130	140 – 270	1.0	5.0
Lubricating Oil	<10-3	150 – 225	300 – 450	— [4]	— [4]
Used Oil	<10-3	>100	>200	— [4]	— [4]

[1] Values vary slightly depending on grade.
[2] Value is for m–xylene.
[3] Similar to gasoline.
[4] Relatively nonflammable, NFPA = 2.
[5] LFL is Lower Flammability Limit; UFL is Upper Flammability Limit.

Table 3–4. Solubility and Viscosity Data for Representative Petroleum Products. Solubility and viscosity data are listed for several representative petroleum products. Solubilities are give in ppm, viscosities are in centistokes. Because petroleum products are mixtures (actually solutions, the distinction is not relevant here) and because gasoline is the most frequently encountered contaminant, data for some of the more hazardous individual components of gasoline are listed. The most soluble components are also the most toxic.

Product	Solubility in cold water [1] (at 20°C in ppm)	Viscosity (in centistokes)
Gasoline [2]	50 – 100	0.5 – 0.6
1-Pentene	150	n/a
Benzene	1,791	0.5
Toluene	515	0.5
Ethylbenzene	75	0.6
Xylenes [3]	150	0.6
n-Hexane	12	0.4
Cyclohexane	210	n/a
i-Octane	8 ppb	n/a
JP–4 Jet Fuel	<1	0.8 – 1.2
Kerosene	<1	1.5 – 2
Diesel	<1	2 – 4
Light Fuel Oil #1 and #2	<1	1.4 – 3.6
Heavy Fuel Oil #4, #5, and #6	<1	5.8 – 194
Lubricating Oil	<1 ppb	400 – 600
Used Oil	<1 ppb	40 – 600
Methanol [4]	>100,000	< 0.1

Data from Riddick, et al., 1986.

[1] 20°C = 70°F; ppm = mg(hydrocarbon)/L(water).

[2] Solubilities of individual components may be higher.

[3] Average data for a mixture of all isomers.

[4] Data for methanol added for comparison.

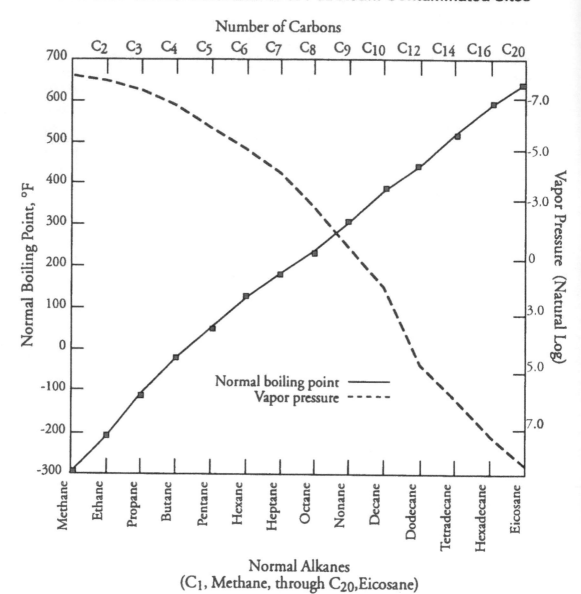

Figure III–19. Volatility of Selected Alkanes. The normal (at 1 atmosphere) boiling point (solid line) and vapor pressure (at 20°C, dotted line) for several representative hydrocarbons are shown. Boiling points increase as the vapor pressure decreases. A higher vapor pressure corresponds to higher volatility and indicates an increased tendency to exist in the vapor phase in porous soils. For compounds having molecular weights greater than decane, C_{10}, the vapor pressure is too low for significant vapors to exist at ambient temperatures.

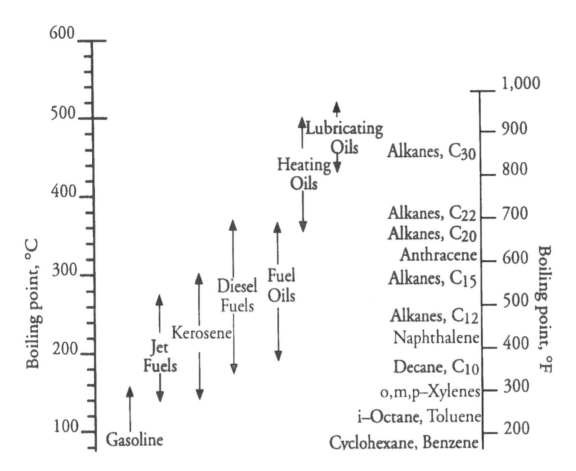

Figure III–20. Boiling Point Distribution of Petroleum Products. Boiling point ranges for representative petroleum products are shown. Since boiling points are inversely proportional to vapor pressures (volatility), the ranges also reflect relative volatilities. Gasoline is in a class by itself. Jet fuel, diesel, and other kerosene derivatives form a group; similarly lubricating and fuel oils have extremely low vapor pressures.

Solubility. From a chemical point of view petroleum hydrocarbons are virtually insoluble in water. Indeed, the *CRC Handbook of Chemistry and Physics*, Table of Physical Properties of Organic Compounds, notes that benzene is, "insol cold H_2O." A more detailed view finds that the maximum solubility of benzene in water is 1,750 µg/liter of water, or about 1,750 ppm (**Figure III–22**). Unfortunately, that amount is sufficient to be harmful to human health and therefore, is unacceptable in groundwater.

Solubility is approximately inversely proportional to molecular weight; lighter hydrocarbons are more soluble in water than higher molecular weight compounds. Lighter hydrocarbons, C_4 to C_8, including the aromatics are relatively soluble, up to about 2,000 ppm; heavier hydrocarbons are nearly insoluble. Benzene, for example, has a maximum solubility of about 100,000 ppm in water, whereas the maximum solubility for toluene is about 1,000 ppm. For comparison, the solubility of methanol is 1,000,000 ppm.

Using a solubility of 1 – 5 ppm (part per million, 1 – 5 mg/L) as a cutoff point, gasoline is the only petroleum product in common use that contains constituents that are soluble enough in water to cause health problems — and then it is only the aromatic compounds, benzene and alkyl benzenes, that are a concern. Solubilities of representative petroleum hydrocarbons are shown in **Table 3–4**.

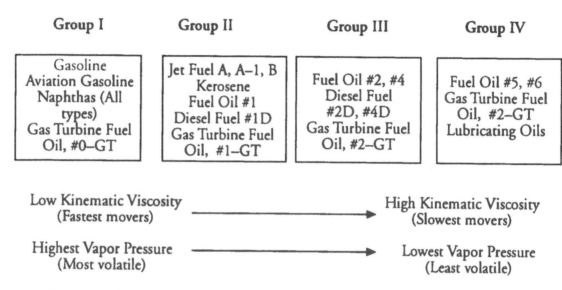

Group I	Group II	Group III	Group IV
Gasoline Aviation Gasoline Naphthas (All types) Gas Turbine Fuel Oil, #0–GT	Jet Fuel A, A–1, B Kerosene Fuel Oil #1 Diesel Fuel #1D Gas Turbine Fuel Oil, #1–GT	Fuel Oil #2, #4 Diesel Fuel #2D, #4D Gas Turbine Fuel Oil, #2–GT	Fuel Oil #5, #6 Gas Turbine Fuel Oil, #2–GT Lubricating Oils

Low Kinematic Viscosity (Fastest movers) ⟶ High Kinematic Viscosity (Slowest movers)

Highest Vapor Pressure (Most volatile) ⟶ Lowest Vapor Pressure (Least volatile)

Figure III–21. Kinematic Viscosity. Schematic representation of the relative kinematic viscosity of representative petroleum products. The higher the kinematic viscosity, the faster the product can be expected to move through soils. Only the products of Groups I and II migrate rapidly enough to be considered "free flowing." Group I products can migrate rapidly enough to warrant aggressive response. The products of Groups III and IV are essentially immobile in all soil types.

Kinematic Viscosity. Kinematic viscosity is a measure of the product's resistance to gravity flow; that is, a measure of the relative ease with which hydrocarbons flow through soils. Kinematic viscosity approximately parallels vapor pressure and is measured in centistokes, cST. Values vary from less than 1 cST for gasoline to a maximum of 638 cST for No. 4–GT gas turbine fuel oil. Viscosities for representative petroleum products and compounds are listed in **Table 3–4**. Petroleum products may be sorted by kinematic viscosity as shown in **Figure III–21**. Only gasoline has sufficiently low viscosity to migrate rapidly in most soils. Diesel and jet fuel will migrate with time, but more slowly than gasoline. The cutoff point for significant migration is about 2 – 3 centistokes. A fuel oil release can thus be treated much less aggressively than, say, a gasoline release because the fuel oil is not going to migrate away from the site.

The health and toxicological properties of petroleum hydrocarbons are well documented. Once again it is worth noting that, despite the fact that petroleum products are legally only regulated products, not hazardous, several petroleum products are, in fact, quite hazardous. The hazardous characteristics are often overlooked because petroleum hydrocarbons are legally regulated materials, not hazardous substances under RCRA.

Petroleum hydrocarbons are hazardous from three different perspectives:

1. Fire or explosive hazards— gasoline is a Class A flammable liquid;

2. Toxicity to human health;

3. Environmental damage, especially to drinking water.

Fire/Explosion Hazards. Only gasoline and the "middle distillates", the kerosene derived products, are acute fire hazards under normal conditions encountered during remedial activities, but, as noted previously, gasoline is the most commonly encountered contaminant. It is not unusual to find raw gasoline in the vadose zone or to excavate soil from which gasoline can be squeezed. It is possible to obtain vapor concentrations high enough to be a fire hazard during soil vacuuming operations.

Products having a high vapor pressure, such as gasoline, have the potential to create serious problems in subsurface structures such as basements, sewers, and parking structures. The vapor concentration required for fire or explosion is generally in the percent (parts per hundred) range relative to air. For example, the lower explosive limit for benzene is 13 mL of benzene per liter of air (1.3% by volume in air). Similarly for volatile hydrocarbons the concentration for acute inhalation toxicity is in the ppm range. For example, the workplace 8 hour averaged threshold exposure limit for benzene is 10 ppm (mg/L) in air (NIOSH).

When working with gasoline or any of the kerosene–derived, middle distillate products, precautions against fire hazards must be observed. Activities include tank removals, excavations, sampling, drilling, and equipment maintenance and monitoring.

For workers routinely exposed to hydrocarbon vapors the most common route of entry is inhalation. Long–term exposure to gasoline has been associated with chronic kidney disease and other effects. Therefore, workers involved in tank removals and remediation should have annual blood tests to monitor benzene levels.

Workers without respiratory gear in excavations heavily contaminated with gasoline must be observed continuously for signs of hydrocarbon intoxication: dizziness, loss of coordination, and unconsciousness. A lifeline should be attached upon entering the confined space. Self–contained breathing apparatus and a lifeline must be readily available and workers trained in their use.

> *Workers should never enter a gasoline contaminated excavation alone and unsupervised.*

Toxicity. Hydrocarbon toxicity can be either *acute* or *chronic.* Acute toxicity is defined as short–term, high–level exposure; chronic toxicity is defined as long–term, low–level exposure to a hazardous substance. Acute effects are usually apparent within minutes to hours or at most a few days. Chronic effects take years to appear and often are frequently very difficult to correlate with the original exposure.

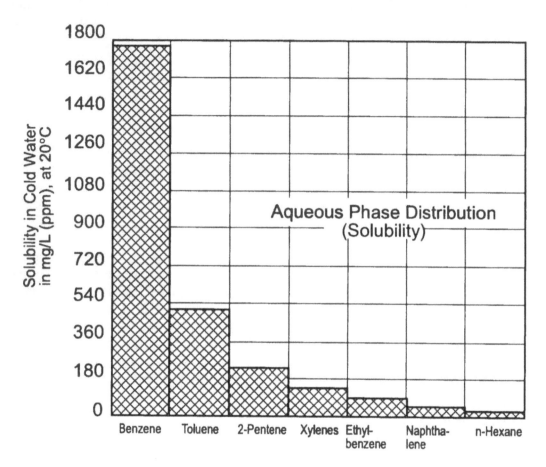

Figure III–22. Solubilities of Selected Hydrocarbons. Most hydrocarbons are not very soluble in water. In fact most petroleum hydrocarbons are almost entirely insoluble in water. Only the gasoline constituents are significantly soluble — that is, soluble enough to be significant from a regulatory standpoint. The most significant solubilities are those listed for the aromatic compounds. Solubilities are listed in Table 3–4. (Adapted from Riddick, et al., 1986)

The effects and symptoms of hydrocarbon exposure are difficult to quantify on a clinical basis. First, there is no sharp boundary between acute and chronic effects. "Short–term and "long–term" are somewhat subjective. Second, individual sensitivity to hydrocarbons varies greatly. Third, it can be difficult to distinguish the symptoms of hydrocarbon exposure from other non–hydrocarbon related symptoms. Therefore, it is particularly important to monitor workers routinely exposed to hydrocarbons. Monitoring may consist of periodic medical checkups for adverse health effects or simply estimating weekly exposure values.

Health effects of petroleum hydrocarbons are summarized in **Table 3–5**. The column, *Potential Health Effects*, is intended to be a summary of symptoms and effects, not a list of diagnostic features. Many of the symptoms listed are, by themselves, indistinguishable from the symptoms of stress, flu, colds, and other ailments, including hangovers. The symptoms must be evaluated in the context of a worker's activities, exposure times, regularity of exposure, and other activities that might have possible health effects. If adverse health effects due to exposure to petroleum hydrocarbons are suspected, the worker should see a physician with experience in environmental health immediately.

The lighter hydrocarbons, including benzene and the alkylbenzenes, are central nervous system, CNS, depressants at relatively moderate concentrations. Cyclopropane was once used as an operating room anesthetic. It was never used widely due to the obvious risk of fire or explosion. General symptoms of CNS depression include loss of coordination, drowsiness, vomiting, and dizziness. More severe short–term effects include unconsciousness, hemorrhaging of lung and intestinal organs, and in rare cases, death due to circulatory failure. Some components of gasoline, especially benzene, are associated with acute (short–term) and chronic (long–term) problems and are discussed separately below.

Hydrocarbons are good solvents for fatty, non–aqueous substances which makes them useful additives in commercial products, such as degreasers. Unfortunately, fatty, non–aqueous substances also include the lipids that make up the cell membranes in tissues. Cells of the skin are well-protected by an impervious layer of keratin, but epithelial tissues such as the lining of the throat and lungs are much more sensitive. That is why inhalation of hydrocarbon vapors can cause damage to lung tissue. And that is the reason why commercial and household products containing petroleum distillates carry a warning,

> "If ingested, do NOT induce vomiting!"

The stomach lining is relatively impervious to hydrocarbons. If ingesting heavier hydrocarbons resulted in metabolic toxicity, few children would have survived the doses of mineral oil administered for years by well–meaning mothers. Therefore, ingesting hydrocarbons, even gasoline, usually results in little absorption. However, vomiting increases the opportunity to introduce hydrocarbons and hydrocarbon vapors into the lungs. Lung tissue is not resistant to the solvent effect (defatting) of hydrocarbons and lung damage results.

A major difficulty with diagnosing acute health problems associated with hydrocarbon exposure is that many of the symptoms are essentially the same as those of stress or other relatively minor ailments: loss of coordination, dizziness, nausea, vomiting, diarrhea, irritability, loss of appetite are all fairly common symptoms that may reflect health effects unrelated to hydrocarbon exposure. Therefore, it is essential that workers be carefully monitored for exposure levels, and have periodic health checkups.

Aromatic compounds. *Benzene* and its alkylbenzene relatives are generally considered to be the most toxic components of gasoline (except for methanol, which is a non–hydrocarbon additive). These compounds, which may be present at concentrations as high as 50% by weight in premium unleaded gasolines, exhibit both acute (prompt) and chronic (long-term) toxicity in humans. The principal difficulty with benzene is that the small, flat ring is very difficult to oxidize biochemically.

All foreign molecules that are ingested or inhaled go through the liver where an enzyme system attempts to oxidize the molecules to more water–soluble forms. More highly oxidized molecules are generally more soluble in the blood, which is essentially just an aqueous solution. The more water–soluble a molecule is, the more readily it (or metabolic by-products) enters the blood stream where it is eventually filtered by the kidneys and excreted with urine. This is why urine tests are used in sports. Benzene does not combine readily with the liver enzymes. This means that the liver cannot easily rid itself of excess benzene, which is stored in fatty tissues. The resistance to oxidation may account for the ability of benzene to cause liver damage over a period of time.

Acute toxic effects include typical symptoms of CNS depression resulting from hydrocarbon inhalation: headache, nausea, and light-headedness. Death has resulted from a single 5– to 10–minute exposure to airborne benzene at a concentration of 20,000 ppm. Concentrations of 3,000 to 7,500 ppm may result in toxic symptoms within one hour. Exposures to 50 to 250 ppm can result in headache and dizziness; recovery is rapid on exposure to fresh air. Concentrations of 25 ppm have no reported effects.

Evidence for skin absorption is readily apparent to anyone who has ever spilled gasoline on his or her hands; the odor may linger for some time, even after washing due to absorption. In elevated concentrations direct contact of benzene with skin may cause redness and dermatitis. Skin absorption for short periods of time is not considered a major route of entry.

Chronic effects of benzene exposure over long periods of time include bone marrow damage and cancer. Early symptoms of chronic benzene exposure are reversible leukopenia (decrease in white blood cells), anemia (decrease in red blood cells), or thrombocytopenia (decrease in platelet count). Lowest air levels capable of producing these effects are 40 to 50 ppm over a period of years.

The most important chronic effect of benzene exposure is destruction of bone marrow cells. With continued exposure there is severe bone marrow damage which results in pancytopenia, a deficiency of all elements of the blood and results in increased susceptibility to infection and hemorrhagic conditions. Eventually, continued exposure leads to the development of a distinctive type of leukemia called acute myelogenous leukemia.

OSHA has set definite limits on workplace benzene exposure; the time weighted average (skin) exposure for a 40–hour workweek for benzene is 10 ppm (NIOSH). Workers who are exposed to hydrocarbon vapors on a routine basis must be monitored for benzene exposure. For UST operations this includes tanker operators, tank removers, reliners, tank and line tightness testers, inspectors, and maintenance workers.

Toluene makes up approximately 4 to 7% of gasoline. The primary hazard associated with acute inhalation exposure to high levels of toluene is CNS depression. It appears that toluene is much less toxic than benzene because liver enzymes can oxidize toluene more efficiently. The extra methyl group provides a "hook" whereby toluene is oxidized to benzoate. Controlled exposure of human subjects to 200 ppm for 8 hours produced mild fatigue, weakness, confusion, lacrimation, and tingling of the skin. At 600 ppm additional effects, including euphoria, headache, dizziness, dilated pupils, and nausea became evident. At 800 ppm the symptoms were more severe and required several days to disappear. Impaired reaction times have been observed in humans after 20 minutes exposure at 300 ppm and after 7 hours at 200 ppm. Eye irritation is noticeable at vapor levels of 300 to 400 ppm.

Xylenes make up about 6 to 8% by weight of gasolines. Data on the effects of long-term human exposure to xylenes are primarily high level occupational inhalation exposures in industrial settings. The primary effects are similar to the effects of exposure to other alkylbenzenes: CNS depression, loss of coordination, nausea, vomiting, and abdominal pain. Xylenes have a narcotic effect on the CNS; exposures at 90 ppm cause loss of reaction time. Xylenes have variable effects on the liver and kidneys and an irritant effect on the gastrointestinal tract. Both liquid and vapor are irritating to the skin and eyes.

The chronic effects of xylene exposure resemble those from acute exposure, but are more severe. Headache, irritability, fatigue, digestive disorders, and sleep disorders have been reported. Inhalation of high concentrations may cause tremors, impaired memory, weakness, vertigo, headache, and anorexia.

Ethylbenzene irritates the skin, eyes, and upper respiratory tract. Systemic absorption causes CNS depression. Limited data are available on human exposure levels. At 200 ppm the vapor irritates the eyes; at 2,000 ppm, eye irritation and lacrimation are immediate.

It is worth noting once again that the Department of Labor and the Occupational Safety and Health Administration have determined that petroleum products *are* hazardous with respect to health and exposure regardless of the CERCLA exclusion.

Alkanes. Aliphatic hydrocarbons in general, and alkanes in particular, have a narcotic or depressant effect on the human nervous system. Propane has been used as an operating room anesthetic. Absorption may be through the skin or lungs, but is much more rapid through the lungs. Symptoms of aliphatic hydrocarbon toxicity are similar to those of the aromatics; dizziness, nausea, and loss of coordination. The symptoms are reversible on exposure to fresh air. Alkanes are good solvents for lipids and can dissolve tissue lipids and cell membranes if aspirated into the lungs.

n–Hexane is the most toxic of the alkanes. It comprises 11 to 13% of gasoline by weight. Most exposure data on humans are derived from workplace exposure to solvent vapors and may not be directly comparable with exposure to contaminated soil or water.

Acute exposure occurs through inhalation. Vertigo, headaches and nausea are the first symptoms of exposure to be noticed. At high concentrations central nervous system depression results in a narcotic–like state. Pre–narcotic effects occur at vapor concentrations of 1,500 to 2,500 ppm.

Table 3–5. Health Effects of Hydrocarbons. A summary of the toxicity and health effects of representative hydrocarbons. Exposure limits are average values for threshold exposure (see note 1 below).

Product	Compound	TLV[1]	Potential Health Effects
Gasoline	Benzene	10	Nausea, vomiting, known human carcinogen
	Toluene	100	CNS depression, fatigue, weakness, eye irritant
	Xylenes	100	CNS depression, nausea, vomiting, skin irritant
	Ethylbenzene	100	CNS depression, nausea, vomiting, liver/kidney damage
	n–Hexane	50	Nausea, dizziness, vomiting, severe but reversible paralysis
	Other hexane isomers	500	Skin, mucous membrane irritant
	Octane	300	Weakness, fatigue, headache, nausea, vomiting, anorexia, diarrhea
	Ethylene dibromide, EDB	0[6]	Probable human carcinogens
	Ethylene dichloride, EDC[2]	10	
	Tetraethyllead, TEL	0.1[4]	Weakness, fatigue, headache, nausea,
Middle	Tetramethyllead, TML [3]	0.15[4]	vomiting, anorexia, diarrhea
Distillates	PAHs		
	Naphthalene	10	Weakness, tremors, dizziness, vomiting
	Benzo(a)anthracene	0	Probable human carcinogens
	Benzo(a)pyrene	0	Probable human carcinogens
	Cresols and Phenols	5	Skin, eye, mucous membrane irritant
	N,N–dimethylformamide	10	Skin, eye, mucous membrane irritant
	Manganese compounds[5]	0.1	CNS affected, non–specific
Residual			
Fuels	PAHs		
	Benzo(a)anthracene	0	Probable human carcinogens
	Benzo(a)pyrene	0	Probable human carcinogens
	Crysene	0	Possible human carcinogens

[1] Threshold Limit Values; Values are time–weighted average exposures in ppm for vapors. STELs, Short Term Exposure Limits, are usually higher.

[2] Additives in leaded regular gasolines.

[3] Legal limit for lead in gasoline is 13 mg/liter of gasoline.

[4] In mg/m^3 for skin absorption.

[5] For the compound manganese cyclopentadienyl tricarbonyl which has been used as an additive in regular gasolines.

[6] A value of zero means there is no acceptable exposure level.

Chronic exposure to n–hexane vapors causes nerve damage. The first clinical sign of nerve damage is a feeling of numbness in the fingers and toes. Further exposure results in increased numbness and loss of muscular stretching reflexes. Reversible paralysis develops with varying degrees of impairment. Recovery begins 6 to 12 months after exposure ceases. The most severe cases take years to recover.

Octane, if inhaled into the lungs, can cause rapid death due to cardiac arrest respiratory paralysis, and asphyxia. It has a narcotic effect similar to hexane. Prolonged skin exposure causes burning and blistering.

Isopentane is a CNS depressant. Acute toxicity effects are similar to the effects of alcohol intoxication: exhilaration, dizziness, headache, nausea, confusion, and loss of coordination. Unconsciousness results from prolonged exposure.

Other Additives

Lead is added to some gasolines to improve octane number in the form of two compounds: tetramethyllead, TML, $(CH_3)_4Pb$, and tetraethyllead, TEL, $(CH_3CH_2)_4Pb$. Two more compounds, ethylene dichloride, EDC, $C_2H_4Cl_2$, and ethylene dibromide, EDB, $C_2H_4Br_2$, are added to scavenge lead to prevent buildup of deposits on valves and cylinder walls.

TEL is absorbed through the skin, inhaled as a vapor, or ingested. Exposure levels of 100 mg/m^3 for 1 hour causes acute symptoms. Lower levels of exposure require longer times to produce symptoms. TEL exposure symptoms are often vague and difficult to diagnose. Mild exposure leads to weakness, fatigue, headache, nausea, vomiting, and diarrhea — not dissimilar to the results of writing a book about petroleum contaminated soils. Prolonged exposure leads to confusion, delirium, manic excitement, and catatonia. Loss of consciousness and death follow.

EDB has been identified as an experimental animal carcinogen; EDC has similar properties, although its potency is much lower. Acute exposure symptoms to EDB include vomiting, diarrhea, abdominal pain and CNS depression.

MTBE, methyl tertiary–butyl ether added to so–called oxygenated fuels, so far appears to have no adverse human health effects. There is little reason to assume that this situation will change since similar ethers have been in commercial and industrial use for many years. Pure MTBE is highly flammable and should be treated appropriately. MTBE is volatile and very soluble in water. During soil venting operations MTBE is recovered first. It is also useful through soil gas surveys as the first indication of a plume.

Middle Distillates

The middle distillates include kerosene, aviation fuels, diesel fuels, and fuel oil # 1 and 2. These fuels contain paraffins (alkanes), cycloparaffins (cycloalkanes), aromatics, and olefins from approximately C_9 to C_{20}. Aromatic compounds of concern include alkylbenzenes, toluene, naphthalenes, and PAHs. Compositions range from avgas and

JP–4, which are similar to gasoline (see Appendix A), to Jet A and JP–8, which are kerosene–based fuels.

JP–4 and JP–5 are volatile, complex mixtures of aliphatic and aromatic hydrocarbons and are principally used in military aircraft. The volatility means that inhalation exposure is a potential problem near fueling facilities, either from spills or leaks. Once the soil has become saturated, remedial activities create both fire and inhalation hazards. Toxic effects are similar to those described for gasoline.

Kerosene has the least amount of aromatic hydrocarbons with alkylbenzenes, indanes, and naphthalenes being the major aromatic components. Kerosene is irritating to the skin and mucous membranes.

Diesel fuels include high percentages of PAHs: naphthalenes, acenaphthalenes, anthracenes, and phenanthrenes. Dermal (skin) exposure to diesel oil is toxic to the kidneys.

Fuel Oil No. 2 (heating oil) contains a higher percentage by volume of benzenes and naphthalenes relative to kerosene and diesel fuels. Most middle distillates contain some benzene, alkylbenzenes, toluene, ethylbenzene, xylenes, and cumenes, but in much lower percentage than in gasoline.

All these products have similar routes of entry: ingestion, absorption, and inhalation. Like the lighter hydrocarbons, inhalation exposure produces the most serious effects. High levels of vapors are irritating to eyes, mucous membranes, and lung tissue. The greatest risk of exposure is to people working in fueling operations or handling contaminated soils from remedial activities.

Acute toxic effects are similar to the effects of exposure to aliphatic hydrocarbons: nausea, headache, confusion, and irritation of the respiratory system. Vapors are irritating to eyes, mucous membranes, and lung tissue and can have a defatting effect on skin due to a solvent effect.

Chronic effects are mainly due to exposure to aromatic compounds, which are found primarily in JP–4 and JP–5, used mainly in military aircraft, and in Fuel Oil #2.

4 Soils and Subsurface Characteristics

One of the more important physical parameters required in the planning phase of a remediation is the nature and type of the soil or soils in which the contamination is located. An open porous soil will offer more options for remediation, whereas a tight, compact soil will limit the available options. On the other hand, an open, porous soil may also require a much more aggressive remedial action since the contamination will migrate more rapidly, while a contaminated clay allows a more deliberate approach. During the planning phase as part of the initial site investigation, it is essential to determine the type of soils and subsurface structures that are present. The information is necessary in order to predict the phases and probable distribution of contaminants.

In this chapter we will look at soil types, classification, and subsurface characteristics from the point of view of remedial choices and planning. Because there are infinitely varied combinations of subsurface conditions, it is impossible to discuss all possibilities. However, we can look at some of the more common structures.

In all of the subsequent discussion in this chapter we assume that groundwater is a factor to be considered. That is, groundwater lies reasonably close to the surface — an arbitrary depth is less that 50 feet. If groundwater is hundreds of feet deep, it is probably not worth worrying about. Groundwater presents several difficulties that must be considered during the planning phase. One difficulty arises when a potable aquifer lies beneath an impervious stratum. It is essential to be aware of the depth to stratum to avoid drilling through to contaminate otherwise pristine groundwater (see, for example, Hoffman and Bell, 1993). A second difficulty occurs with highly fractured bedrock, which provides a virtually unmanageable pathway for migration.

There is a distinct slant toward gasoline as the contaminant. Gasoline accounts for approximately 90% of all remedial sites and it is the most volatile and most mobile of all contaminants. In most circumstances any product heavier than the kerosene derivatives, diesel or jet fuel, will not migrate rapidly enough to create a problem, at least in the short term. Diesel and jet fuel are intermediate in volatility and mobility. Diesel and jet fuel contamination accounts for many of the very large sites, railroad fueling yards and commercial and military airports. Heavier hydrocarbons are found as contaminants at sites such as heavy industrial plants (hydraulic oils, solvents, and lubricants) and refineries.

Soil Characteristics

Soils that are contaminated by petroleum products include the entire range of soil types from sands, glacial tills, and gravels, to sediments, silts, and clays. Contaminated soil types also include artificial fills, tailings and even abandoned landfills. Two physical properties that may affect fluid and contaminant movement through the subsurface environment are:

- porosity, which is the ratio of column of void (pore) space to the total volume of material;

- and permeability, which is a measure of a soil's ability to transmit fluids.

Another characteristic important to residual, adsorbed contamination and to biodegradation processes is particle size as measured by grain surface area. First, the greater the surface area per gram of soil particles, the greater the amount of adsorbed hydrocarbons. Second, since most microbial oxidations occur at the soil–water interface on the particle surface, microbial degradations are more efficient on more finely divided particles. Adsorbed hydrocarbons are difficult to remove and are the reason that soil remediation guidelines are flexible. Residual contamination remaining after closure is in the form of adsorbed hydrocarbons.

Characteristics of soil water and degree of saturation are also important. Shallow aquifers are frequently briny and non–potable. Therefore, contamination in a non-potable aquifer requires a less aggressive approach than contamination in an aquifer used or potentially used for drinking water. Very dry soils are unsuitable for bioremediation since microorganisms cannot survive without water.

Soil pH actually refers to the relative acidity of soil water. The aqueous layer on the surface of soil particles is the "life zone" in soils; the region of most abundant microorganisms. Soil pH should not deviate from 7 by more than 1 pH unit for optimum bacterial activity. As a broad generalization, western soils tend to be mildly alkaline; eastern soils tend to be mildly to moderately acidic.

Before a petroleum release occurs, air and water are distributed throughout the interstitial spaces of soil zones in a ratio roughly proportional to depth — saturation increases as depth increases. The distribution of air and water in the subsurface is shown schematically in **Figure IV**–1. After a release occurs petroleum hydrocarbons compete with air and water for pore spaces. Hydrocarbon vapors will readily displace air but bulk liquid hydrocarbons compete poorly with water for available interstitial space.

Immediately below the organic horizon, the upper surface of the soil, there is an unsaturated or "vadose" zone. Interstitial spaces in this zone are partially filled with water (pore water) and partially filled with air. Depending on soil type, this zone is moderately well aerated and can support abundant microbial life.

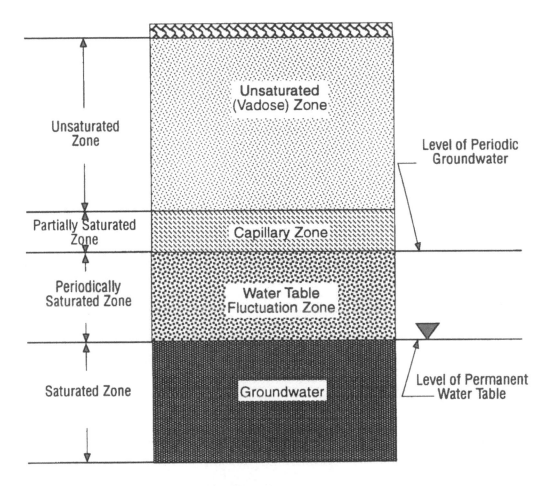

Figure IV-1. Generalized Soil Column. The figure shows a generalized, but nonetheless useful, soil column. The upper layer is the organic horizon, a relatively thin layer no more than a few inches to a few feet thick. The next layer is the unsaturated or vadose zone. This zone includes everything above the highest level of groundwater. Next is the capillary zone, followed by the water table fluctuation zone. The lowest region is permanently saturated, and the top of this level is designated as the water table, the highest level of permanent groundwater. The vadose zone is normally unsaturated, except for brief periods of heavy runoff or snow melt. The vadose zone may be a few inches to thousands of feet thick. The capillary fringe has bulk liquid water as the major phase and the fluctuation zone is saturated periodically as the water table rises and falls on a seasonal or occasional basis. In these intermediate zones hydrocarbon mobility decreases as saturation increases, since hydrocarbons compete poorly with water for available pore spaces. Fluctuation in groundwater levels contributes to "smearing" of hydrocarbons. The saturated zone is the highest level of permanent groundwater. Only dissolved (soluble) hydrocarbons appear in this zone.

Below the vadose zone the soil becomes increasingly saturated and anaerobic. Saturation increases until it reaches 100%, which defines the water table. The intermediate zones that underlie the vadose zone have pore spaces partially to almost completely saturated with water. Because the degree of saturation may vary seasonally and is derived from different mechanisms, it is customary to separate the intermediate region into two zones.

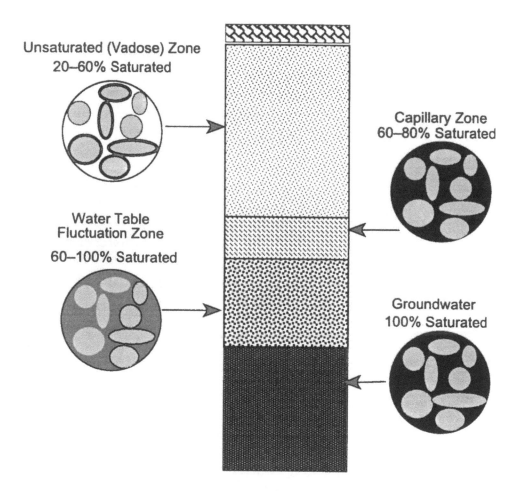

Figure IV-2. Generalized Soil Column in Microview. A microscopic view of soil particles and phases indicates that soil moisture in the vadose zone is largely confined to an aqueous layer surrounding the particles. The interstitial pore spaces in this zone are filled with air. The aqueous layer and the vapor spaces are important since the majority of microbial action is carried out in the region. The capillary zone is partially saturated with bulk water mostly on the surface of soil particles. The fluctuation zone is mostly saturated with bulk liquid phase interstitial water. There are no interstitial vapor spaces left in the saturated zone. Consequently this region is largely anaerobic.

Immediately below the vadose zone is a *capillary* zone. The name derives from the fact that bulk water in interstitial spaces between soil particles is drawn upward by capillary action. Water molecules are attracted to the surfaces of soil particles the same way as water is attracted to a glass surface. The surface tension acts to draw water up into normally unsaturated interstitial spaces and increases the degree of saturation in soils above the water table.

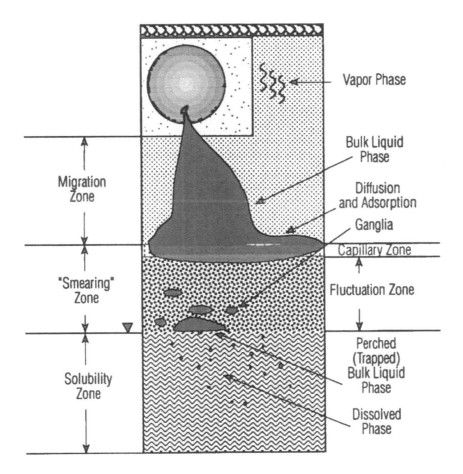

Figure IV-3. Distribution of Hydrocarbon Phases in Soils. Petroleum hydrocarbons tend to partition among soil particles, vapor, the aqueous phase, and a bulk hydrocarbon phase. In the unsaturated or vadose zone the open pore spaces among soil particles allow volatile contaminants to vaporize. Vapors will migrate through a loose, porous soil, but remain trapped in a tighter, more dense soil. In the saturated zone water is the primary bulk phase and contamination is normally limited to dissolved hydrocarbons or to trapped, dispersed bulk hydrocarbons. In the intermediate zones hydrocarbons can migrate during dryer periods when the zone is drained and less than 100% saturated. During periods when the water table rises hydrocarbons become trapped since water migrates much faster than bulk liquid phase hydrocarbons.

The next lower zone is the *fluctuation* zone. Typically the highest level of saturation, the water table, is not a static boundary, but fluctuates in cyclical manner as the water table rises and falls. Many coastal areas experience groundwater fluctuations related to tidal movements. In other areas the water table rises during a rainy season or as snow melt arrives in the early summer. The fluctuation zone becomes virtually saturated as groundwater rises during periods of heavy surface runoff.

There are two considerations for Light Non–Aqueous Phase Liquids, or LNAPL contamination in the intermediate zones. First, the mobility decreases as saturation increases. Hydrocarbons are competing with water for available interstitial space and water is much more mobile through pore spaces.

Second, transfer of contaminant molecules from bulk hydrocarbon phases to the aqueous phase increases. The reason for this latter observation is the following: As the water table rises, bulk liquid hydrocarbons are more viscous and migrate much less readily than water. The result is to trap bulk liquid phase hydrocarbons within a bulk liquid aqueous phase. As the saturated level retreats, liquid hydrocarbons are extracted into the aqueous phase. The net result is an increase in dissolved phase hydrocarbons and an increase in contaminant levels in the groundwater. The other effect, that of "smearing," is discussed below.

Hydrocarbon Phases

Subsurface petroleum contamination is distributed in a complex mixture consisting of different physical states. The complicating factor is the insolubility of petroleum. In contrast methanol, for example, is 100% miscible with water, and simply dissolves in groundwater. The resulting contamination may be more difficult to remediate, but the concept is relatively simple.

Hydrocarbons, on the other hand, can be present in the subsurface in any of the following phases:

- The vapor phase;

- A bulk liquid phase;

- Adsorbed phases;

  and

- An aqueous or dissolved phase.

In the vapor phase hydrocarbon vapor exists in the vadose zone in interstitial spaces that are not already occupied by water or bulk liquid hydrocarbons.

Hydrocarbon contaminants move downward by force of gravity; some horizontal spreading will also occur because of capillary forces between the migrating liquid and soil particles. In the dissolved phase soluble hydrocarbons are transferred to water until an equilibrium is achieved. Transport of the dissolved phase depends on groundwater velocity and migration rates. Migration and spread of dissolved phase hydrocarbons depend on relative adsorption of the hydrocarbon to soil particles, rates of volatilization to the vadose zone, rates of biodegradation by bacteria. **Figure IV–3** illustrates the vertical distribution and degrees of mobility of hydrocarbon phases in soils. **Figure IV–4** shows a similar scheme but from a microscopic view. This view is more important for remediation purposes.

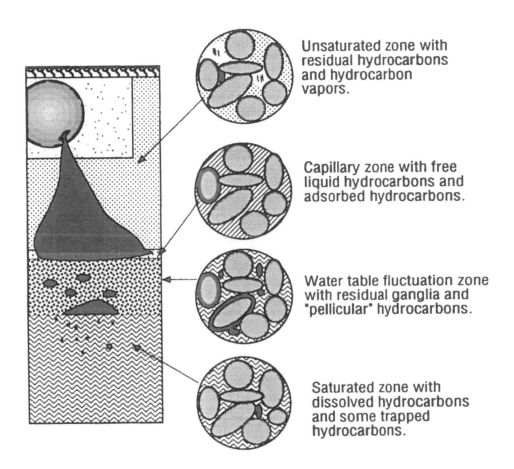

Unsaturated zone with residual hydrocarbons and hydrocarbon vapors.

Capillary zone with free liquid hydrocarbons and adsorbed hydrocarbons.

Water table fluctuation zone with residual ganglia and "pellicular" hydrocarbons.

Saturated zone with dissolved hydrocarbons and some trapped hydrocarbons.

Figure IV–4. Microview of Petroleum Contaminated Soils. Expanded view of contaminant phase distribution in petroleum contaminated soils. The vadose zone contains partially saturated soils in which the aqueous layer is distributed over the surface of the particles and open "pore spaces" containing vapor phase hydrocarbons. An appreciable distribution in the vapor phase is limited to gasoline hydrocarbons. The plume migrates relatively freely through the vadose zone, but, because migration occurs through interstitial pore space, it cannot compete with water in more saturated zones. Therefore, mobility decreases. As groundwater levels fluctuate, bulk–phase hydrocarbons, ganglia, are "smeared" across soil particles, and more soluble compounds are extracted into the aqueous phase. The extraction process is facilitated by the increased surface area created as the ganglia are smeared across the soil particles. As the molecular weight of contaminant hydrocarbons increases, mobility and solubility decrease.

Vertical movement of water through periodically saturated soils extracts adsorbed hydrocarbons and hydrocarbons trapped in dispersed bulk phases. **Figure IV–5** illustrates the effect of saturating and draining contaminated soils in which the hydrocarbons become *pellicular,* or adsorbed onto soil particles. Adsorption is more of a factor in silts and clays than in sands and gravels.

The bulk hydrocarbon phase therefore becomes trapped by the rising groundwater. As the water level rises, liquid phase hydrocarbon residuals tend to break up into globules sometimes called "ganglia," which are persistent, long term sources of hydrocarbon contamination, both in soils and in groundwater. (Mackay and Cherry, 1989, Mackay, et al. 1985). Fluctuations in water table and mixing between the two bulk phases accelerates movement of hydrocarbons into the aqueous phase.

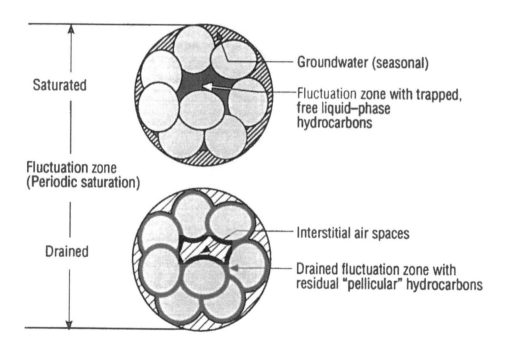

Figure IV–5. Seasonally Saturated Soils. A portion of the bulk phase hydrocarbons become "pellicular," or adsorbed onto soil particles. Pellicular hydrocarbons can be difficult to remove from soils and account for the fact that following remediation efforts, benzene levels can return to high levels after a period of time. (Adapted from Friesen, 1990.)

A common observation in vapor extraction and groundwater pump and treat systems is an initial decrease in hydrocarbon levels, especially benzene levels, as measured by BTEX analyses. After a period of time in which water levels rise and fall, benzene levels are observed to rise again as residual hydrocarbons are flushed from soil particles. This is one of the reasons in situ treatment methods must run for long periods of time.

From Table 4.1 petroleum constituents can be arranged into the following groups:
- Those that adsorb strongly onto soil particles;
- Those that volatilize readily;
- Those that migrate readily into groundwater;
- Those for which multiple pathways exist.

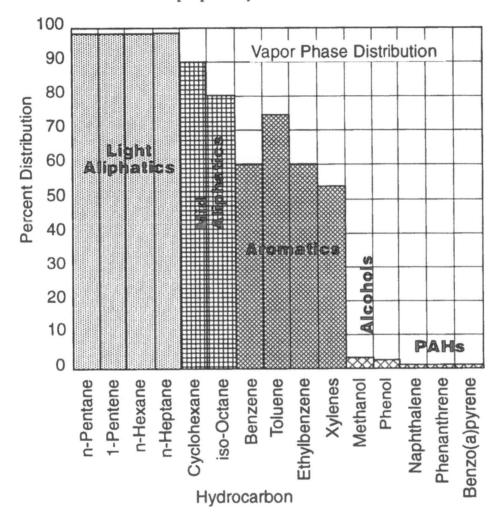

Figure IV-6. Importance of Volatilization as a Migration Pathway. Volatilization is an important migration pathway for lighter aliphatic hydrocarbons. It is moderately important for aromatic compounds and not important at all for PAHs. Data for PAHs are representative of all heavier hydrocarbons in the middle distillate range. Methanol and phenol are alcohols, not pure hydrocarbons. They are included for comparison and since they are found in some gasoline blends.

   The actual distribution and mobility of hydrocarbons in the soil depends heavily on the nature of the particular hydrocarbon and, hence, on the nature of the original product (Friesen, 1990). Organic compounds typically found in petroleum can be divided into four groups according to their ability to partition through a soil column. The results are shown in Table 4–2 and in Figure IV–6.

The implications of these data are that before remediation proceeds, careful attention should be paid to the nature of the soil layer or layers in which the contamination resides and the type of product which was released. Thus, soil venting is not the first choice for recovering a diesel or fuel oil spill or release. On the other hand, soil venting may be an excellent choice for remediating a gasoline release (McKee, et al., 1972). Similarly, in situ bioremediation is an unlikely choice in more compact soils since necessary nutrients must be supplied through the soil.

Less obvious is the presence of lenses and corridors. Clay deposits can radically alter the course of migration of vapors and liquids and thus reduce the effectiveness of remedial methods that rely on contaminant mobility. Likewise, the presence of ancient or artificially filled streams, lakebeds, and other drainages can change the direction and speed of migration.

Because of this complexity the term "radius of influence" as applied to extraction wells is a misnomer because it implies that there is a smooth spherical volume from which air or water is extracted. Likewise, injection wells rarely have a sphere of constant radius into which water or air is injected.

Soil Permeability

Migration of contaminants also depends on the type and porosity of the soil. A measure of porosity is coefficient of permeability, measured in cm/sec. As product leaves a leaking tank, the backfill is essentially transparent unless the tank was installed prior to 1988 and has native soil as backfill. Movement through native soil depends on the coefficient of permeability. A factor of 10^{-3} cm/sec translates as nearly 3 feet per day.

The viscosity of the product is also important. Migration rates used in this chapter are for gasoline, the most common contaminant and the least viscous. Petroleum products heavier than diesel have very low flow rates due to the high viscosity. Table 4–3 gives an approximate correlation between hydraulic conductivity and soil types.

Table 4–1. (Facing page) Distribution of Subsurface Contaminants. Hydrocarbon contaminants partition among several phases depending on the age and location of the plume. The principal phases are vapor, bulk liquid phase, adsorbed to soil particles, and dissolved in groundwater. Data are listed for representative constituents of gasoline. The columns give the average percent distribution for each compound across three phases. Thus, benzene can be expected to partition approximately 3% adsorbed onto soil particles, 65% in the vapor phase, and 35% dissolved in groundwater. Numbers will vary according to specific conditions, especially soil porosity.

Compound	Adsorbed to Soil Particles	Volatilization	Soluble in Groundwater
Aromatics			
Benzene	5	60	35
Toluene	5	75	20
Ethylbenzene	20	60	20
Xylenes [1]	15	55	30
Aliphatics			
n–Pentane	<1	95	4
iso–Pentane	<1	95	4
n-Hexane	<1	95	4
n–Heptane	<1	95	4
Cyclohexane	1	90	4
PAHs			
Naphthalene [2]	60	10	30
Benzo(a)pyrene	100	0	0
Anthracene	100	0	0
Phenanthrene	90	3	7
Alcohols			
Methanol	0	2	98
Ethanol	0	2	98
Phenol	9	>1	90

[1] Value is for o–xylene.
[2] Naphthalene is chemically an aromatic compound more similar to benzene than pyrene or phenanthrene. It is included with PAHs since it is found in diesel fuels, kerosene, and other middle distillates.

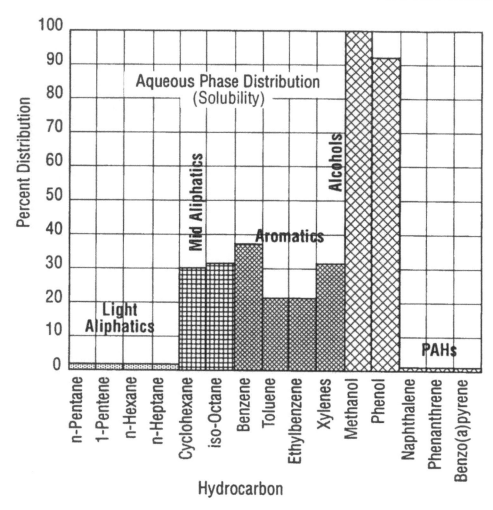

Figure IV–7. Distribution of Contaminant Phases for Typical Soils. The chart shows the distribution of representative petroleum hydrocarbons, particularly those commonly found in gasolines, in the aqueous phase. The percent of a given contaminant that will be found dissolved in the aqueous phase, either as ground-water or as contaminated soil water. The distribution is based on Table 3–4. The lighter hydrocarbons show strong tendencies to vaporize and are less soluble than aromatics. More than 90% of 1–pentene is found in the vapor phase. Aromatic compounds are the most soluble of the hydrocarbon components of gasoline, with benzene being the most soluble of the aromatics. Benzene and the alkylbenzenes are about equally distributed across vapor, adsorbed, and aqueous phases. Mid–range alkanes and cycloalkanes and cycloalkenes are intermediate in solubility and are about equally distributed across vapor, adsorbed, and aqueous phases. PAHs and related compounds are strongly adsorbed to soil particles and relatively insoluble. Aromatics are also volatile and vaporize readily. While methanol and phenol are not pure hydrocarbons, they are included for comparison. Both are found in typical gasoline blends and are very soluble.

Petroleum hydrocarbons will normally partition between an immobile phase and a mobile phase. The mobile phase flows in the direction of groundwater flow. This is always true for LNAPLs. Just this one fact can save time, effort, and money relative to a remediation involving DNAPLs, which can actually flow upgradient if bedrock happens to slope in the right direction. DNAPLs flow in the direction of bedrock slope rather than with groundwater. Usually bedrock slope and groundwater flow are in the same direction, but not always. Because of this, if DNAPLs are involved, a substantial number of monitoring wells must be placed upgradient in order to completely characterize contaminant distribution. If only LNAPLs are involved, upgradient wells are necessary only to insure that contamination is coming from the primary source and not migrating from offsite.

Table 4–2. Phase Migration Routes. The most important migration phases for several representative hydrocarbons are summarized. The table is derived from data in Tables 3–4 and Figure IV–7. Few hydrocarbons are present exclusively in a single phase. Very light gasoline constituents, such as 1–pentene and iso–pentene, migrate mostly as vapor. Naphthalenic hydrocarbons and PAHs are strongly adsorbed to soil particles. Alcohols, such as methanol, ethanol, and iso–propanol (added as a water scavenger) are very soluble in water. Aromatic compounds are distributed across several phases.

Adsorbed	Volatile	Soluble	Mixed Phases
Naphthalene	n–Pentane	Methanol	Benzene
Benz(a)pyrene	iso–Pentane	Ethanol	Toluene
Anthracene	n-Hexane	Phenol	Ethylbenzene
Phenanthrene	n–Heptane		Xylenes
	Cyclohexane		

Hydrocarbon flow velocities are proportional to viscosities, particle density, and soil conductivity. Migrating hydrocarbons tend to partition between an immobile phase and a mobile phase. The mobile phase will contain those hydrocarbons with lower viscosities and higher volatility. The mobile phase consists of the lighter hydrocarbon components, including benzene, toluene, and the lighter aliphatic hydrocarbons, butanes, pentanes, and hexanes. The immobile phase will consist of hydrocarbons with higher viscosities and lower volatility. The immobile phase consists of less soluble components, mostly heavier aliphatics in the C_{10} and above range, and naphthalenic hydrocarbons (PAHs).

Table 4–3. Classification of Soil Types by Coefficient of Permeability. Soils vary widely in their ability to transmit contamination. Flow rates for common soil types are listed. Flow rates are for water or petroleum dissolved in water. Corresponding flow rates for hydrocarbons are proportional, but lower. A flow rate of 1,000 cm per day means the soil is essentially transparent to the contamination. A flow rate of 1 cm per week means the soil is essentially impervious to migration.

Flow rates in cm/sec	10^2	1	10-2	10-4	10-6	10-8
Flow rates in ft/day	30,000	3,000	30	0.3	3x10-3	3x10-6
Soil Type		Gravels	Sands	Silts	Clays	
	←				→	
		More permeable		Less permeable		

Contaminant Migration

Hydrocarbon contaminants migrate by a variety of routes, some of which may be unexpected as shown in **Figure IV–8**. As noted above, when a UST system is releasing product, the excavation backfill is sufficiently porous that it is essentially transparent to contaminant flow. The contaminant plume is drawn down by gravity into the native soil where it begins to spread horizontally due to capillary action. The dispersion of the hydrocarbon plume is similar to the dispersion of light passing from one medium to another.

Figure IV–8. (Facing page) Potential Routes of Migration of Hydrocarbon Contaminants. Hydrocarbons are spread through the subsurface by various routes of migration. Gravity tends to draw released hydrocarbons down through the vadose zone, which also diffuses and spreads the plume ①. Flow rates decrease through the capillary and fluctuation zones as saturation increases ②. Bulk hydrocarbons perch on the water table and may become trapped as seasonal fluctuations in groundwater occur ③. Seasonal fluctuations also lead to "smearing" as hydrocarbons are adsorbed onto soil particles ④. Fractured bedrock or outcroppings provide a path for contaminants to flow in unexpected directions ⑤. More soluble components, including aromatics dissolve in and flow with groundwater ⑥. Poorly-planned well borings that penetrate otherwise impervious subsurface layers can provide a path between aquifers ⑦. A major route in urban areas is utility corridors which can spread contamination over long distances ⑧.

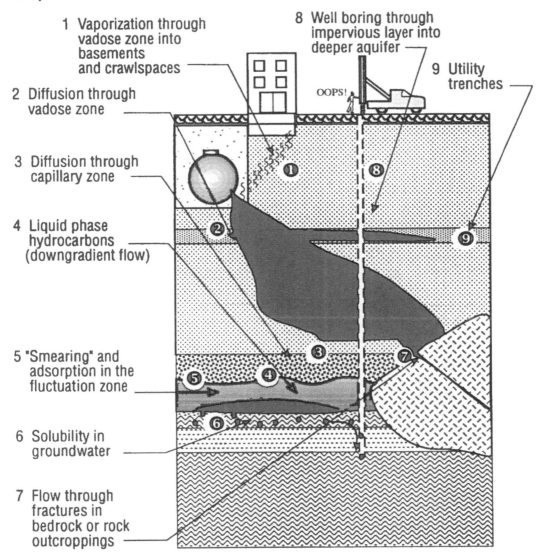

1 Vaporization through vadose zone into basements and crawlspaces

2 Diffusion through vadose zone

3 Diffusion through capillary zone

4 Liquid phase hydrocarbons (downgradient flow)

5 "Smearing" and adsorption in the fluctuation zone

6 Solubility in groundwater

7 Flow through fractures in bedrock or rock outcroppings

8 Well boring through impervious layer into deeper aquifer

9 Utility trenches

OOPS!

Releases into heavy, dense clays will not migrate rapidly; releases into gravels, glacial tills, or other loose soils can travel at a rate of feet per day. Therefore it is necessary to determine the soil type from the beginning since that alone can dictate response time.

Contaminants may migrate in completely unexpected directions and contaminate remote sites if utility trenches or corridors are present. The backfill in utility trenches is nearly always less dense than the surrounding soil, even if native soil has been used as backfill. The lower density provides the path of least resistance and thus the direction of travel. Examples abound in which gasoline leaks have migrated miles from the actual source by traveling through utility trenches. Another unsuspected mode of migration is filled drainages, which can be ancient or man-made.

One example included a filled lake. The lake had partially filled with debris in ancient times. The process was completed to accommodate a small shopping center which now stands where the lake once stood. A service station complete with leaking gasoline storage tank was adjacent the shopping center. The tanks failed a tightness test, but wells failed to reveal the presence of gasoline. Meanwhile, gasoline migrated freely through the fill.

Contaminants may also migrate in unexpected directions if fractured subsurface country rock or rock outcroppings occur. More mobile contaminants will migrate rapidly through granites and sedimentary formations. These cases can be a real challenge for local regulatory authorities to sort out.

Soil Classification System

Soils are classified according to whether they are coarse–grained or fine–grained. The latter category includes silts and clays, the former includes sands and gravels. Soil type is determined primarily by local geology; that is, by whether the soil was deposited as stream–borne sediments, wind–deposited loess, glacial tills, etc. Secondarily, soil type is determined by grading and artificial fills. Real soils are mixtures and may not fall exactly into a single, well–defined category. Soil types usually grade from one type into another. However, the fifteen categories listed in **Figure C–1** in Appendix C work remarkably well for most soil types.

Soil descriptions are generally taken from well borings. Hollow stem augers are especially convenient for noting soil types and correlating with depth. Unfortunately hollow stem augers are not always suitable. In describing soils from well borings it is important to note the appearance and odor of fresh samples.

5 Environmental Assessments

The Legal Environment

Remediation of petroleum contaminated sites should be simply a matter of getting the right engineering to the right place at the right time. In practice, remediation efforts are frequently complicated by disputes among PRPs, and are more often than not driven by liability — or the avoidance of liability — considerations. Strictly speaking, owners and operators of petroleum facilities are not exposed to CERCLA liability because of the petroleum exclusion. However, the petroleum exclusion does not offer protection from conventional environmental impairment liability and third party damage costs; nor does the petroleum exclusion afford protection in commercial property transactions. Further, court decisions involving petroleum contamination have tended to follow CERCLA reasoning. (Bulletin, 1992)

In this chapter we will look at CERCLA liability issues, including recent court decisions, and the concerns that have led to the rise in Phase I environmental assessments. The focus is not on Phase I assessments as a stand–alone activity. That is a book in itself. Rather, most of the features of the Phase I assessment are relevant to an initial site characterization and to the overall site evaluation. The document and historical review need not be as detailed as in a Phase I, but, at least, the preparer must cover most of the same ground.

State regulatory agencies require specific information about the site and surrounding land use. In many, if not most, petroleum contaminated sites, there exists the possibility of third–party contamination, either as the result of plume migration onto adjacent property, or as a result of an unrelated plume migrating onto the site. Other potential problems that must be addressed include possible contamination from hazardous wastes and public health threats such as groundwater contamination impacting drinking water wells or offsite vapor migration.

In this chapter a distinction is made between a site assessment, which is further divided into an initial site assessment and a second level assessment, and Phase I and Phase II assessments[‡]. All of these are defined later in this chapter. There is a real distinction between an initial site assessment and a Phase I; the distinction between a Phase II and a second level assessment is less distinct. Nevertheless, the terms are retained since they are in common use.

Phase I assessments are used as a framework for the document and records review necessary in evaluating the potential risk and problems associated with a site. Much of what is required in a Phase I is also required in an initial site assessment. Further, it is likely that a Phase I has already been done on a site so that duplication becomes unnecessary.

Overview of CERCLA

The "big three" pieces of environmental legislation are RCRA, Clean Water Act, and Clean Air Act. Each requires the EPA administrator to set regulations for a segment of the regulated community. Within that scope, RCRA and CERCLA are companion acts. At the risk of oversimplification, RCRA sets technical standards for hazardous substances; CERCLA sets legal standards for hazardous substances. Together they have had an immense influence on the regulatory and legal communities, on the economy, and on the means of doing business in this country.

The requirements of RCRA have changed forever the way companies use and handle hazardous materials and wastes. Waste minimization is now a routine activity for businesses that handle hazardous substances. CERCLA has irreversibly altered usual and customary standards for everything from laboratory data to the legal basis for negligence and liability. It is probably too soon to properly assess the long–term influence of CERCLA and RCRA, but it is certain that neither can be ignored.

Congress enacted CERCLA in 1980 in response to perceived environmental and public health hazards caused by the improper disposal of hazardous waste. The act was later reauthorized and amended by the Superfund Amendments and Reauthorization Act, SARA, in 1986. At the time of enactment a major concern was to provide a mechanism whereby contaminated sites could be restored.

The long–term result has been to establish a new standard for liability exposure in the U.S. Prior to CERCLA demonstrating liability normally required the demonstration of negligence; after CERCLA liability could result from nothing more than ownership or being in the wrong place at the wrong time. One commentator has stated

---

[‡]   A note on terminology: It has been customary to use the phrase *Environmental Audit*; however, the new ASTM standard uses the term *Assessment* instead. Therefore, in this chapter we will adhere to the revised usage, Phase I environmental *Assessment*. In this context an *Audit* is an environmental inspection to determine whether an operation is in compliance with applicable standards.

that there are only two types of environmental liability, CERCLA and everything else — and everything else only contributes about 10%. (Sweeney, 1993)

The underlying philosophy of CERCLA is to place ultimate responsibility for cleaning up hazardous wastes on those "Potentially Responsible Parties," PRPs, that have been identified as having caused the contamination in the first place (Krendl and Gibson, 1992). The intent of Congress is clear: remediation is of paramount importance and the party responsible for contamination must pay all costs of remediation, including costs of response, damages for injury or loss of natural resources, and any costs for health assessment.

Definitions in CERCLA

CERCLA contains a number of definitions that are essentially the same as those found in RCRA. The definitions summarized here are only those required for the ensuing discussion. The list is by no means complete.

The first significant term to be defined is "facility." A "facility" means [CERCLA §101, (9)]

"(A) any building, structure, installation, equipment, pipe or pipeline (including any pipe into a sewer or publicly owned treatment works), well, pit, pond, lagoon, impoundment, ditch, landfill, storage container, motor vehicle, rolling stock, or aircraft, or (B) any site or area where a hazardous substance has been deposited, stored, disposed of, or placed, or otherwise come to be located; but does not include any consumer product in consumer use or any vessel."

CERCLA defines "hazardous substances" as any substances having characteristics defined in RCRA and Solid Waste Disposal Act; that is, any substance that has certain characteristics of ignitability, corrosivity, reactivity, or toxic characteristics. The definition of hazardous substances in CERCLA contains a notable exclusion by excepting petroleum and petroleum products from the definition of hazardous substances, the so–called "petroleum exclusion" [CERCLA 42 U.S.C.A. §101(14)].

"The term {hazardous substance} does not include petroleum, including crude oil or any fraction thereof which is not otherwise specifically listed or designated as a hazardous substance under {other definitions in this act}, and the term does not include natural gas, natural gas liquids, liquefied natural gas, or synthetic gas usable for fuel (or mixtures of natural gas and such synthetic gas)."

RCRA and CERCLA definitions are consistent. The same petroleum exclusion is found in the definition of hazardous substances in RCRA. The exclusion was deemed necessary due to the enormous volume of stored petroleum products in the U.S. A substantial number of owners and operators of petroleum storage systems would have been unable to financially comply with regulations requiring gasoline to be stored as a hazardous substance. Strictly speaking, owners and operators of petroleum storage facilities or petroleum contaminated sites should not need to be concerned about

CERCLA because of the petroleum exemption. However, as explained below, reality has proved somewhat different.

At the heart of CERCLA definitions and CERCLA liability is the term "owner or operator." The term as defined in CERCLA is sufficiently imprecise to have caused great difficulty. "Owner or operator" means [CERCLA §101 (20) (A)]

"(i) in the case of a vessel, any person owning, operating, or chartering by demise, such vessel, (ii) in the case of an onshore facility or offshore facility, any person owning or operating such facility, and (iii) in the case of any facility, title or control of which was conveyed due to bankruptcy, foreclosure, tax delinquency, abandonment, or similar means to a unit of State or local government, any person who owned, operated, or otherwise controlled activities at such facility immediately beforehand."

In other words, an owner or operator owns or operates a facility. But this apparently simple definition has proved difficult to apply. The courts have attempted to provide a degree of specificity to these terms, but have succeeded only in interpreting them broadly. (Krendl and Gibson, 1992) At the core of the problem is CERCLA's failure to define "operate" and "management."

Potentially responsible parties, PRPs, which may include any or all of the following [CERCLA, §107(a)];

- the owner and operator of a vessel or a facility,

- any person who at the time of disposal of any hazardous substance owned or operated any facility at which such hazardous substances were disposed of,

- any person who (agrees to transport) for disposal or treatment, of hazardous substances…, and

- any person who accepts or accepted any hazardous substances for transport to disposal or treatment facilities, incineration vessels or sites selected by such person, from which there is a release, or threatened release, which causes the incurrence of response costs…

shall be liabile for all costs of removal or remedial action, any other costs incurred by any other person, damages for injury to, destruction of, or loss of natural resources, and the costs of any health assessment or health effects study.

There is a two–fold potential difficulty for PRPs handling or storing petroleum contaminated soils at permitted landfills:

- Gasoline contamination in soils is no longer a product and, therefore, may conceivably fall under TCLP at a future date (see related material in Chapter 6, especially the proposed EPA exclusion); and

- Permitted landfill operations normally take possession of — but not *ownership* of — petroleum contaminated soils.

Potential problems arise because benzene, toluene and other aromatic components of gasoline are listed hazardous materials covered by the definitions of RCRA, Subtitle C. These compounds are permanently excluded from the requirements of Subtitle C under a final rule by the EPA so long as they are components of gasoline or other petroleum products. That is, the compounds remain under the petroleum exclusion even when they are no longer products. However, there are no guarantees and the court tests of EPA's rule have just begun. Therefore, owners of record of the stored contaminated soils remain PRPs indefinitely and would incur liability in the event that the landfill should be found at some future time to be contaminated with hazardous wastes under Subtitle C or to have been operated improperly.

This scenario may not be likely, but it is not remote since the history of hazardous waste management is littered with good intentions that have ended up costing large amounts to rectify.

The Secured Creditor Exemption

In the course of hearings leading to passage of CERCLA, Congress recognized that lenders must be allowed to do business as usual. That is, lenders whose participation in a business consists of lending money to a business and securing the loan with a security deed on the property or some portion of the business such as inventory, must have some protection from liability.

Congress expressly excluded from the definition of owner or operator
"...a person who, without participating in the management of a vessel
or facility, holds indicia of ownership primarily to protect his security
interest in the vessel or facility." [42 U.S.C.A. §101 (20)(A)].
This is called the "secured creditor exemption" or sometimes "the security interest exemption." The term "indicia of ownership" is not defined by CERCLA, but has been interpreted to mean evidence of an interest in real or personal property. (Krendl and Gibson, 1992)

In the business and legal world there is a distinction between *financial management* and *operational management*. The distinction is particularly important with respect to lenders. A lender can take property or other assets as security against a loan and review the business practices of that facility to maximize the ability of the facility to repay the loan. This is distinct from operational management, which generally refers to the day–to–day operation of a facility. Congress clearly intended to exclude lenders from CERCLA liability in cases where they did not participate in operational management. Even if a lender foreclosed on a property that had some contamination, the lender was not exposed to CERCLA liability. However, Congress did not define what action or actions constitute "participation in management" that avoids CERCLA liability. As a result the courts have had to distinguish between financial management and operational management.

The risk to lenders and purchasers of commercial property cannot be overstated. In one case a lender received only $200,000 during a foreclosure sale for a service station valued at $450,000 because the property required $250,000 in remedial costs. In another case lenders wrote off $57 million in loans rather than foreclose on a refinery due to fears of the costs of remediation (Sweeney, 1993). Much of the property now owned by the Resolution Trust Corporation is contaminated, a fact that has contributed to the slowness and complexity of disposal of these properties. Since 1980 the EPA has charged at least forty–four lenders as defendants in CERCLA liability cases (Krendl and Gibson, 1992).

The CERCLA Concept of Liability

Because environmental liability exposure in commercial property transactions is almost exclusively CERCLA liability, it is useful to examine the legal definitions and climate associated with environmental impairment liability. The digression into liability issues is necessary to provide a foundation for Phase I environmental assessments.

Under CERCLA, liability has three components:

*   Strict Liability. A party may be held liable *without* negligence or misconduct. This was a new feature in U.S. common law.

*   Joint and Several Liability. All contributing parties are usually held responsible, starting with the current owner or operator. However, the current owner may be held liable for the full amount of restoration without regard for actual cause. A common scenario is for the regulatory agency to sue the current owner/operator for restoration; the current owner/operator then sues other PRPs in separate actions.

*   Retroactive Liability. Damage caused before passage of CERCLA is still covered because the courts have held that CERCLA liability exposure is for contamination that exists now, not for the release of the contaminant which may have occurred before passage of CERCLA.

CERCLA recognizes only three defenses [CERCLA, §107(b)]:

"There shall be no liability under {CERCLA} for a person otherwise liable who can establish by a preponderance of the evidence that the release or threat of release of a hazardous substance and the damages resulting therefrom were caused solely by
1)   an act of God;
2)   by an act of war;
3)   an act or omission of a third party other than an employee or agent of the defendant, or than one whose act or omission   occurs in connection with a contractual relationship, existing directly or indirectly, with the defendant…, if the defendant establishes by preponderance of the evidence that (a) he exercised due care with respect to the hazardous substance concerned … and (b) he took precautions against

foreseeable acts or omissions of any such third party and the consequences that could foreseeably result from such actions or omissions."

In other words, the innocent purchaser defense is allowable only if the defendant
shows that [CERCLA :

1) The release or threat of release was caused solely by a third party;

2) The third party is not an employee or agent of the defendant;

3) The acts or omissions of the third party did not occur in connection with
   direct or indirect contractual relationship to the defendant, or if there was
   a contractual relationship, the defendant i) acquired the property after
   disposal or placement of the hazardous substance, and ii) at the time the
   defendant acquired the facility the defendant did not know and had no
   reason to know that any hazardous substance (had been disposed of on the
   facility), and

4) The defendant exercised due care with respect to the hazardous substance
   and took precautions against foreseeable acts or omissions of the third
   party.

The Innocent Purchaser Defense

The third defense listed above is the "innocent purchaser" or "innocent landowner"
defense, and was added in 1986 to protect participants in property transactions who
unknowingly acquire contaminated property. The defense, however, requires that the
purchaser exercise all "due diligence" before the fact to determine whether or not the
property is contaminated. Due diligence is not defined precisely or otherwise. CERCLA
leaves it up to the property buyer to determine how extensive an investigation he or she
wishes to do. In past two years the definition of due diligence has increasingly been set
by lenders and insurers.

The type of examination conducted by a potential buyer or required by lenders and
insurers to avoid acquiring contaminated property is frequently referred to as a *Phase
I Environmental Assessment*. Indeed "Phase I's" have become so pervasive that they will
be discussed in a separate section.

Prior Case Law

Until 1990 courts made clear distinctions between financial management, such as
controlling accounts receivable, which was permissible, and operational management,
which was not, and which resulted in CERCLA liability (*United States v. Mirabile,
1985*). Lenders could avoid CERCLA liability by not becoming involved in day-to-day
operations, but could participate more or less freely in the financial management.

In the case of foreclosure, lenders who foreclosed to protect the lender's security
interest escaped CERCLA liability (*Mirabile, 1985*). *Those* who foreclosed and became

actual owners of record and operated the facility incurred CERCLA liability (*Guidice v. BFG Electroplating & Mfg.. Co., Inc., 1989*), (*United States v. Maryland Bank & Trust Co., 1986*).

The present climate is confused by conflicting cases in 1990. When CERCLA was passed in 1980, Rep. Dannenmeyer declared that its epitaph would one day read, "Noble of purpose, but notorious in operation." This statement will certainly prove true if *Fleet Factors* ultimately turns out to be the litmus test for CERCLA liability.

The "Black Hole" of Lender Liability

The lender in *Fleet Factors*, The Fleet Factors Corp., held a security interest in the equipment, inventory and fixtures of Swainboro Print Works, Inc., a print company in southern Georgia. As additional collateral, the lender was granted a security interest in the real property where the equipment was located. The company went out of business. Being aware of previous court decisions, Fleet foreclosed on the equipment and inventory, but *not* on the property. Fleet then hired an auctioneer to sell the equipment and inventory collateral.

The EPA had inspected the facility in 1984 and found 700 55-gallon drums containing toxic chemicals and 44 truck loads of material containing asbestos. The government sued Fleet to recover cleanup costs. The government argued that, in preparation for the auction, Fleet had caused the drums to be moved and had released asbestos by disconnecting pipes from equipment. Such action constituted "management" of the facility.

The district court (Southern District of Georgia) determined that Fleet's preforeclosure conduct did not constitute management and was consistent with the CERCLA exemption. However, the court denied a motion for summary judgment, stating that there was genuine issue of fact as to whether Fleet operated the facility after the foreclosure and during the auction, and whether the activity exposed the company to CERCLA liability.

On appeal, the Eleventh Circuit Court of Appeals rejected the arguments in *Mirabile*, and other cases, distinguishing between permissible financial and impermissible operational management. The Court, driven by the "overwhelmingly remedial" goals of CERCLA, held that,

"a secured creditor may incur [CERCLA] liability, without being an
operator, by participating in the financial management of a facility to
a degree indicating a *capacity to influence* the corporation's treatment of
hazardous wastes." (Emphasis added).

The court left no doubt that it was making policy by severely limiting the secured lender exemption, and reasoned that its ruling would encourage creditors to investigate more thoroughly the waste treatment systems of potential borrowers. To qualify as a borrower, waste handling systems must meet lender–mandated standards. The Court clearly felt that this would provide strong incentive for borrowers to improve their waste handling systems.

Conflicting Case Law

A similar case before the 9th Circuit Court of Appeals in August, 1990 resulted in a decision that appears to contradict *Fleet Factors*. In *Bergsoe vs Port of St. Helens*, there were a series of interlocking transactions in which the revenue bonds issued by the Port of St. Helens were held in trust by the U.S National Bank of Oregon, and revenue from the bond sales went to Bergsoe to purchase land and construct a secondary lead recycling plant. Bergsoe ended up in involuntary bankruptcy. The bank and the bankruptcy trustee sued Bergsoe's stockholders to recover company debts, including costs of cleaning up the site's environmental pollution.

Bergsoe countersued, alleging that the bank and Port were liable for cleanup costs. The Port's motion for summary judgment that it was not an "owner" of the plant under CERCLA was granted by the district court. Bergsoe appealed.

The appeals court's analysis indicates that in order for the secured interest exemption to apply the Port had to show that:

(1) it held indicia of ownership primarily to protect its security interest in the plant; and

(2) did not participate in the management of the plant.

The court was aware of *Fleet Factors*, but rejected the arguments in that case stating,

> "whatever the precise parameters of 'participating,' there must be some actual management of the facility before a secured creditor will fall outside the exemption."

Even though the Port was using Bergsoe's revenue to cover its own indebtedness, this did not change the analysis since there was no evidence that the Port had exercised any control over Bergsoe. The Port was, however, in a position to *influence* the operation of the plant, and so, under Fleet Factors, would have participated in the operation of the plant. Thus, *Bergsoe* appears to directly conflict with the decision in *Fleet Factors*. Such conflicts are normally resolved by the Supreme Court.

Legal Outcomes

Currently, most lenders insert environmental covenants into loan agreements, requiring environmental assessments, warranting compliance with environmental laws, and indemnifying the lender from environmental liability. The very presence of these covenants might suggest that a lender has the "capacity to influence" hazardous material handling and operations, and, therefore, be exposed to CERCLA liability under *Fleet Factors*.

It was widely assumed among the commentators and the legal community that the decision in *Fleet Factors* would be reversed by the Supreme Court. Unfortunately, in March, 1991 the Court refused, without comment, to hear the appeal (*certiorari* denied). Since the appeal was on a motion for summary judgment, the case returns to district court for trial.

The impact of *Fleet Factors* has clearly exceeded the impact of the actual decision itself. Since the appellate decision was on a defense motion for summary judgment, the direct legal result of the decision was merely that the appellate court agreed that the government had sufficient facts to warrant a trial. However, because of the strong wording and because the court obviously intended to send a message, the lending community has reacted strongly to defend its interests.

In testimony before Congress, a spokesman for the Petroleum Marketers Association of America stated that:

> "The court has clearly shifted more of the burden for non–compliance with hazardous waste laws to lenders, having decided, 'Lending institutions are especially well equipped for this function.' ...We think the decision in Fleet Factors may represent an important extension of the trend in case law toward expanded lender liability for environmental contamination." (Bulletin, 1992)

Thus, the immediate effect of *Fleet Factors* has been to severely limit the availability of capital to businesses just at a time when it is most needed. Although petroleum USTs do not fall under the jurisdiction of CERCLA, *Fleet Factors* has still had an impact on petroleum marketers who are in the process of complying with EPA–mandated UST upgrade deadlines. Few owner–operated petroleum outlets can afford the cost of tank removal and replacement without financial assistance, which is unavailable.[‡]

Financial Outcomes

The outcome of *Fleet Factors* then, has been to make it difficult or impossible for companies with existing or potential waste problems, including those with USTs, to borrow money. Lenders will refuse to become insurers against the possibility of environmental liability. Indeed, at the present time many lenders simply will not loan money to any business with a potential for environmental risk, including the presence of USTs.

The President of a bank in Ohio told a House subcommittee hearing on lender liability,

> "...our bank has had to carefully consider what types of businesses in our trade area could present liability problems. As a result of this analysis, we have amended our loan policy by adding a category of loans deemed 'undesirable.' Included in the bank's list is 'any business with an underground storage tank.' We simply cannot afford to risk the

---

[‡]   The *Fleet Factors* case has been resolved. A summary of the decision is reported in Appendix E. Several other related and relevant court decisions are also summarized in Appendix E.

potential expense that, ..., these businesses present to my [bank]."
(Clark, 1991)

Companies are in a "Catch 22" situation. They are unable to borrow to upgrade UST facilities or they are able to borrow only after upgrading or replacement. The financial crisis means loss of business for lenders, insurers, and the companies that are forced out of business. Increasing numbers of UST sites are being abandoned. The inability of companies to borrow is increasing the number of orphan sites which are left for state governments to clean up — at taxpayer expense.

Proposed Legislation

At the time of writing, two bills have been introduced in the Senate and the House to reverse the effect of *Fleet Factors* by defining actions lenders can take without losing their exemption. The Senate bill, S. 651, introduced by Jake Garn (D-Utah), allows liability to be imposed if a lender caused the release, or threatened release, or failed to take reasonable action to prevent the release of a hazardous substance.

The House bill, H.R. 1450, introduced by John LaFalce (D-N.Y.), amends the "secured creditor" exemption to provide incentives to lenders to conduct environmental assessments and remedy environmental damage "rather than walk away from their collateral." The bill imposes liability on lenders that are directly responsible for environmental damage. The bill goes beyond CERCLA and addresses liability under RCRA, Subtitle I, which includes properties containing petroleum storage tanks.

The EPA Rule on Lender Liability

The EPA has opposed further legislation to clarify potential exposure to CERCLA liability in the belief that the problem could be handled by regulation. Consequently on April 29, 1992 the EPA published an interpretive rule which would establish a "safe harbor" allowing a lender to engage in normal lending activities without incurring CERCLA liability (FR, 1992; and 40 CFR §300). The rule specifically defines what constitutes "business as usual"; i.e., those activities in which lenders may engage without incurring CERCLA liability (see relevant case law in Appendix E). Lenders may

- provide financial advice to distressed borrowers;

- require cleanup of a facility prior to or during the life of a loan;

- require a facility owner or operator to comply with applicable federal, state, and local environmental or other laws during the life of the loan;

- monitor or inspect a facility and the owner or operator's business or financial condition;

- restructure or renegotiate the terms of a loan;

- liquidate assets; and,

- foreclose on a security.

Critics have stated that the rule does not provide a safer harbor than any other legislation nor does it reconcile the dilemma raised by *Fleet Factors* (Krendl and Gibson, 1992). It appears likely that some legislation will be passed to amend CERCLA. How quickly is an open question.

Overview of Environmental Assessments

In the present climate as described above, owners and operators (sellers) and buyers and lenders (potential owners) risk liability exposure whenever commercial property changes hands. Under CERCLA in order to establish that an {the owner} had no reason to know of existing contamination —

"...the defendant must have undertaken, at the time of acquisition, all appropriate inquiry into the previous ownership and uses of the property consistent with good commercial or customary practice...." [CERCLA, ¶101 (35)(B)]

This concept has been described above as "due diligence." If due diligence has been exercised and the property is later found to be contaminated, the new owner has recourse to the innocent landowner defense. The question is, "What is due diligence?"

Commercial sites must be carefully evaluated to determine whether contamination is present or absent. However, it is neither necessary nor cost effective to begin corrective action or sampling immediately. In recent years the definition of due diligence has increasingly been determined by lending institutions and insurance companies, exactly the organizations that by training and inclination are least competent to evaluate environmental conditions.

Therefore, environmental investigations have actually been carried out by environmental companies. Still, there is no guarantee that the investigation carried on a particular property was adequate. In short, there are no national or industry–wide accepted standards for due diligence.

Because it is not necessary to evaluate most properties by intrusive sampling, environmental investigations have customarily been carried out in stages closely analogous to the stages in a site assessment. Indeed, there is so much overlap that distinctions are somewhat superficial.

The stages of environmental assessment are called Phase I, Phase II, and, sometimes, Phase III assessments. These phases are described briefly as follows:

Phase I: The first Phase of an environmental or site assessment consists of a physical inspection of the site and an appropriate documentary review. Phase I assessments normally include the following:

- physical inspection;

- historical (or document) review;

- record review.

A physical site inspection means just that; visit the site, walk around, and visually inspect the site for potential problems. The physical site visit is as important in an initial site assessment as it is in a Phase I environmental assessment. The historical review is often called a document review. It includes a check of as many documents relevant to the site as possible. A checklist is given below. The record review includes local or state records relevant to the site. In an initial site assessment the historical review is necessary to discover prior uses and potential problems and risks.

Phase II: A second Phase of an environmental assessment may be required to determine the nature and extent of suspected contamination inferred from the Phase I assessment. Phase II assessments normally include the following:

- Installation of monitoring and sampling wells;

- Intrusive sampling and laboratory analyses;

- Data evaluation.

The number and location of monitoring wells and preliminary sampling plans should be determined in consultation with the appropriate regulatory agency.

Phase III: the third Phase is the actual remediation and/or stabilization of the contamination. Phase III of an environmental assessment consists of the following:

- Detailed sampling and analyses;

- Initiation of a corrective action plan;

- Stabilization and/or isolation, if required;

- The remediation project.

Property Risk Analysis

To expedite the environmental assessment, land and structures must be categorized according to relative risk from low probability of contamination to high probability. Risk analysis is a cost effective first step to determine the depth of succeeding procedures. Risk analysis is not an exact science and should not be treated as such. There are numerous examples of apparently pristine farm land that have been used for illegal dumping or previously residential land contaminated from a distant source. Nevertheless, it is a necessary part of any site evaluation.

Note that in this context "risk analysis" is not to be confused with "risk assessment" that is required in connection with a corrective action plan. Risk analysis is simply the evaluation of the potential of land and property for existing or future contamination. Risk assessment is the evaluation of the potential threat to human health and the environment due to known or suspected contamination.

Classification of the relative risk of a property is subjective and should be viewed with caution. Examples of low probability of environmental risk include the following:

- Rural areas with no evidence of previous contamination;
- Most single family residential areas, multifamily areas in most but not all states;
- Projects involving no change in profile grade or significant excavation.
- Areas that have been previously studied[‡] and involving no new grade changes or excavations.

Exceptions abound: single–family dwellings and rural properties may be contaminated with agricultural residues, single family dwellings may be located in the plume from a nearby leaking underground storage tank. Agricultural properties have been used for illegal hazardous waste disposal and as unregulated landfills.

One feature to consider in evaluating the relative risk of a property is whether or not excavation is planned. Projects for which new excavation, grading, and/or trenching are planned require more care and consideration than those for which no new excavation is planned. The idea is that contaminants left undisturbed often pose little risk, whereas contamination that will be excavated poses a high degree of risk. High risk properties and surroundings are listed in **Table 5–1**.

Table 5–1. Examples of High Environmental Risk. Examples of properties and surroundings having a high probability for environmental risk are listed. (Characteristic hazards are in parentheses):

- Unregulated municipal or private dumps and landfills;

  (All kinds of hazardous wastes were landfilled prior to SWDA in 1965)
- Waste accumulation sites;

  (Everything from legitimate wastes to illegal storage)
- Treatment plants;

  (Solvents, corrosive and reactive chemicals, PCBs, etc.)
- Manufacturing plants;

  (Solvents, corrosive and reactive chemicals, PCBs, etc.)
- Photo processing or printing operations;

  (Silver, corrosive and reactive chemicals, solvents)
- Paint and plating operations;

  (Solvents, corrosive and reactive chemicals)

[‡]   ASTM specifies a six month time limitation for using previous studies. There is no corresponding limitation for an ISC; however, data older than six months should be used only with care.

- Battery recycling plants;

  (Lead)

- Automotive salvage yards;

  (Lead, heavy metals, ethylene glycol, acids, petroleum)

- Metals and paper processing plants;

  (Solvents, corrosive and reactive chemicals, heavy metals)

- Agricultural operations;

  (Pesticides, fertilizers, unregulated dumps)

- Medical supply facilities;

  (Biohazards, radionuclides)

- Service stations;

  (USTs, petroleum, ethylene glycol, other hazardous wastes if used oil was collected)

- Dry cleaners;

  (Solvents, USTs, if a converted gas station)

- Older (asbestos–containing) buildings;

  (Asbestos, lead)

- Mining (particularly abandoned milling) operations;

  (Heavy metals, other inorganics)

- Abandoned tailings piles

  (Heavy metals, other inorganics, low pH water).

Phase I Environmental Assessments

Due in part to the uncertainty created by *Fleet Factors*, and in part to the substantial expense involved in remediation, the business of Phase I environmental assessments has boomed since about 1990. Because of the "innocent purchaser" or "innocent land-owner" defense the process of identifying "Recognized Environmental Conditions" (ASTM E50.02, 1993) has assumed a very high priority in commercial property transactions. As stated in the ASTM Standard Practice, "Recognized Environmental Conditions" means

"... the presence or likely presence of any Hazardous Substances or Petroleum Products on a property under conditions that indicate an existing release, a past release, or a material threat of a release of any Hazardous Substances or Petroleum Products into structures on the Property or into the ground, groundwater or surface water of the

Property. The term includes Hazardous Substances or Petroleum Products even under the conditions in compliance with the laws. The term is not intended to include *de minimus* conditions that generally do not present a material risk of harm to public health or the environment and that generally would not be the subject of an enforcement action if brought to the attention of appropriate governmental agencies." (E.50.02, 1993, ¶1.1.1).

Petroleum products are included because 1) they are a source of concern, 2) courts have tended to extend CERCLA liability to petroleum contaminated sites, and 3) current and customary practice is to include them.

There are three purposes underlying an environmental assessment:

- The Transaction Screen (E50.01, 1993) can be a relatively inexpensive means of deciding that no further investigation is required or that it is necessary to proceed directly to Phase II sampling;
- Virtually all commercial transactions must now be accompanied by a Phase I environmental assessment ;
- Information from a Phase I assessment facilitates the planning process for sampling, analysis, and remediation, which is crucial to all but the simplest remediation project.

Together the data provided by these steps not only provide the environmental consultant with information on which to base decisions but also provide the owner with the necessary documentation to satisfy the requirements of lenders and insurers.

The basic requirements for Phase I assessments or initial site characterization are:
- To identify and avoid potential environmental liability on any property or transaction;

    This is useful information in an initial level assessment. Information gathered in this context can reveal potential sources of contamination from offsite that may interfere with subsequent sampling or remediation.

- To ensure against worker and public safety exposure to air, soil or water contamination, particularly in areas where excavation is necessary;

    A principal goal of the site assessment process is to protect workers and the environment from exposure to hazardous substances.

- To identify hazardous materials that might be disturbed and released to the environment during subsequent construction activities;

    Hazardous substances, especially metals, radionuclides, and other inorganics, pose a greatly reduced level of risk if left undisturbed. Early identification allows alternative plans for excavation and grade changes which should be avoided if possible.

- To provide information about environmental conditions during planning stages, so that alternative designs can be used to avoid problems and reduce costs;

  Known contaminants are already expensive enough, unknown surprises are guaranteed to strain budgets.

- To demonstrate that all due diligence was exercised in case of unexpected problems and/or litigation;

  This is the bottom line in property transfers. In effect it is essential in an initial site assessment to allow for appropriate planning.

It should be apparent that in each case the focus is on the planning process and the acquisition of information essential to planning. Surprises are always expensive, and can be *very* expensive. The planning process is discussed in more detail in Chapters 7 and 8.

Search Radius

For a Phase I assessment or an initial site assessment the minimum search radius is dependent on the area in which the property is located. The ASTM standard uses the term "search distance" since few properties have circular boundaries. It is certainly necessary to examine property usage outside the immediate property boundaries; the question is, how far out?

Factors to consider in determining the minimum search distance include the following:

- The density of use (urban industrial, urban commercial, suburban commercial, suburban multifamily, suburban residential, rural);

- Potential migration routes;

  and

- Distance to drinking water sources.

In an urban industrial area a Phase I search might extend ½–mile or more. The main criterion for search distance is to try to ensure that potential sources of environmental risk have been covered.

Phase I Site Inspection

It is generally agreed that at a minimum a Phase I assessment should include a physical site inspection, called a "Site Reconnaissance" by the ASTM Standard Practice (ASTM, 1993, §3.3.30). A site reconnaissance should be conducted by an Environmental Professional (EP) familiar with the usual and customary practices of the area.

The ASTM Standard Practice requires that, although the information may be generated by government agencies, third–parties, the User, and present or past owners, the information must be obtained under the supervision of an EP.

In the course of the site visit the EP should have knowledge of and consider the present and past usage of the site, current and past usage of adjacent properties, and present and past usage of surrounding properties. The EP should visually and physically inspect the site for the conditions listed in **Table 5–2**.

Table 5–2. Visual and Physical Site Reconnaissance Checklist. The following is a list of features to check during a physical site inspection that is required in a Phase I assessment, and should be included in an initial site characterization. The list is not exhaustive for all possible situations, but does include the features most commonly associated with contamination.

☐ Evidence of abandoned USTs, including vent stubs or piping with no apparent use;

☐ Stained soil, distressed vegetation, groundwater seeps, and standing pools which may be evidence of leaking USTs or hazardous wastes;

☐ Pits, ponds, or lagoons that may be repositories for hazardous wastes or petroleum products;

☐ Hazardous material or petroleum storage containers such as old drums or barrels;

☐ Unidentified containers;

☐ Drains and sumps;

☐ Significant odors indicating possible contamination;

☐ Electrical transformers and hydraulic equipment indicating possible PCBs;

☐ Artificial fill and altered topography indicating possible abandoned landfills;

☐ Friable asbestos, including wall and ceiling insulation, pipe wraps, etc. (especially in older buildings);

☐ Good housekeeping (poor housekeeping may be evidence of other significant problems);

☐ Utility trenches and corridors;

☐ Septic systems and drain fields, and waste water trenches;

☐ Adjacent property use;

☐ Possibilities for radon gas seeps (in appropriate areas).

A review of historical documents and records is an essential feature of any Phase I assessment because the effort can yield information obtainable by no other means. The search can save time and money since sampling and laboratory time are expensive and may be inconclusive. Information gathered in document searches is usually beneficial in the planning process.

The "appropriate minimum search distance" includes areas outside the immediate property. Typical search radii range from ¼ mile to 1 mile. Search radius is often a factor in distinguishing one bid from another. Information is available from a variety of sources including public documents, published resources, large data bases, and third–party resources (see **Table 5–3**).

Table 5–3. Resources Available for Historical and Records Review. This table includes those documents most frequently used in document, historical, and official record searches. The list may not be complete for all possible search criteria, but does include the documents and resources most often required. Not all resources will be required in every case.

Federal Documents

☐ National Priorities List (NPL) site list

☐ Comprehensive Environmental Response, Compensation and Liability Information System, CERCLIS

☐ RCRA treatment, storage and disposal, TSD, facilities

☐ RCRA generators list

☐ Emergency Response Notification System, ERNS, list

State Documents

☐ State NPL equivalent site list

☐ Hazardous waste site list

☐ Landfill or solid waste site list

☐ LUST lists

☐ Registered UST lists

☐ SARA Title III (§304) Reports (Emergency response reports)

Local (County) Documents

☐ Historical maps and plats

☐ Tax records

☐ Surveys

- ☐   Landfill or solid waste disposal sites list
- ☐   Building permits
- ☐   Permitted well records

Other Local Agencies

- ☐   Fire department hazardous materials responses
- ☐   Fire department MSDS files
- ☐   Sanitation department records
- ☐   Air quality/pollution control agency records
- ☐   Water quality agency records

Published and Commercial Sources

- ☐   Aerial photos
- ☐   Fire Insurance directories (Sanborn Directories)
- ☐   City directory listings
- ☐   USGS topographical maps
- ☐   State or USGS groundwater survey maps
- ☐   State or USGS subsurface geology maps

Other Sources

- ☐   Interviews with current and past owners/operators
- ☐   Interview with Key Site Manager
- ☐   Interviews with previous employees
- ☐   Interviews with neighbors

Phase II and Phase III

The terms "Phase II" and "Phase III" are sometimes encountered in RFPs and RI/ FS documents, as well as in general use. Phase II assessments consist of intrusive sampling to determine the nature and extent of the contamination. Phase II is indicated by a "flag," or indicator of contamination, in a Phase I assessment or after confirmation of contamination. Phase III, the third Phase of the environmental assessment, is the actual remediation or stabilization itself. Phase III is normally initiated after the corrective action plan has been approved and approval from the reimbursement fund obtained. Phase II and Phase III are the subjects of Chapters 7 through 10.

6 Site Assessments

In this chapter we will look at specific site characterizations, often called collectively a Phase II environmental assessment or a second–level assessment. Site assessment at this level requires intrusive sampling (monitoring well installation) in order to completely characterize subsurface features and detail the vertical and horizontal extent of the contaminant plume. The characterization should proceed in steps from initial planning to detailed lab analyses. The sequence of steps follows a logical, geostatistical approach in which subsequent steps are dictated by the current information.

Initial Site Characterization

Understanding the conditions at the site is essential to selection of an appropriate soil or water treatment technology. A complete site assessment need not, and should not, be done all at once. Rather, the assessment should proceed in steps beginning with an initial site characterization. This step must include all of the formal data required by the primary regulatory agency.

Release of a regulated product may be indicated by one of the following conditions (see Appendix B):

- A failed tank tightness test;

- Leak detector alarm;

- Inventory control overage or shortage by 1% plus 130 gallons for 2 months in a row;

- Municipal sewer alarms;

    or

- Other indications, such as reports of gasoline odors from nearby residents, sheen on surface waters, seeps, etc.

Federal regulations require an initial corrective action plan to be submitted within 20 days of the confirmation of a contaminated site (see Appendix C). All state implementing agencies therefore have the same or more stringent requirements for initial site characterization. Most states have more stringent requirements. In Colorado and many other states, for example, if a release of a regulated substance is suspected, the appropriate agency must be notified within 24 hours. Presence or absence of the release must be confirmed within 7 days. These time limits will vary from state to state.

Regardless of time requirements, the first consideration must always be, "Is emergency response required?" Emergency response may consist of spill cleanup or containment or diversion trenches for free product. After

- confirmation of a release and stabilizing or containing the release as appropriate;

  or

- establishing that emergency response is not required;

the cause of the release must be established and corrected.

Next an initial site characterization report must be submitted. An initial site characterization will include systematic information useful to planners and regulatory agency personnel (see Table 6–1).

The data submitted should include clear and concise information, referenced where necessary. Among the more important data that may be difficult to establish is ownership. CERCLA establishes that direct liability for environmental damage lies with the *owner*. In many cases the property owner and facility owner are different. The property owner, the facility owner, and the facility operator may all be different. In preparing the initial site characterization, all PRPs should be listed, if possible. If not, the more important information is the facility (from which the release originated) owner.

The amount of product released is an important piece of information that is useful in planning further actions and estimating completion. It is also a difficult piece of information to establish. Estimating the amount of product is most readily done from existing inventory records. It is, unfortunately, the case that releases seldom happen at facilities with clear, accurate inventory records. Examples abound of service stations that have leaked 10,000 gallons of gasoline without the operator being aware that anything was wrong. Amounts of product released can be estimated from monitoring well data; however, that method requires extensive sampling, a well–characterized plume, and is expensive.

Along with estimates of amount of product released, the preliminary assessment information should address whether the USTs or fuel containers contained or have contained leaded gasoline. Also if waste oil was collected on–site, that information should be noted since it impacts the type of sampling that will be required. Eventually samples collected from the vicinity of the used oil collection tanks will have to be analyzed for the presence of ethylene glycol and/or other hazardous waste (see TCLP in Chapter 7).

The extent of the contaminant plume is the next information that must be established. In the initial stage of site assessment, the extent can estimated using reliable criteria for the estimate. Examples include preliminary sampling data, soil vapor probe data, or soil vapor surveys. In establishing the extent of contamination both horizontal and vertical extent are important. The potential for off–site impact should be determined at the same time.

The potential for migration should be estimated based on reliable criteria. Examples include public utility maps, preliminary sampling, and records review. If the plume has reached, or will reach, the water table, flow rates should be determined. At least some sampling data from upgradient points are necessary before assuming that the plume is all downgradient.

The document review described in the preceding chapter will have determined the locations of drinking water wells, surrounding populations that may be affected, and potential health and safety concerns. Routes of migration other than through normal subsurface pathways should be identified. These include utility corridors, artificially filled topographical features such as streams, wetlands, gullies, drainages, etc., and any other features that result in subsurface densities being different from the ambient conditions.

Table 6–1. Initial Site Characterization Report. Information required in an initial site characterization report is listed. The form will certainly vary from state to state, but the data are required by virtually all.

General Background Information

☐ Cause of the release

☐ Type(s) of regulated product(s) released

☐ Amount of product released. (Inventory records if available)

☐ Period during which the release continued

☐ Contaminant extent at the present time

☐ Current condition of the tank(s) (if not removed)

☐ Dimensions of any excavations, estimated volume of material excavated

☐ Treatment or disposition of excavated material

☐ Current condition of the piping network (if not removed)

☐ Visual evidence of contamination

General Facility History

☐ The responsible owner‡ or operator, business information, physical address, phone, etc.

---

‡   This information may not be easily established. Actual ownership may be hidden in dummy companies. Abandoned tanks discovered in a Phase I assessment or as the result of a release also present problems of actual owner identification.

- ☐ Relevant history of the site (including former owners/operators)
- ☐ List of all regulated substances that are currently stored, or have been stored, on the site
- ☐ Any hazardous substances that may be located or stored at the site, such as ethylene glycol
- ☐ Size and volume of each UST
- ☐ Tank tightness test history

Detailed Site Description
- ☐ Site location (include site and surrounding area maps)
- ☐ All on–site buildings and structures
- ☐ Present and former tank locations
- ☐ Present and former piping locations
- ☐ Ground cover
- ☐ Extent of excavations and location of stockpiles of contaminated soils
- ☐ Relevant subsurface features (include utility corridors, basements, sewers, crawlspaces, etc.)
- ☐ Soil sampling locations
- ☐ Soil vapor monitoring locations
- ☐ Monitoring and sampling well location with identifying labels

Health and Safety Information
- ☐ Surrounding populations that may be affected
- ☐ Adjacent land uses and structures that may be affected by the release
- ☐ Locations of water wells that may be affected
- ☐ Probable health and safety risks
- ☐ Probable routes of exposure

Contamination Information
- ☐ Time of release
- ☐ Contaminant phases present
- ☐ Mobility of the product constituents
- ☐ Likely migration direction and rate

Subsurface Information
- ☐ Types of soils and subsurface formations
- ☐ Depth to permanent saturated zone (water table)

❑ Depth and extent of the unsaturated (vadose) zone

❑ Seasonal or periodic fluctuations in depth to groundwater

Regional Geology of the Site

❑ Description of the geology of the site

❑ Description of the soil and/or rock types at the site

❑ Estimated hydraulic conductivity typical for soil types

Hydrogeology of the Site

❑ Depth to the uppermost saturated zone beneath the site and method (or records) of determination

❑ Regional and site groundwater flow direction

❑ Seasonal or periodic groundwater fluctuation

❑ Climatological information, including average monthly precipitation, annual precipitation, forms of precipitation, and periods of runoff

Sampling and Analysis Plan

❑ Sampling methods and sampling points

❑ Matrix to be sampled

❑ Laboratory procedures and results

Summary of Initial Site Characterization

An initial site characterization contains information useful in planning and to regulatory agency personnel. It should contain general information adequate to decide on the specific steps and information required in subsequent stages. The initial site characterization breaks down into the following information sections:

• Ownership;

• UST system condition and history;

• Amount and type of contamination;

• Local depth to groundwater;

• Groundwater flow direction;

• Locations of private and/or municipal drinking water wells;

• Hydraulic conductivity;

• Subsurface layering and permeability;

• Rainfall, runoff, and infiltration rate;

- Monitoring well locations;
- Initial sampling plan and analysis results.

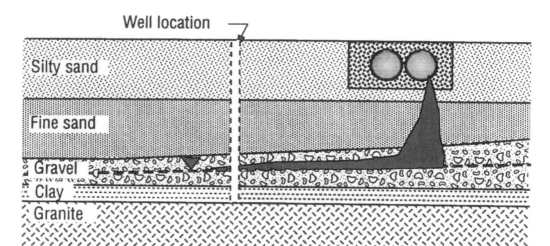

A. Preliminary subsurface profile.

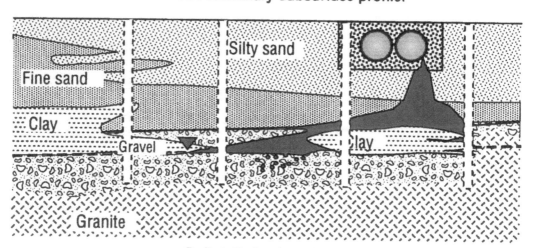

B. Detailed subsurface profile.

Figure VI–1. Subsurface Profiles. (A) A preliminary subsurface profile based on few monitoring wells and limited data. A profile at this level of detail may be suitable for an initial site characterization report, but not for final remedial action. (B) Detailed profile based on an adequate number of sampling wells. A subsurface profile at this level of detail is required for a second stage assessment report and is suitable for remediation decisions. The detailed profile should show plume extent as accurately as possible.

Detailed Site Characterization

The second level of a Phase II site assessment consists of more extensive sampling to determine the full areal extent of contamination, both vertically and horizontally. As the number of borings increase, lithologic logs should provide a more concise picture of subsurface features (**Figure VI–1**).

A principal focus of this level is reporting results of laboratory analyses. Chain–of–custody, QA/QC, and field procedures should be reported along with the lab results. The second stage should also confirm that the site in question is indeed the primary source. In many cases there may be multiple sources of petroleum contamination. Deciding on a single source or matching plume to source may require a field of sampling wells.

Installation of sampling wells treads a fine line between being detailed enough to adequately characterize the contaminant plume and bankruptcy. Well installation is expensive. There is no magic formula for placement of monitoring wells. Preliminary soil vapor surveys can provide a starting point. They are feasible and recommended in porous soils that allow vapors to migrate. A soil vapor survey point consists of driving a steel probe into the ground. Air is withdrawn from the soil and sampled. The usual method is to use a photoionization detector, PID, or flame ionization detector, FID. PIDs are easier to use, but limited in detection range. FIDs are more difficult to use but capable of analyzing a wider range of hydrocarbons. Either instrument must be used with great care to avoid erroneous data.

Wells are then placed so as to define the full extent of a plume or discriminate among several plumes (see Chapter 8). It is normal to install an initial set of wells, then a second or third set or fourth set as needed. Clients need to understand that initial well siting is usually followed by additional wells as needed to completely define the plume extent.

Further information required includes detailed site geology and hydrology, and extent of contamination. Specific information requirements are listed in **Table 6–2**

Table 6–2. Detailed Site Assessment Checklist. The following table summarizes the most important items covered in a second level site assessment. Not every site will require all of the information; a few could require more information.

Site Geology
☐ Specific site subsurface lithology from borings, excavations, and sampling

☐ Well logs and excavation cross section

☐ Geological cross section if the local geology is complex

Soil Characteristics

- ☐ Soil moisture content
- ☐ Soil temperature
- ☐ Soil pH
- ☐ Soil surface area
- ☐ Particle density
- ☐ Bulk density
- ☐ Field capacity

Site Hydrogeology

- ☐ Depth to the uppermost saturated zone
- ☐ Local groundwater depth, flow rate, and gradient
- ☐ Soil characteristics in the capillary and fluctuation zones as determined by sampling

Extent of Contamination

- ☐ Full areal extent of contaminant plume
- ☐ Soil sampling and appropriate analyses
- ☐ Protocols for sampling
- ☐ Groundwater sampling, if necessary
- ☐ Map of sampling wells

Sampling Protocols
- ☐ Field documentation
- ☐ Site sampling map with reference grid
- ☐ Chain–of–custody records
- ☐ Field QA/QC

Well Documentation

- ☐ Design schematics
- ☐ Map with well reference grid or labels
- ☐ Well logs

Laboratory Results

- ☐ Documentation of appropriateness of laboratory tests
- ☐ Analysis results
- ☐ Interpretation

Other sampling as appropriate for gasoline additives that may affect cleanup operations

- ☐ Tetraethyllead and tetramethyllead
- ☐ Ethylene dibromide (EDB) and ethylene dichloride (EDC)
- ☐ Methanol
- ☐ Dimethylamine
- ☐ Methyl *tert*-butyl ether (MTBE)
- ☐ 1,2-Dichloroethane (DCE)
- ☐ Ethylene glycol (antifreeze)

Other sampling as appropriate for possible mixed waste contamination that may affect cleanup operations
- ☐ PCBs
- ☐ Radionuclides (Ra, U)
- ☐ Inorganics (Pb, As, Cd, Hg, Ag, Se)
- ☐ Metals (V, Cr, Fe, Co, Ni, Cu, Mo)
- ☐ Chloro– or bromo– hydrocarbons
    - tri–, tetra–, and perchloroethane
    - tri–, and tetrachloroethylene
    - Halothane®
- ☐ Organic solvents
    - Methyl ethyl ketone and organic alcohols
    - Aromatic hydrocarbons (not covered by the petroleum exemption)
- ☐ Pesticides
- ☐ Asbestos
- ☐ Radon

Offsite sampling
- ☐ Contamination migration
- ☐ Baseline data (point of compliance)
- ☐ Third party sources
- ☐ Offsite risk assessment

Sampling off–site, on property belonging to a third party, to determine the extent of a plume is required by regulatory agencies. Even if it is not required or there is no reason to believe the plume has migrated beyond property boundaries, off–site sampling is necessary to establish baseline data and to establish liability defense in the event of the inevitable lawsuit. Permission to sample off–site may not be easy to obtain.

Some states have enacted legistation giving the implementing agency power to order off–site sampling. Most rely on persuasion. A third party property owner who denies permission to sample risks incurring liability for the contamination even though it migrated on to the property from elsewhere.

Sampling and Sampling Protocols

The most neglected aspects of environmental data collection operations have been in the areas of sampling design and sample collection (Johnson and Haeberer, 1991). The quality of environmental data has traditionally been regarded as having been determined by the quality of laboratory analyses. As noted in the next chapter, extensive protocols have been developed to standardize laboratory environmental measurement methods, develop QA/QC, improve documentation and automation, install LIMS, and provide for periodic audits. In fact, the pendulum has shifted too far on the opposite direction; currently, laboratory analyses are sometimes subordinate to good science. Documented procedures must be followed even when the procedure does give the most useful results. The conclusion is that the laboratory is rarely if ever the principal source of error in the information flow.

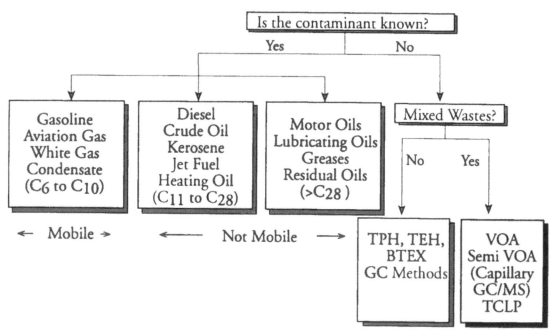

Figure VI–2. Decision Tree for Contaminant Determination. If the contaminant is known to be a petroleum product (only), it is necessary only to establish the exact identification of the product: gasoline (leaded or unleaded), diesel, fuel oil, etc. If the contaminant is unknown, it is necessary to run more extensive tests to identify the contaminant precisely and distinguish between regulated products and products regulated under Subtitle C of RCRA.

A significant potential exists for environmental data to be flawed due to errors in sampling design and sample collection; problems that often go entirely undetected in the laboratory. The old expression in software design, "Garbage In, Garbage Out," certainly holds true in laboratory analyses. The reliability of lab data is crucial since decisions based on this data affect risk assessment and management, remediation technologies, evaluation, and monitoring. It is relevant to note that sampling occurs high up in the flow charts of Chapter 8. Sampling and analyses are expensive, and sampling errors can significantly increase the cost of a remedial effort.

Effective sampling is achieved by defining what is to be accomplished, planning how to accomplish the defined goal, and assessing the effectiveness of the field work and results. A major obstacle to these actions is summarized in the following wonderful quote:

> ...(I)n technocracies there are predominately two types of individuals: those who understand what they do not manage, and those who manage what they do not understand (Johnson and Haeberer, 1991).

The statement is referred to as "Putt's First Law." Effective planning then consists largely of establishing effective communication between the two groups.

Planning

The function of the planning stage is to set protocols, establish priorities, define the potential for problems, and assess the risk to humans and the environment. This stage should pose questions relevant to identifying and obtaining data required to determine the extent and physical and chemical nature of the contaminant and the physical characteristics of the subsurface environment. This information in turn is crucial to support remediation decisions and to establish criteria for monitoring and evaluation.

One of the first steps in designing sampling protocols is to establish the error level acceptable to the data user. Error level is built in to a defensible statistical–based sampling plan, and determines the frequency and distribution of samples. Poor design at this level results in too many, too few or unrepresentative samples. Constraints on acceptable error level affect the level of quality control.

Since the ultimate goal is cleanup and closure, successful planning rests on understanding regulatory guidelines, and on negotiating practical cleanup levels. A successful remediation plan requires thorough documentation. Each primary decision must be documented along with the input associated with the decision, and the data required to support the decision.

Both sampling and planning will be discussed again in the next chapter. The topics have been included here because they are important to the site assessment process. It is impossible to overstate the relevance of proper planning and sampling protocols in a successful remediation project. Surprises are always expensive!

Table 6–3. A Summary of Project Planning. The categories required in planning sampling protocols are listed. Not all sites will require all categories. A few sites may require more (see also Chapter 7). (Clarke, 1991)

- ☐ Selected chemical compounds
- ☐ Selected environmental media
- ☐ Acceptable error levels
- ☐ Required detection limits
- ☐ Anticipated concentration levels
- ☐ Anticipated difficulties
- ☐ Need for special analytical procedures
- ☐ Required turnaround times
- ☐ Field QA/QC
- ☐ Laboratory QA
- ☐ Report format
- ☐ QA/QC deliverables
- ☐ Number of samples
- ☐ Location of samples
- ☐ Frequency of sampling
- ☐ Laboratory response times and required holding times
- ☐ Special reporting requirements
- ☐ Required documentation
- ☐ Logistics
- ☐ Sampling collection and handling procedures
- ☐ Chain–of–custody
- ☐ Decontamination and cross contamination avoidance procedures

7 Environmental Sampling and Laboratory Analysis

The process of generating information describing the nature and extent of contamination at a site is obviously a crucial step for any remediation project. If the vertical and horizontal extent of contamination is not defined clearly and completely, well design and placement are only guesswork, a remedial technology cannot be selected, and the corrective action plan will not be approved. Obtaining usable information is the result of generating data and interpreting the data. Data are just numbers, information tells the user something about the site.

The only way to obtain the necessary data is through sampling and laboratory analysis. Initial planning, therefore, must include sampling protocols and data quality objectives, DQOs. One criterion that is often overlooked is appropriateness of both sampling and analyses. Appropriateness includes sampling sites, number of samples, and type of analyses requested. Data are only useful in the context of the overall site assessment scheme, applicable regulatory guidelines, and the needs of the end data user.

In recent years an interesting phenomenon has occurred. The emphasis on generating and using experimental data in the course of a project has shifted from the *scientific* quality to the *legal* quality of the data. The interesting part is that from the viewpoint of a scientist accustomed to the peer review process, this is an extraordinary transition. It means that sample collection and generation and interpretation of data has been taken out of the hands of scientists and engineers (where it belongs) and placed in the hands of lawyers (where it does not belong). Lawyers are good at law, not at analytical chemistry.

The course of events was probably inevitable for two reasons. First, the EPA has published standards and protocols for sampling and laboratory analyses, as for example, SW–846 for drinking water analysis and other protocol documents that are listed in the References section. The original purpose of the standards was laudable; to set baseline standards for routine environmental sampling and analysis. The standards have evolved into a set of rigid guidelines that must be followed at all costs.

The second reason is that most contaminated sites — petroleum or chemical — involve litigation, or at least attorney–mediated negotiation to some degree. During this process defendant's attorneys *always* attack plaintiff's data. The entire process is questioned; collection, analyses, evaluation, and interpretation. If, for example, an analysis was not carried out according to EPA Method 8240 or 602 or some other protocol, the data are suspect or invalid. The consequences are serious and expensive.

One unfortunate result is that laboratories are constrained to follow mandated procedures that may not always give the best data. There is no flexibility left for operators to use their best judgment in performing a given analysis. It must be done "by the book" and the situation arises that valid data and useful information are not necessarily the same thing. The conclusion is that development of sampling and analysis protocols during the planning stage is as often driven by considerations of data *defensibility* as by the information content contained in the data.

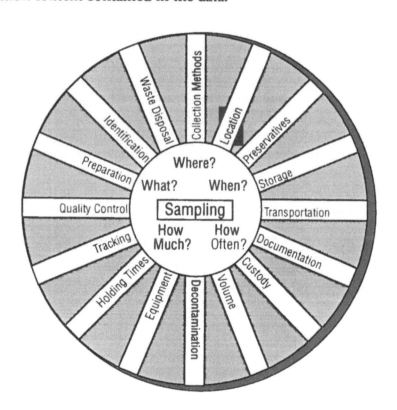

Figure VII–1. Sampling Wheel. Even relatively simple sites require appropriate sampling techniques. The sampling wheel shows the 16 factors that are important to collecting valid data. The factors are arranged in no particular order. Each one must be taken into account when setting up sampling protocols.

All this having been said, in this chapter it is assumed that *valid* data and *useful* data are concomitant goals. The treatment begins with sampling requirements and continues to laboratory analyses. The chapter does not attempt to provide a complete guide to environmental sampling and analytical procedures (see Keith, 1988). Rather, issues are listed that should be considered in collecting and analyzing samples and developing information about the site.

Sampling Procedures

Sampling in the environment for chemical analyses is a complex subject with many different aspects (**Figure VII–1**). The development of meaningful sampling protocols demands careful planning of the actual procedures used in sample collection, handling, and transfer. Preliminary sampling and a well–conceived sampling experiment can provide the validation and experience necessary to design efficient sampling protocols that will meet program needs (Barcelona, 1988). It is a common misconception that one can simply collect a sample and take it to the laboratory to find out what is in it. This approach, which is frequently noted at petroleum contaminated sites, can lead to wasted effort, excessive laboratory costs, and flawed data (Bryden and Smith, 1989). It is especially important to use appropriate protocols and exercise care when sampling groundwater since contaminant values are often near detection limits.

It is necessary to plan sampling protocols that will generate valid data and useful information (see **Figures VII–2 and VII–3**). Requirements include the following:

- Regulatory jurisdictions;

- Regulatory compliance guidelines;

- Proper documentation;

- Statistical validity;

- Relevance to the needs of the site;

- Appropriate sampling techniques.

Sampling must be carried out with regard to the appropriate regulatory jurisdiction. Examples include city ordinances, sewage and storm drainage districts, landfills, water districts, health or environmental departments, and U.S. EPA. Fortunately most areas have straightforward jurisdictions; California excepted, of course. Sampling protocols are based on which regulatory requirements are being enforced.

Regulatory compliance guidelines must be considered at the earliest possible stage. Sampling procedures vary depending on the particular matrix and location that are being sampled. Compliance levels for contaminated soils must be negotiated.

All sampling must be properly documented. Sampling personnel must set up, document, and follow Standard Operating Procedures, SOPs, and Field Operating Procedures, FOPs. Log books must be used and maintained. Changes in procedures must be referenced in an appendix or addendum.

Sampling must have statistical validity. Representative samples of the contaminated site must be collected. A representative sample is one that exhibits average properties of the whole waste. Sufficient samples must be taken to represent the variability of the contaminated site. If groundwater is being sampled, four replicates (subsamples) should be taken for analysis and the mean contaminant concentration level be calculated.

There are many standard sampling protocols available (see References). Early in the planning stage it must be decided which concerns must be addressed that meet the needs of the specific site. These considerations should include relevant peripheral information, such as history, soil gas surveys, possible offsite sources of contamination, and initial composite or exploratory sampling results. One of the key factors here is that sampling and analysis are expensive. Therefore, an early focus is needed to stay within budget, yet not be so restrictive as to overlook significant contamination.

Appropriate sampling techniques address scientific and legal quality assurance objectives. Field equipment rinses and travel blanks test whether contamination has been introduced during sampling and transportation. Acceptable precision and accuracy ranges (control limits) dictate sampling and laboratory QC procedures.

Planning Sampling Protocols

Sampling plans and protocols must be set up in advance — either by adopting existing procedures of the environmental company or by designing custom protocols in the case of a complex or unusual site. Sampling plans may be intuitive- or statistically–based or a hybrid of the two methods. Statistically–based plans are generally more defensible, but intuitive–based plans may yield more information. The choice depends on the needs of the end data user.

Sampling protocols should include data quality objectives, DQOs, sometimes called, *quality assurance project plans,* QAPjPs; that is, a detailed site specific quality assurance program that may be part of a work plan or a standalone document. DQOs include the following:

- Project description;
- Monitoring network design;
- Monitoring parameters;
- Collection frequency;
- Analytical references;
- Project organization;
- Data quality requirements and assessments;
- Data documentation, reporting and validation;
- Program auditing;
- Corrective action.

Project description should include a summary of the project goals, organization, and responsible personnel. The site description should include a detailed site map, including topography and surroundings, and an appropriate reference grid for documenting samples. If the client needs to contact the laboratory regarding a specific sample or set of samples, there must be some common ground for communication between client and laboratory personnel. (See section on **Operating Procedures** below.) Both parties need to refer to samples using a common language. The simplest way to accomplish this is to use a grid referenced to the site or to unambiguously label samples with reference numbers referred to a log book.

The monitoring network design and layout should accurately describe the contaminant plume. If preliminary investigation indicates that the contaminant plume has migrated offsite, this section should include plans for monitoring contaminant levels offsite.

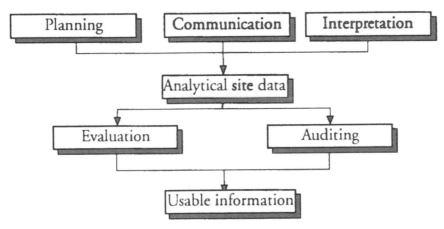

Figure VII–2. Obtaining Usable Data. The chart shows the factors important in obtaining usable data. Planning and communication begin the process and establish the parameters within which data collection is carried out; Interpretation converts data into information and auditing guarantees validity of the information. Evaluation relates the information to the site.

Monitoring and evaluation parameters are introduced in Chapter 8. Monitoring parameters may be simply measuring benzene levels in contaminated strata, CO_2 evolution, or there may be more complex requirements such as monitoring bacteria populations or VOC levels in groundwater. Whatever the choice, monitoring parameters should be cost–effective and compatible with regulatory compliance guidelines.

Operating Procedures

Sampling sites should have a reference grid superimposed on a detailed site map to document the exact location of each sample or set of samples. Ideally the grid size should be fine enough to allow one sample per grid section. A sampling grid provides a convenient and exact device for communication between supervisors and field personnel and with laboratories. Laboratory personnel frequently need to communicate with the user. A sample reference system simplifies communication.

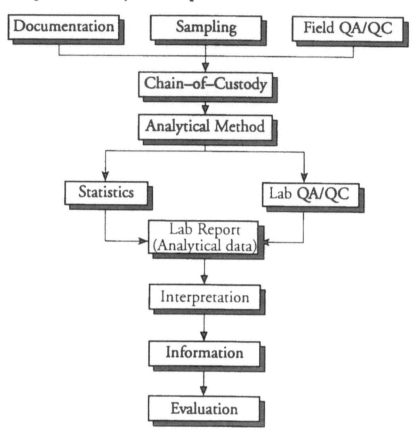

Figure VII–3. Data Flow in Sampling and Analyses. The flow of data through sampling and laboratory analysis includes documentation, field QA/QC, and chain–of–custody. The choice of analytical method(s) depends on the level of information required. The lab report must be interpreted carefully with regard to the analytical method, accuracy, and detection and error limits.

All references to samples, descriptions of sampling activities, site references, and other day–to–day documentation must be entered permanently in log books. The first requirement is a Master Logbook that is hardbound (no loose pages) with permanently numbered pages. The Master Logbook never goes into the field. The Master Logbook must be keep in a secure area with limited access. Purely as a matter of responsibility, access should be restricted to a few known individuals. That is, only certain authorized personnel may have access to the Master Logbook, and that access is always documented. Extreme security measures may be necessary only in sensitive or legally contested sites; nevertheless, access to master logs should be monitored. The problem is always one of authenticity. It may be necessary in court or in a deposition to be able to testify that the references in the Master Logbook are authentic and unaltered.

Field Logbooks are referenced to the Master Logbook, and then checked out as needed, noting the responsible person, departure time, date, destination, and site. (See also **Table 7–1**). Information in the Master Logbook should include the following for each incidence:

- Date(s) of sampling;
- Site address (location);
- Sample collection location (or grid index);
- Sample reference number for identification;
- Sampler
- Deviations, if any, from prescribed protocols, with reasons and references.

Logbooks are legal documents and must be treated as such. The information is the basis for client reports and corrective action plans and may need to be accessed years after data entry. Eventually, the information gathered in logbooks will be tested in court, or at least in affidavits. Invalid data are worthless and a very expensive waste of time. The following are some rules of the road for using logbooks.

- Never tear out pages. Pages are permanently numbered and omissions or skips are immediately obvious and can invalidate data.

- All entries are made in ink; corrections are made by crossing out the error and reentering the data with the responsible person's initials.

- Pages are signed and dated as they are used. At regular intervals, logbooks should be notarized.

- Deviations from FOPs are necessary on occasion. When this happens the deviations must be noted in an appendix together with reasons and circumstances.

Table 7–1. Field Logbook Checklist. These items should be included in a field logbook protocol to accompany other sampling protocols. The list is not necessarily complete for all cases, but does include the items most frequently required. (Adapted from CDPHE, Groundwater Monitoring Guidance, 1991).

- ☐ Sampling area grid identification
- ☐ Sampling site location
- ☐ Trip blank (Yes/No)
- ☐ Identification and/or location of well
- ☐ Time (to the minute, military) and date of activities
- ☐ Well depth
- ☐ Static water level depth and measurement technique

☐ Purge volume and pumping rate

☐ Time well purged

☐ Well evacuation procedure and equipment

☐ Sample withdrawal procedure and equipment

☐ Sample blank (Yes/No)

☐ Date and time (to the minute, military) of collection

☐ Frequency of collection

☐ Well sampling sequence

☐ Decontamination procedures

☐ Types of sample containers used and sample identification numbers

☐ Matrix spike (Yes/No)

☐ Duplicates (Yes/No)

☐ Preservatives used

☐ Reagent blank (Yes/NO)

☐ Parameters requested for analysis

☐ Field analysis data and method

☐ Sample distribution and transporter

☐ Field observation of sampling

☐ Name of collection

☐ Climatic conditions including air temperature

☐ Internal temperatures of field sample containers with time of measurement

☐ Internal temperatures of shipping containers with time of measurement

Field Sampling

The Heisenberg Uncertainty Principle, a fundamental concept in quantum physics, states that the position and momentum of an elementary particle can never be known with precision simultaneously. The reason is that the very process of observing changes the state of the particle. The principle has application in field sampling: The process of sampling can alter the values being measured. Therefore, it is important to formulate field sampling plans that minimize the possibility of error, maximize the usable information, and minimize costs.

The total variance (or uncertainty) in measurement data is

$$s^2_{total} = s^2_{measurement} + s^2_{sample}$$

where s^2_{total} is the total variance, $s^2_{measurement}$ is the variance due to the process of measurement, and s^2_{sample} is the variance due to the sample itself (Taylor, 1988). Protocols are designed to minimize the value of each parameter. The variance in measurement is controlled and evaluated through an appropriate QA/QC program. Replicates, splits, duplicates, and blanks are straightforward to design so that measurement variance can be known with confidence.

Sample uncertainty is a combination of systematic and random errors. Sample uncertainty can be difficult to estimate especially when analyses near the limits of detection are required. Sample variance arises because of a lack of randomness in sampling sites (the sampling does not result in data representative of the site), device errors (errors due to the sampling device itself), and discriminatory sampling (using personal bias in selecting sampling sites). Discriminatory sampling arises most often in an effort to keep costs at a minimum. It can be a positive factor, but can also be difficult to defend from a legal point of view.

Field protocols must begin with the level of information required. Decide in advance exactly what information is needed and how it can be obtained. For example, if soil cleanup levels are established at 100 ppm benzene, it is unnecessary to use "clean room" decontamination procedures. Likewise, sampling procedures appropriate for, say TEPH, are not appropriate for volatile analytes.

"Grab samples" or composite samples may be appropriate for initial site characterization, but not for a second stage report. The accuracy of collection is mirrored in the accuracy of the information. At each level consult with regulatory personnel — often a single individual; he/she can be helpful in negotiating specific requirements and constraints. Other questions include the necessity of TCLP sampling, procedures for groundwater sampling, and sampling for other contaminants.

All sampling personnel must be familiar with the goals of the project so that the purpose of specific sampling protocols is understood and appropriate field decisions can be made.

Contamination of samples and sampling equipment is a common source of error, particularly at petroleum contaminated sites. Contamination avoidance and decontamination procedures should be built into field protocols to avoid systematic errors propagating through the data. Contamination from outside sources can be controlled or estimated by obtaining background or baseline data for the site.

Always use appropriate containers designed for specific sampling needs. The customer service representative at the destination lab can be very helpful in recommending appropriate sample containers.

Samples must be preserved according to specifications and are typically stored at 4°C (40°F) using either ordinary ice or "blue ice." Ordinary ice should not be used with water sensitive samples. Internal temperatures must be measured before, during and after sample collection, and before and after shipping.

Chain–of–Custody

Chain–of–custody rules and protocols must be observed absolutely (Table 7–2). Chain–of–custody is an apparently simple concept, but it is one of the most frequent points on which data are thrown out of court. (Glade, 1993). The following points should be noted on chain–of–custody.

- Never leave samples unattended or unobserved. Samples or the sample container must be under positive observation at all times until relinquished to the next person in the chain.

- Always follow field operating procedures wherever possible. If it is necessary to deviate or modify procedures to adapt to local conditions, document and justify the changes.

- Always document transfer of custody. Sample chain of custody documents are shown in Appendix D.

- It is appropriate to video tape sampling activities to clearly document procedures, especially at sensitive sites.

Table 7–2. Chain–of–Custody Record Checklist These items should be included in a chain–of–custody record. The list may be used as a checklist or the items are included in the sample document shown in Appendix D. The list is not necessarily complete for all cases, but does include the items most frequently required.

☐ Sample number

☐ Grid reference (if applicable)

☐ Signature of collector

☐ Date and time (to the minute, military) of collection

☐ Identification of well

☐ Number of containers

☐ Parameters requested for analysis

☐ Signatory of person(s) involved in chain of possession

☐ Inclusive dates of possession

☐ Internal temperature of shopping container

☐ Maximum temperature recorded during shipment

☐ Internal temperature of shipping container upon opening at the laboratory

☐ Disposal of wastes

Decontamination and Cross–Contamination

Data validity and accuracy in field sampling depend heavily on avoiding contamination and cross–contamination while sampling. In one sense the fact that petroleum products are abundant and common in commercial use can create problems and false contamination levels. It is nonproductive to go to the expense of putting in a well to sample groundwater only to find that the laboratory values reflected someone's cleaning solvent spilled on the ground beside the well (**Figure VII–4**).

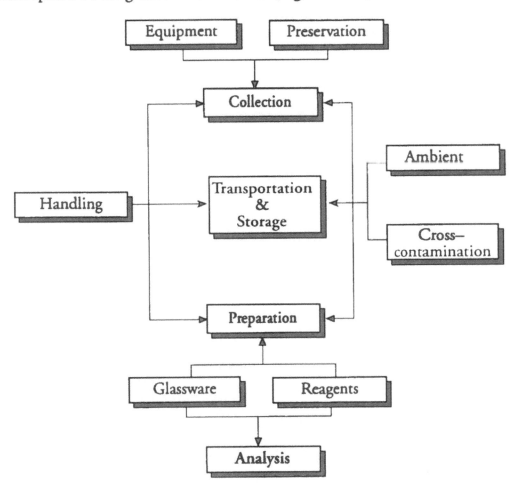

Figure VII–4. Sample Contamination Sources. Contamination is a common source of error in all types of environmental measurements. Contamination can be introduced at several points: from ambient sources, in the field, during transportation, or at the laboratory. Sample handling is a major source of contamination. Ambient contamination is a serious problem due to the ubiquitous nature of petroleum products. Sampling tools are another source. Other sources include glassware and contaminated reagents. Contamination is controlled and estimated by the use of blanks.

Precision of measurement is estimated from the standard deviation of the random errors that affect the measurement. The square of the standard deviation is the total variance, which can be represented as the sum of the variances of individual operations. Thus, the total variance can be written as:

$$s^2_T = s^2_x + s^2_t + s^2_s + s^2_r + s^2_g + s^2_p + s^2_e + s^2_m$$

where s^2_T is the total variance, s^2_x is spatial, s^2_t is time, s^2_s is sampling, s^2_r is transportation, s^2_g is storage, s^2_p is preparation, s^2_e is chemical treatment, and s^2_m is measurement. The magnitude of these factors is estimated and evaluated by the use of blanks and controls as shown in **Figure VII–5** (Black, 1988 and Lewis, 1988). Cross–contamination can be a serious problem at a site with multiple wells since the same bailer or spoon is used to collect repeated samples.

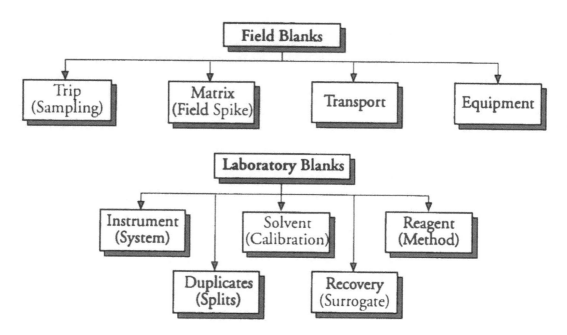

Figure VII–5. Blanks and Controls. Blanks and controls are two versions of the same process; namely, to estimate the degree of contamination, to evaluate contamination sources, and to authenticate laboratory results. In this chapter blanks and controls are used synonymously as quality control and assurance methods. Field blanks and laboratory blanks are necessary to ensure quality data.

Decontamination procedures are basically efforts to guarantee that sampling personnel are using very clean equipment and containers and to remove a source of error due to equipment contamination. The procedures are similar to the practices required in a quantitative analysis laboratory or in a bacteriological laboratory. Glassware is straightforward to clean since it does not dissolve in anything and is resistant to heat; plastic requires more careful selection of reagents and treatments.

Whenever possible, equipment should be cleaned in the laboratory or preparation facility rather than in the field. Cleaning is usually accomplished with hot water and laboratory grade detergent followed by rinsing, first with hot water then distilled or deionized water, DI. This works well for spoons, bailers, and other containers, but pumps can require disassembly, cleaning, and reassembly. At sites with heavy contamination from hazardous wastes, heavy equipment will need decontamination.

Never use organic solvents such as acetone on equipment intended for VOA sampling. Containers to be used to collect samples for metals analyses should be rinsed with 6M nitric acid and rinsed with DI water. During cleaning, attention must be paid to the waste generated as a result of washing and to appropriate health and safety procedures. Alternatively, use commercial precleaned, disposable containers. Wraps and protective covers on commercial equipment should be left on until the equipment is ready for use in the field.

In the field, equipment should be laid on a protective groundcloth to ensure against accidental contamination. Equipment must be cleaned between each well or sampling point to avoid cross–contamination. Rinsate from cleaning should be collected even if it is not hazardous to avoid contaminating wells. Field decontamination is less thorough than laboratory cleaning, but may be dictated by circumstances. DI water and detergent must be sufficient for field decontamination. Therefore, equipment blanks are used to detect inadvertent contamination.

Sample Storage and Holding Times

Field–transportable lab instrumentation is becoming more popular, more affordable, and is cost–effective. A truck outfitted with sample handing facilities, a GC and an IR can make data available in a matter of hours instead of the weeks normally associated with laboratory analyses. Field–transportable laboratory instruments can reduce the number of laboratory samples from; for example, 100 to 10, with a concomitant savings in the cost of the analyses. However, field labs are expensive, require qualified technicians to operate, and the QA/QC may not be sufficient for the data to be acceptable to the regulatory agency.

Thus, in normal circumstances environmental sampling procedures require that samples be transported to laboratory facilities for analysis. Because samples can deteriorate with time and become nonrepresentative, standard methods dictate the amount of time that can elapse between sample collection and analysis or extraction. Allowance for holding times must be built in to sampling protocols (Maskarinec and Moody, 1988).

Typical holding times are 7 to 14 days. It is important to note that holding times are measured from the moment of collection to actual analysis or extraction. If holding times are exceeded, the data that are generated may not be usable; that is, the data are neither representative nor defensible. If holding times have been exceeded, laboratory customer service representatives should make note of that fact and not proceed with an analytical procedure that results in unusable data.

For volatile analyses a typical holding time is 7 days from collection to analysis —
not delivery to the lab. A typical holding time for semivolatiles is 14 days (**Table 7–3**).
For some samples that must go through an extraction procedure, the holding time
applies through the extraction. After extraction there is an additional time period until
actual analysis. For short holding times it is prudent to reserve time in advance at the
laboratory for analysis. Otherwise, the analyses become very expensive.

*Table 7–3. Laboratory Methods. Standard laboratory methods are listed by EPA
reference number. The second column,* Analysis, *gives the type of compound
analyzed for, and the third column,* Technique, *is the instrumental method used.
The fourth column,* Container, *is the normal recommended field sample collection
container. The fifth column,* Preservative, *lists the recommended or required
preservative. If no preservative is used, holding times are usually cut in half. The
sixth and seventh columns,* Holding Times, *list the maximum time between
sample collection and analysis (volatiles) or extraction. The last column,* Sample
Preparation, *lists the standard method of preparation for the corresponding
method. These may vary depending on specific analytical needs. Addition of
preservative is based on the assumption of 18M HCl, which is the standard 1:1
dilution of concentrated 36M HCl.*

Notes for Table 7–3.

Detection methods
   FID = Flame ionization detection
   PID = Photo ionization detection
   ECD = Electron capture detection
   HPLC = High pressure liquid chromotography

Sample containers
   G = Glass container only; Teflon septum
   GA = Amber glass container only; Teflon septum
   ZH = Zero headspace VOA container only; Teflon septum

Preservatives
(Required in hydrocarbon samples to prevent biological degradation.)

   a Approximate volumes for pH < 2.
   b d = drops (20 drops = 1 mL);
      based on 1:1 dilution of 36M HCl to 18M HCl
   c 14 days preserved; 7 days unpreserved

Sample preparation
   P&T = Purge and trap

Holding times
   d 7/40 = 7 days to extraction; analysis 40 days after extraction

Method	Analysis	Technique	Container Size/Type	Preservative [a] (Chill to 4°C)	Holding Times Water	Holding Times Soil	Sample Preparation
418.1	TPH	IR	1L/G	2 mL H_2SO_4	28 days	28 days	Extraction
503	Aromatics	GC/MS	120 mL/ZH	2 d HCl [b]	14 days [c]	14 days	P&T
601	Halogenated	GC/PID	120 mL/ZH	2 d HCl	14 days	14 days	P&T
602	Aromatics	GC/PID	120 mL/ZH	2 d HCl	14 days [c]	14 days	P&T
604	Phenols	GC/FID	120 mL/GA	2 d HCl	7/40 days [d]	7/40 days	Extraction
608	Pesticides, PCBs	GC/ECD	1 L/GA	2 d HCl	7/40 days	7/40 days	Extraction
610	PAHs	HPLC	1 L/GA	2 d HCl	7/40 days	7/40 days	Extraction
624	VOCs	GC/MS	120 mL/ZH	2 d HCl	14 days	14 days	P&T
8010	VOCs	GC/FID	120 mL/ZH	2 d HCl	14 days [c]	14 days	P&T
8015	TPH (diesel)	GC/FID	120 mL/ZH	4 d HCl	14/40 days	14/40 days	Extraction
8020	Aromatics	GC/PID	120 mL/ZH	2 d HCl	14 days [c]	14 days	P&T
8040	Phenols	GC/FID	120 mL/GA	None	14 days [c]	14 days	Extraction
8100	PAHs	GC/FID	120 mL/GA	None	7/40 days	7/40 days	Extraction
8240	Aromatics	GC/MS	120 mL/ZH	None	14 days [c]	14 days [c]	P&T
8260	Aromatics	GC/MS	120 mL/ZH	None	14 days [c]	14 days [c]	P&T

During the time of transportation, samples must be stored such that the samples that arrive at the laboratory are still representative of the site conditions. For analyses appropriate to petroleum contaminated sites this normally means storage at 4°C (40°F); that is, average refrigerator conditions. Samples should be packed in portable, well–insulated coolers in which the temperature is maintained with ice or "blue ice." Normal sampling protocol calls for the field operator to measure the internal temperature of the container at the time of storage and when the container arrives at the laboratory.

Care must be taken in the following conditions:

- Do not to use ordinary water ice if the samples are water sensitive;

- Do not freeze aqueous samples;

- Soil samples may be frozen, but not if being analyzed for VOCs since freezing causes loss of volatile components;

- Samples that have been allowed to warm, then recooled are no longer representative;

- Field and trip blanks, replicates, and spikes must be maintained under exactly the same conditions in the same container as field samples.

Problems with Sampling

In all the steps from planning to information there are many opportunities to invalidate the information. The most common place for mistakes and oversights is in sampling; failure to follow sampling protocols. At least part of the reason for this has to be the use of inadequately or improperly trained personnel. Sampling is an exacting activity that requires attention to detail and intelligent thought.

The following is a list of the most frequently encountered problems in collecting field samples.

1. Field operators are not aware of the objectives of the project or of any requirements specific to the site. In these circumstances trained technicians are unable to make intelligent decisions on the spot. The results are the collection of invalid samples, loss of time, and increased costs.

2. Field operators do not follow written plans for collection. Field operators must follow FOPs whenever and wherever possible. Situations do arise when it is not possible to follow prescribed protocols. In these cases trained operators must be able to make decisions on the spot. All deviations from FOPs must be documented in appendices.

3. Inappropriate containers are used. The most blatant example is to use inappropriate containers when collecting VOA samples. Volatile means that a compound will vaporize easily. Analyzing soils for VOAs after half

the volatile compounds have vaporized is an expensive waste of time and effort.

4. Samples are not identified or identified improperly. This is a very common problem for laboratory personnel. The solution is to develop a consistent site location system and site identification grid.

5. Chain–of–custody sheets are not filled out properly. Without a valid chain–of–custody, record samples are not valid and, therefore, the information derived from the samples is not usable. Care in observing chain–of–custody can prevent expensive resampling.

6. Samples are not packaged or sealed properly. Samples usually must have tamper–proof seals placed at the time of collection. VOA samples must be carefully sealed.

7. Actual sampling activities are not properly documented. See comments following item 2.

8. Field personnel have no knowledge of analytical procedures with respect to volume or amounts of sample required. Volumes collected should be 3 or 4 times in excess of what is required for laboratory analyses.

9. Samples are contaminated by outside sources or cross–contaminated from other wells.

10. Inadequate field QA/QC, such as field blanks, duplicates, splits, recovery blanks, etc. This is usually due to efforts to economize or stay within an inadequate budget. If the data are never challenged or questioned by regulatory agencies, there would be no problem. Unfortunately, that is seldom the case.

11. No backup equipment or spare parts available. Field personnel should go out properly prepared and equipped.

Laboratory Methods

Laboratory analyses generate potentially vast amounts of information, but are only one part of the process of generating *usable* information (see, for example, **Figures VII–2 and VII–3**). Data and information are not synonymous; information is data that have been interpreted into a useful form. Therefore, it is important to know what analyses to ask for in order to generate appropriate information.

At petroleum contaminated sites soil and groundwater samples usually require semivolatile or volatile analyses. If the contamination is gasoline, required analyses will include TPH and BTEX. If the contamination includes diesel, TPH and TEPH will be necessary.

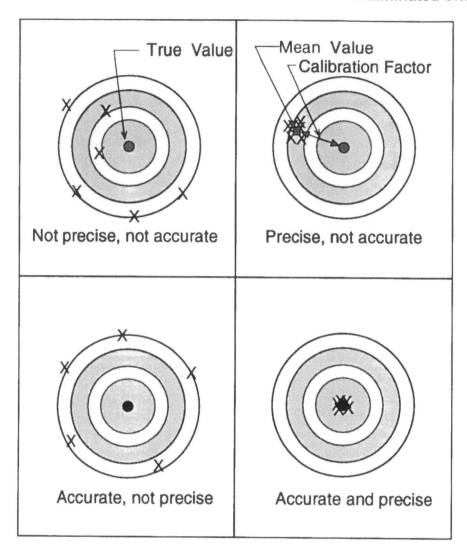

Figure VII–6. Precision and Accuracy. In evaluating analytical results it is important to remember that precision and accuracy are not the same thing. A common mistake in using field instruments is to confuse accuracy and precision. Users are typically looking for accuracy when they should be looking for precision. Precision refers to the standard deviation of a set of values; it is a measure of the reliability of the analytical method. Accuracy refers to the approach of a measurement of some "true" value. Most good instruments are precise, but not necessarily accurate. The correction to the "true" value is called a calibration factor and is common in analytical methodology.

Identification and quantitation of contaminant compounds is a sequence of separation and identification. For petroleum hydrocarbons separation is carried out by extraction and/or gas or liquid chromatography, which may be followed by mass spectrometry. Identification is carried out by spectral analysis, usually by comparing the observed spectrum with known library spectra. Library spectra offer quick, inexpensive identification of routine compounds, but must be used with caution when analyzing pesticides, PCBs, and unusual mixtures.

Separation of volatile compounds is done by gas chromatography (GC) (**Figure VII–7**). This is a separation technique in which the sample is vaporized, mixed with an inert carrier gas, usually helium, and passed through a small diameter column containing a stationary phase. To create the stationary phase the column is packed with a silica–based material that retards diffusion of the volatile compounds to a greater or lesser degree. As the vapor diffuses through the column, the lighter, more volatile components diffuse faster while heavier, less volatile components are retained. The separation is thus a function of molecular weight.

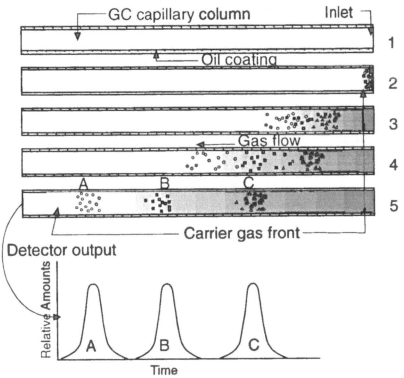

Figure VII–7. Chromatography Separations. A liquid chromatography column that would be used, for example, to separate the component of diesel fuel, consists of packing called the stationary phase. The mixture to be analyzed is injected into the column together with an appropriate solvent. As the solvent migrates, the compounds present in the mixture migrate at varying rates depending on how strongly they are attracted to the stationary phase. In step 2 the mixture is beginning to be separated. By step 5 the three components are essentially separated. At the bottom is an illustration of the appearance of the detector output. In this mixture compound A is attracted least strongly to the stationary phase; compound C is attracted most strongly.

For low resolution GC, a packed column is used (Method 8240 or 8250); for high resolution GC a capillary column is used (Method 8260). A capillary column has inherently higher resolution than the larger diameter columns, but is associated with somewhat higher costs per sample.

Separation of semivolatiles is also done by GC, but must be preceeded by physical manipulations — extraction and separation procedures. Extraction means that the sample is washed with a solvent or series of solvents to dissolve and separate the component(s) of a mixture of contaminants. The analyte(s) of interest are concentrated and analyzed. Thus, analysis of "semivols" is usually a more involved procedure than analysis of volatile contaminants.

In the analysis of either vols or semivols, individual compounds are separated in time by passing through the column. As the molecules leave the end of the column a detector is activated and records a chromatogram. The pattern of bands is then interpreted by reference to library or calibration chromatograms. In **Figure VII–7** compound C is most strongly attracted to the stationary phase and hence moves along more slowly. Compound A is attracted least strongly and moves more rapidly. In real hydrocarbon mixtures the large number of compounds result in overlapping peaks on the chromatogram which makes interpretation and identification difficult. Computer analysis is necessary to resolve peaks into individual compounds.

One difficulty with GC is that the method does not distinguish well between molecues with similar molecular weights and structures, toluene and methyl cyclohexane, for example. Further separation must be carried out by mass spectroscopy (MS). This is a method in which molecules are converted to ions by bombardment with electrons. The ions are passed through an electromagnet where they are deflected and separated according to mass. Individual ions are detected and counted. The pattern that results is not simple, even for a pure compound. To be useful as a routine analytical tool, patterns must be catalogued and stored in automated libraries. When a field sample is analyzed, the observed pattern is compared to a library pattern and specific, individual compounds are identified.

Data Reporting

There is considerable confusion surrounding the units used in reporting amounts of contaminants. Regulatory guidelines may state maximum contaminant levels (MCLs) or practical quantifiable limits (PQLs) and remedial guidelines for soils or water in weight of contaminant per kilogram or liter of sample.

$$\frac{\text{Weight in grams of contaminant}}{\text{Total weight of soil sample in grams}} = \text{A fraction in units of grams of contaminant per g of soil}$$

$$\frac{\text{Weight in grams of contaminant}}{\text{Total volume of water sample in milliliters}} = \text{A fraction in units of grams of contaminant per mL of water}$$

Since amounts of contaminants are usually very small relative to the total amount of sample the fraction that results is very small, on the order of one millionth, 10^{-6}, or one billionth, 10^{-9}, of a gram of contaminant per kilogram or liter of sample. These numbers are too small to use conveniently. It is customary to convert these very small numbers to numbers that are more convenient by multiplying the fraction above by a multiplier. The multiplier is arbitrary depending on the relative size of the fraction. The process is familiar in everyday life in the use of percentages. The notation "10%" is more convenient to use than the fraction 0.01.

Table 7–4. Conversion Multipliers for Common Fractions. To convert very small contaminant levels into more conveniently sized numbers, the fraction, [g of contaminant/gram of sample], is multiplied by arbitrary multipliers such that the result is close to one. If the multiplier is 100, 10^2, the result is called percent; if the multiplier is one billion, 10^9, the result is parts per billion, ppb. The multiplier is always applied to the fraction g/g. Customary practice uses kilograms, Kg, of soil or liters, L, of water. Therefore, equivalent units for soil and water samples are listed. Abbreviations used are as follows: dg = decigram, mg = milligram (10^{-3} g), µg = microgram (10^{-6} g), ng = nanogram (10^{-9} g), ppm = parts per million, ppb = parts per billion, ppt = parts per trillion.

			Equivalent Units	
Factor	Multiplier	Name	For Soils	For Water
10^0	One	Unity	1 g/g	1 g/mL
10^2	Hundred	Percent	1 dg/Kg	1 dg/L
10^3	Thousand	Per mille	1 g/Kg	1 g/L
10^6	Million	ppm	1 mg/Kg	1 mg/L
10^9	Billion	ppb	1 µg/Kg	1 µg/L
10^{12}	Trillion	ppt	1 ng/Kg	1 ng/L

For example suppose that a soil sample is brought to a lab for analysis. The soil sample is weighed and found to be 0.50 Kg (1.1 lbs). The sample is divided into smaller portions and analyzed for benzene, toluene, ethyl benzene, and xylenes. The actual amount of contaminant found is converted back to the original weight of soil and found to be 23 mg (micrograms or 10-6 grams) of benzene in the soil sample.

$$\frac{23 \ mg \text{ of benzene}}{0.05 \ Kg \text{ of soil sample}} = 460 \text{ mg of benzene per } Kg \text{ of soil}$$

$$or \quad \frac{23 \times 10^{-3} \ g \text{ of benzene}}{0.05 \times 10^3 \ g \text{ of soil sample}} = 4.6 \times 10^{-4} \text{ g of benzene per g of soil}$$

$$or \quad \frac{23 \ mg \text{ of benzene}}{0.05 \ Kg \text{ of soil sample}} \times 10^6 = 460 \text{ ppm benzene}$$

Table 7–5. Laboratory Methods for Petroleum Contaminated Soils. Appropriate laboratory methods and costs per sample are tabulated. Methods refer to EPA drinking and waste water standard methods (see EPA, 1983). Soil analyses refer to EPA methods for solid waste analysis (see EPA, 1984). Costs, which are typical at the time of writing, are listed only for comparison among different methods.

Laboratory Methods for Petroleum Contaminated Sites

Analysis	Method/Cost		
	Potable Water [a]	Waste Water [b]	Soils [c]
Group Methods			
TPH (IR)	418.1/$85	418.1/$85	418.1/$65
TVH (GC)	524.1/$245	624/$245	8015/$65
TVH (GC/MS)			8240/$225
TVH (Capillary)			8260/$245
BTEX (GC)	503.1/$150	602/$85	8020/$85
BTEX (GC/MS)		624/$245	8240/$225
BTEX (Capillary)			8260/$255
TEH		625/$255	8250/$85
TRPH			80154/$85
VOCs [d]	524.1/$245	624/$245	8015/$85

[a] EPA method; see Reference X, p. 251.

[b] SWDA methods; see Reference X, p. 251.

[c] SW-846; see Reference X, p. 251.

[d] Modified method; see Reference X, p. 251.

[e] TVH, BTEX, and VOAs are all quantified by basically similar methods.

Methods Appropriate to Petroleum Contaminated Soils

Laboratory analyses range from relatively inexpensive screening methods that detect broad classes of hydrocarbons to specific methods that detect individual compounds or groups of compounds. Other choices include method of detection and high or low resolution depending on site requirements, level of inquiry, and needs of the end user.

Screening methods include the following: Total Petroleum Hydrocarbons, TPH; Volatile Organic Compounds, VOC, to screen for the presence of mixed wastes, and, possibly, Total Extractable (Petroleum) Hydrocarbons (TEH or TEPH). TPH is basically an infrared (IR) screen for petroleum contamination without regard to type. VOC (or a VOA) will identify organic compounds that may be present. There is no reason for this analysis unless there is suspicion of mixed wastes. TEH is specific for diesel and related hydrocarbons. There is also a Total Recoverable Hydrocarbons (TRH) in use. TRH is an oil and grease analysis and detects heavier oils associated with lubricating and fuel oils.

Methods for specific compounds or groups of compounds include the following: Benzene, Toluene, Ethylbenzene, and Xylenes (BTEX) is a specific method for aromatic gasoline components. This is one of the most important analytical methods for gasoline contaminated soils and groundwater since the aromatics are the most toxic contaminants. A GC run to identify all hydrocarbons through about C_{12} can be done. This procedure is called a Total Volatile Petroleum Hydrocarbons (TVPH) (see Tables 7–5, 7–6, and 7–7). If it is known or suspected that the contamination includes leaded gasoline, soil samples should be analyzed for ethylene dibromide (EDB) and ethylene dichloride (EDC). Also analyses for tetramethyllead (TML) and tetraethyllead (TEL) may be required.

Methods of analysis for benzene, toluene, and ethylbenzene are specified under 40 CFR Part 136. The methods include (see Tables 7–5, 7–6, and 7–7) EPA Methods 602, 624, and 1624 (Standard Methods). No method is specified (as of March 1, 1993) for the analysis of xylenes. EPA Region VIII recommends that the total concentration of the three isomers of xylene be obtained by adding the values for *o*–xylene to the single value for *m*– and *p*–xylene, which coelute and are not resolved by normal commercial GC/MS techniques.

If the presence of mixed waste is suspected, other analyses will be necessary. If waste oil is included in the contamination, it will be necessary to analyze for ethylene glycol, the principal component of antifreeze. Ethylene glycol is frequently found in used oil collection facilities and, as a result, is a common soil contaminant where waste oil tanks have leaked. Waste oils are often found to be contaminated with ethylene glycol, PCBs, PCP (pentachlorophenol), and inorganics, including lead, chromium, and arsenic (FR, 1991).

The Toxic Characteristic Leach Procedures (TCLP) a screen for a number of hazardous wastes, may be required for contaminated soils containing mixed wastes. The TCLP is explained below. Volatile organic analyses (VOAs) may be required to detect the presence of volatile organic compounds (VOCs). Other specific methods that may be needed as required include analyses for phenol, cumenes (additives in unleaded gasolines), and creosotes (wood preservatives).

Table 7–6. Analytical Methods for Quantifying Petroleum Hydrocarbons. The first column is the quantity measured by the method listed in column 2; the second column is the standard method used. Numbers refer to EPA standard methods (see references given in Table 7–5). The third column gives lower detection limits and significant figure limits.

Laboratory Methods for Petroleum Contaminated Sites

Analysis	Method		Detection Limits [a]		MCL [b]	
	Water	Soils	Water	Soils	Water	Soils
BTEX (GC/MS packed column)	602	8020/ 5030	10 ng/L (1 ppb)	100 ng/Kg (0.1 ppm)	10 ng/Lc (10 ppb)	70 ng/Kg (70 ppb)
BTEX (GC/MS capillary)	602	8260/ 5030	1 ng/L (1 ppb)	10 ng/Kg (10 ppb)		14.4 mg/L d (14.4 ppm)
TPH	418.1	8015	0.2 µg/L (200 ppb)	10 µg/Kg (10 ppm)		
TPH by IR	413.2					
TEPH (diesel)		8015 (Modified)		1 µg/L (1 ppm)		
VOCs	602 624	8015 8020 8240	1 ppb 5 ppm	10 ppb 10 ppb		

a Estimated average values.
b Maximum Contaminant Levels for hazardous wastes under Subtitle C of RCRA.
c For benzene.
d For toluene.

Table 7–7. Analytical Methods for Petroleum Contaminated Soils and Ground-water. This table lists analytical methods for groups of compounds and individual compounds according to whether the sample is for contaminated soils or ground-water. Abbreviations are defined in the text. TPH is total petroleum hydrocarbons; modifiers include V = volatile, E = extractable, and R = recoverable. Costs are average and may vary depending on specific circumstances.

Analysis	Contaminant	Matrix	By	Method	Cost
Group Methods					
TPH	Petroleum	Soil	IR	418.1	$65
TVPH	Gasoline	Soil	GC	8015 M	65
TEPH	Diesel	Soil	GC	8020	45
TRPH	Heavy oils	Soil	IR	413.1	45
BTEX	Aromatics	Soil	GC	8015	85
TPH	Petroleum	Water	GC/MS	620	90
TVPH	Gasoline	Water	IR	8270	90
TEPH	Diesel	Water	GC	625	90
TRPH	Heavy oils	Water	GC	8015	90
BTEX	Aromatics	Water	GC/MS	624	90
VOCs	Volatiles	Soil	GC/MS	8240	225
VOCs	Volatiles	Water	GC/MS	624	225
Individual compounds					
MTBE	In gasoline	Water	GC/MS	8240	125
Ethylene glycol	Waste oil	Soil	GC	8240	75
EDB, EDC	In leaded gasoline	Soil	GC/MS	8240	125
TML, TEL	In leaded gasoline	Soil	GC/MS	8240	125

Table 7–8. TCLP Hazardous Levels for Inorganic Elements

Element	ppm
Arsenic	5
Barium	100
Cadmium	1
Chromium	5
Lead	5
Mercury	0.2
Selenium	1
Silver	5

Table 7–9. TCLP Hazardous Levels for Organic Compounds

Compound	mg/L
Benzene	0.5
Carbon tetrachloride	0.5
Chlordane	0.03
Chlorobenzene	100
Chloroform	6
o–Cresol[a]	200
m–Cresol[a]	200
p–Cresol[a]	200
2,4–D	10
1,4–Dichlorobenzene	7.5
1,2–Dichloroethane	0.5
1,1–Dichloroethylene	0.7
2,4–Dinitrotoluene	0.13
Endrin	0.02
Heptachlor	0.008
Hexachlorobenzene	0.13
Hexachlorobutadiene	0.5
Hexachloroethane	3
Lindane	0.4
Methoxychlor	10

Methyl ethyl ketone	200
Nitrobenzene	2
Pentachloroethane	0.7
Pyridine	5
Tetrachloroethylene	0.7
Toxaphene	0.5
2,4,5–Trichlorophenol	400
2,4,6–Trichlorophenol	2
Vinyl chloride	0.2

---

[a] If cresol isomers cannot be separated, the total cresol level is 200 mg/L.

Toxic Characteristic Leaching Procedure

On March 29, 1990 (FR, 1990a) the EPA published the Toxicity Characteristic Final Rule, which included test method 1311, the Toxicity Characteristic Leaching Procedure or TCLP. The procedure was revised on June 29, 1990 (FR, 1990b). The TCLP is a series of extractions designed to simulate climatic leaching action expected to occur in landfills — including those used to warehouse petroleum contaminated soils — and is frequently required by operators of permitted landfills accepting petroleum contaminated soils for storage. The TCLP test may also be required by regulatory agencies on contaminated soils before the soil can be removed offsite or otherwise treated.

The TCLP identifies and quantifies 8 metals (**Table 7–8**) and 25 organic compounds, such as pesticides, herbicides, etc. (**Table 7–9**) having a potential to leach into groundwater. The test expands and refines the previous EPTOX extractions. TCLP may be performed on either solid (soils) or liquid (groundwater) samples. For liquid samples with <0.5% solids, the waste, after filtration through a 0.6 to 0.8 μm glass fiber filter, is defined as the TCLP extract. For samples containing > 0.5% solids, the liquid and solid portions are separated and analyzed independently. The liquid portions may contain either volatiles or semivolatiles.

The EPA excluded until January 25, 1993 injected groundwater that is hazardous only because it exhibits the toxicity characteristic and is injected through an underground injection well in free–phase hydrocarbon recovery operations performed at refineries and spill or release sites. After this date operations are subject to written state agreements. (Biedry and Martin, 1991).

The EPA originally deferred indefinitely a final decision on petroleum contaminated media and debris subject to UST corrective action. Strictly speaking these wastes are not subject to hazardous waste regulations under Subtitle C of RCRA, provided that the media are contaminated with petroleum wastes only and not with other hazardous wastes.

Unacceptably high levels of inorganic elements, especially inorganic lead, are often found in used oil tanks at collection and recycling facilities. In recent years the U.S. EPA has been under intense pressure to designate used oil as a hazardous waste for just this reason (FR, 1991). The elements come from machinery wear and from inorganic compounds that occur naturally in crude petroleum. A final rule on used oil was issued May 20, 1992 (FR, 1992). Used oil destined for disposal is not listed as a hazardous waste unless toxicity characteristic contaminants exceed specified maximum levels. Used oil destined for recycling must be managed in accordance with new management standards codified in 40 CFR Part 279.

On February 12, 1993 an EPA proposed rule would permanently exempt petroleum contaminated media and debris from the provisions of the TC rule. (FR, 1993). The proposed rule includes only the 25 organic chemicals listed in **Table 7-3**. At the time of writing the Hazardous Waste Treatment Council and Natural Resources Defense Council have sued EPA on the grounds that EPA lacked authority to defer petroleum contaminated media. Current estimates indicate that soils at most UST petroleum contaminated sites would fail TCLP, at least with respect to benzene levels.

The EPA has argued that removing the deferral would significantly hinder corrective action and cleanup procedures at UST sites, and increase remediation costs. Moving cleanup of petroleum contaminated sites under Subtitle C would result in seriously increased permitting requirements — at a time when permitting requirements already account for a significant fraction of remediation costs — and result in greatly increased remediation and disposal costs. The EPA estimates that approximately 50,000 new UST sites are identified each year and that the states are in the best position to oversee disposal and treatment of contaminated media from these sites.

Data Validity and Authenticity

In order for information to be useful and usable, the data upon which the information is based must be reliable; that is, the accuracy of the data must not be in question. Put another way, the accuracy of the data must be demonstrable. In technical terms the data must have validity and authenticity (**Figure VII-6**). Validity and authenticity are the result of the quality control and quality assurance, QA/QC. Various pieces of QA/QC protocols and procedures have been mentioned in previous chapters and sections in this chapter, particularly the use of blanks. In this section data quality, validity, and authenticity will be summarized.

Data validity has two components which are not necessarily mutually exclusive: scientific validity and legal validity. Scientific validity refers to how accurately and precisely a value is known. In statistics, this is the confidence placed on a value. Scientific validity is fairly easily established by standard statistical methods discussed earlier in this chapter, as well as by the QA/QC procedures in effect at the laboratory.

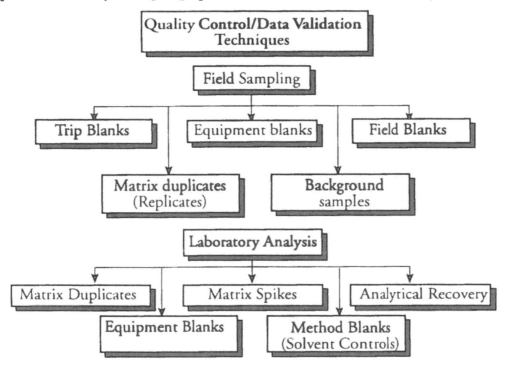

Figure VII–8. Quality Control/Data Validation Techniques. Quality control and data validation are necessary in order for data to be "believable." In the field, trip blanks and field blanks are necessary to ensure that nothing associated with travel or sampling interfered with the analyses. Matrix duplicates or "splits" mean the sample is divided and the splits analyzed separately. Laboratory duplicates or splits are sent to separate laboratories for independent analyses. Laboratory blanks are pure solvent samples that guarantee against laboratory artifacts.

Legal validity refers to the way in which data are generated, which means, in effect, how closely the protocols adhered to recognized standards. Legal validity is more diffuse and difficult to establish than scientific validity. In general, legal validity must be established by courts, on a case–by–case basis. Legal validity has three components:

Admissibility To be admissible the data must have scientific validity. Would the methods of acquiring the data withstand normal scientific scrutiny?

Relevance To be relevant the data must have a bearing on the issue at hand. The extent of contamination may not be relevant to the issue of who caused the contamination.

Weight To have weight the data must be persuasive and convincing. This does not necessarily have anything to do with scientific validity, but rather on the QA/QC methods used.

Quality control and data validation techniques are shown in **Figure VII–6.** "QA/ QC" is largely a matter of documentation and using appropriate blanks and controls both in the field and in the laboratory. **Table 7–10** is a checklist of factors important in QA/QC procedures.

One of the most valuable QA/QC tools is the recovery spike (control) in which a precisely prepared and known amount of contaminant in the appropriate matrix is sent to the laboratory. The laboratory's results are compared with the known value. The results of the recovery spike are an accurate measure of the laboratory's accuracy.

Measurement Integrity

A complete discussion of all the factors affecting the integrity and validity of measurements and sampling is beyond the scope of this book. **Table 7–10** is included to summarize the factors which can affect measurement integrity.

Table 7–10. Summary of Measurement Integrity Factors. The table is a checklist of "QA/QC" factors that impact data validation procedures. Each area and item within each area has the potential to create an error or errors. All of the items in the list may not be required all of the time, but most of the items will be required most of the time.

Analytical
- ☐ Subsampling
- ☐ Preparatory method
- ☐ Analytical method
- ☐ Matrix interferences
- ☐ Detection limits
- ☐ Holding times
- ☐ Turnaround times
- ☐ Contamination
- ☐ QC samples
- ☐ Reagents and supplies
- ☐ Reporting requirements

Project Details
- ☐ History

- ☐ Waste generation
- ☐ Waste handling
- ☐ Contaminants
- ☐ Fate and transportation
- ☐ Sources of contamination
- ☐ Areas to sample
- ☐ Adjacent properties
- ☐ Exposure pathways

Sampling

- ☐ Representativeness
- ☐ Health and safety
- ☐ Logistics
- ☐ Sampling approach
- ☐ Sampling locations
- ☐ Number of samples
- ☐ Type of samples (field QA)
- ☐ Sample volume
- ☐ Composition
- ☐ Containers and equipment

Objectives

- ☐ Need of program
- ☐ Regulations
- ☐ Thresholds or standards
- ☐ Protection of human health
- ☐ Environment protection
- ☐ Liability
- ☐ Data quality objectives
- ☐ Company/agency directives
- ☐ Public relations

☐ End--use of data

Validation and Assessment
☐ Data quality objectives
☐ Documentation of quality
☐ Documentation of activities
☐ Completeness
☐ Representativeness
☐ Accuracy and precision
☐ Audits
☐ Chain--of--custody

Measurable Factors
☐ Quality of reagents
☐ Quality of supplies
☐ Analytical recovery
☐ Sensitivity
☐ Interferences
☐ Contamination
☐ Blank--detectable artifacts

Nonmeasurable Factors
☐ Bias sampling/subsampling
☐ Sampling wrong area or matrix
☐ Sample switching
☐ Mislabeling
☐ Misweighing or misaliquoting
☐ Incorrect dilutions
☐ Incorrect documentation
☐ Matrix--specific artifacts

8 Data Integration and Technology Selection: The Corrective Action Plan

Overview

This chapter provides a systematic approach to the Corrective Action Plan, CAP. The steps followed are planning, sampling, data evaluation and integration, and selection of a remedial technology. Many of the topics have been introduced in previous chapters. This chapter draws the material together and focuses on preparation of the CAP.

Although the discussion that follows almost always goes into more detail than any single remedial project would require, it nevertheless provides a systematic approach to the overall process and a thorough review of all relevant areas that must be considered in choosing the best available, most cost–effective technology. More and more remediation efforts are sufficiently complex projects that do require serious planning and protocols.

One of the problems that can arise in any project is the sheer volume of information available, which can be overwhelming. When this happens, the project can deteriorate into the infamous "black box" approach. Information is processed through a series of black boxes with little or no understanding of real causes and effects. It is crucial to the goal of the project to have a workable scheme for integrating all available data into the decision–making process. Without an integrated approach, the goals of a clean site and closure become difficult and expensive.

In the event of petroleum releases or spills into the subsurface environment, it is necessary to address the problem of cleaning up the vadose zone and/or groundwater by selecting a technology that is likely to be effective. The object is to remove petroleum hydrocarbons from the soil or groundwater and minimize the impact on surrounding properties. Before an appropriate technology can be selected, the nature and extent of the contaminant plume must be completely determined. Therefore, in this chapter we begin at the beginning with planning and the regulatory agencies and move through the process to technology selection.

Because petroleum releases are often sudden and unexpected, before anything else is done, the need for emergency response must be considered. The appropriate agencies must be notified of suspected releases, usually within a week. The agency must be notified again within 24 hours of confirming the presence of a release.

Emergency Response

Unfortunately, petroleum releases from underground storage systems rarely announce themselves at the onset. The first indication of a release is usually indirect evidence: a failed tank tightness test, complaints from neighbors, or leak detectors going off. Gasoline leaking from a service station location has the potential to be a public health hazard because of the confined location and surrounding population. Since the primary concern must be to protect public health and safety, prompt emergency response may be required.

Before any remedial or corrective action the following steps must be taken:

1. Is there a need for temporary containment due to any of the following situations?

    •   A drinking water source, surface water, storm drainage, or wetland is in imminent danger of contamination.

    •   An adjacent property — especially underground structures, basements, crawlspaces, sewer or utility trenches, or storm drainages — is in imminent danger of contamination.

2. Has the source of the contamination been stopped and removed or corrected?

    •   The source of contamination is usually a spill, accidental release, or leaking UST, but may also be heavily contaminated soil.

3. Have the appropriate agencies been notified? It is difficult to notify too many groups. Some of the agencies that should be notified include the following:

    •   The state agency with jurisdiction over remediation;

    •   The agency or group that approves LUST reimbursement;

    •   Any other state or local agencies with possible jurisdiction, including air and water quality districts, sewer districts;

    •   Fire districts or marshals.

An obvious, but sometimes overlooked point, is that the source of the release must be stopped and removed or repaired. Leaking USTs must be removed or repaired, or at least drained. Leaking ASTs must be drained, which can be a protracted process for large storage systems. LUSTs and LASTs are not the only sources of concern. Heavily contaminated soils are also sources of contamination of surrounding areas. Excavation of heavily contaminated soils prior to in situ treatment is recommended to improve remediation efficiency and to remove a source of contamination.

Regulatory Agencies

Notification of appropriate regulatory agencies is the initial step in an usually long, hopefully amicable, relationship with your local health or environmental department. Notification is normally required within 24 hours of confirmation of the release. Owners/operators are sometimes reluctant to report releases because of adverse publicity associated with a leaking UST system; however, nonreporting can lead to serious risks and possible fines.

Satisfying regulatory requirements must be addressed from the beginning since the ultimate goal is closure of the site. Closure by itself does not guarantee that current or future liability will be avoided. Without it the property has greatly reduced value, and there is a high probability of future liability.

"The action or lack of action by the appropriate agency is often the most important factor in determining remedial action or closure requirements for hydrocarbon contaminated sites. The position of the regulatory agency is important because of strong laws and public support for pollution control... Thus, satisfaction of agency requirements becomes the most important factor in addressing contaminated sites." (Daugherty, 1991)

The ability of state regulatory agencies to keep up with the workload has not kept pace with the increasing number and complexity of remediation sites. The result is a bottleneck in starting remediation. There are four factors at work.

1. A long time delay between initial submission of a corrective action plan and final approval has been partially due to an ongoing education process. Environmental companies have had to learn what constitutes an adequate CAP. For most petroleum contaminated sites, preparation of a CAP should be almost routine. In an informal survey of 12 state agencies all reported that sending CAPs back for more information was the norm rather than the exception. Having CAPs returned for more, or more complete, information, or having to repeat sampling is time–consuming and expensive. As environmental companies become better educated, and as state requirements become more routine, these kinds of delays should decrease.

2. The number and complexity of petroleum contaminated sites being restored is increasing and will continue to increase (Bulletin, 1993a). The result is an increasing number of more complicated CAPs being

submitted. That not only means an increasing workload, but also each CAP requires more time for approval. Environmental companies are relying less on "scoop and run" excavation and land filling, and more on onsite or offsite treatment methods. This increase in technological sophistication requires more time and expertise on the part of regulatory personnel. Expertise, in turn, requires training, which requires money, which is always in short supply.

3. There are often two sets of approvals required: the CAP itself, and approval for payment or reimbursement from the state's LUST fund. As the complexity of sites increases, processing delays due to evaluating "appropriate costs" and technical feasibility are increasing and will continue to increase.

4. The average turnover time for state regulatory personnel appears to be on the order of three to five years. Frequently the state regulatory agency becomes the on–the–job apprenticeship to train employees for the environmental companies. The result is a steady influx of new, inexperienced personnel who need time to gain experience.

OUST has announced a program of *streamlining* to cut the time required to approve cleanup activities (Bulletin, 1993b). The streamlining program is aimed directly at applying total quality management (TQM) and innovative regulatory techniques to the CAP approval process.

A simple step to speed up processing time for state regulatory agencies is to offer a "Consultant's Day" seminar.[‡] When the major bottleneck is inadequate, poorly–documented, and inconsistent CAPs submitted by environmental companies in a hurry, an obvious step is to improve CAP preparation. Simply getting the necessary information in the right spaces can speed up processing time, probably by a factor of 2. A Consultant's Day brings together representatives of environmental companies that regularly file CAPs to explain what information is required, when it is needed, how it should be submitted, etc. The participants can be provided with flow charts and guidance documents to facilitate preparation of CAPs that do not need expensive and time–consuming revisions.

Project Planning and Priorities

Remediation of petroleum contaminated sites can range from simple to quite complex; from excavating a neighborhood service station site to a refinery or airport; from spills to mile–long plumes. Contamination may consist of a simple gasoline

---

[‡]   The concept of a "Consultant's Day" is not original. OUST has sponsored Consultant's Days in several states around the country. The Colorado Department of Public Health and Environment and local environmental consultants benefited from one of the first such programs in April 1993.

leak to a mixed waste site, including not only petroleum but also organic solvents, heavy metals, and/or radioactive wastes. Even if the site is very simple and straightforward, the remediation process must be preceded by a planning phase in which all participating parties have input. For a small, uncomplicated site the planning may be a simple procedure, for a complex site planning may take months. It is dangerous to assume a site is simple because even a "simple" site can turn up unexpected complications.

Either way the planning process is a crucial stage in the overall scheme. It is the point where priorities are set and potential problems can, as far as possible, be anticipated. Risk and the potential health threat must be assessed, and remediation objectives and priorities set. Thorough planning will pay off in the avoidance of surprises. Surprises are *always* expensive. Since surprises are occasionally inevitable, it is best to avoid as many as possible.

The steps in the planning phase may be summarized as follows:

- Set priorities and cleanup criteria based on regulatory requirements;

- Identify contaminant characteristics and information that will be required;

- Identify required physical site characteristics and information;

- Identify problems that may arise based on site–specific characteristics;

- Anticipate health and safety concerns that may arise;

- Assess potential risks;

- Arrive at a coherent project strategy;

- Identify the number and location of monitoring wells initially needed;[‡]

- Define sampling protocols.

- Narrowly define lab analysis requirements.

All groups associated with the remediation effort should be involved in the earliest stages of the planning process. For a large complicated site the planning process might include consultants, construction supervisors, design engineers,

---

[‡] Wells are placed so as to define the full extent of a plume or plumes and should include wells outside the plume itself. It is rare that an initial placement of wells proves to be adequate. A second or third set is installed as needed.

sampling personnel, laboratory personnel, company (owner/operator) representatives, insurance company representatives, lawyers, state and local regulators, EPA representatives and local fire authorities, to name a few.

The planning group will evaluate all data requirements consistent with client needs and regulatory compliance and choose the site–specific cleanup criteria (**Figure VIII–1**).

Historical records may need to be consulted to determine past usage of the site. This step, which may have been included as part of a Phase I assessment, can avoid many potential problems and hazards, such as discovery of mixed waste or worker exposure to hazardous compounds. Prior usage may affect decisions to excavate. If hazardous wastes are present, leaving them undisturbed, even temporarily, presents a greatly reduced health hazard than excavated wastes that become airborne.

Review of historical records and identification of surrounding land use may turn up potential hazards or offsite contributors to an existing plume. A gasoline contaminated site frequently has more than one contributor, including PRPs that are no longer in business.

Risk Assessment

Protection of public health and safety has been incorporated into the requirement by regulatory agencies for a risk evaluation as part of the corrective action plan. The potential risks to public health and safety associated with the site and cleanup criteria must be assessed and categorized.

Risk assessment is often difficult to defend due to the perception that any risk is "bad." Further, risk *assessment* is sometimes confused with risk *assurance*. In one study the National Academy of Sciences recommended that regulatory agencies... "maintain a clear conceptual distinction between assessment of risks and consideration of risk management alternatives..." (Wong, et al., 1991).

Risk assessment decisions frequently revolve around the question of "How clean is clean?" Risk assessment always requires some subjective decisions. A plume that is about to contaminate a municipal drinking water well is a clear and imminent threat to public health, but few contaminated sites are that clear–cut. Further, it is generally agreed that soil remediation will not result in completely clean soil; that is, removal of all contaminants to ambient background levels.

(Facing page) Figure VIII–1. Site Assessment and Evaluation. The overall process of site assessment and evaluation must begin with a planning phase to set priorities, needs, and protocols. This phase begins with establishment of an integrated site management plan. Accompanying actions must be included and relevant questions addressed. The figure gives a step–by–step layout of the required actions through design implementation to provide a framework within which protocols can be formulated. Assessment of risks and potential problems comes early in the planning process. Also, regulatory guidelines must be fitted into the planning at an early stage.

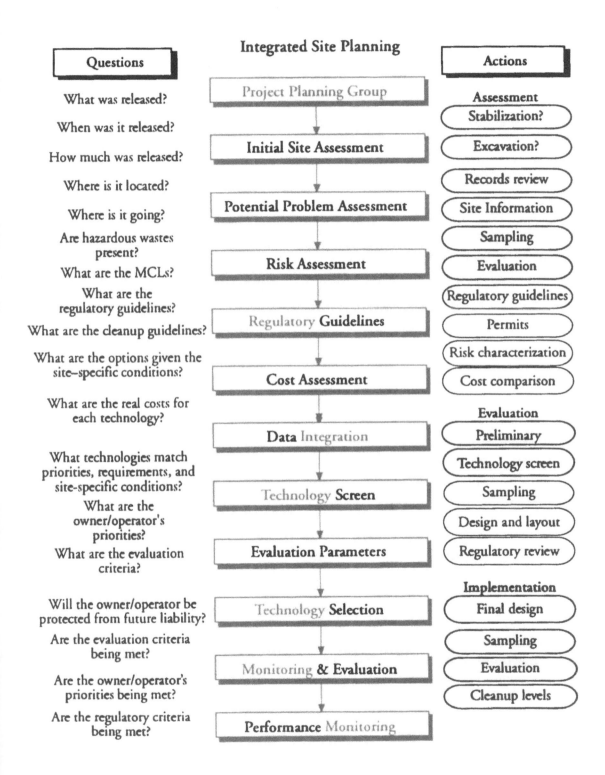

Integrated Site Planning

Questions		Actions

Questions	Integrated Site Planning	Actions
What was released?	Project Planning Group	**Assessment**
When was it released?		Stabilization?
How much was released?	**Initial Site Assessment**	Excavation?
Where is it located?		Records review
Where is it going?	**Potential Problem Assessment**	Site Information
Are hazardous wastes present?		Sampling
What are the MCLs?	**Risk Assessment**	Evaluation
What are the regulatory guidelines?		Regulatory guidelines
What are the cleanup guidelines?	Regulatory **Guidelines**	Permits
What are the options given the site–specific conditions?		Risk characterization
	Cost Assessment	Cost comparison
What are the real costs for each technology?		**Evaluation**
	Data Integration	Preliminary
What technologies match priorities, requirements, and site-specific conditions?		Technology screen
What are the owner/operator's priorities?	Technology **Screen**	Sampling
		Design and layout
What are the evaluation criteria?	**Evaluation Parameters**	Regulatory review
		Implementation
Will the owner/operator be protected from future liability?	Technology **Selection**	Final design
Are the evaluation criteria being met?		Sampling
	Monitoring **& Evaluation**	Evaluation
Are the owner/operator's priorities being met?		Cleanup levels
Are the regulatory criteria being met?	**Performance** Monitoring	

The following steps are normally included in the risk analysis:

- Hazard identification;

- Dose/response assessment;

- Exposure assessment;

- Risk characterization;

- Risk management.

Hazard identification includes determining the potential toxicity of the contaminants and the potential for negative impact on the environment. For petroleum contaminated sites, hazard identification will always include determination of concentrations of aromatic constituents, benzene, toluene, ethylbenzene, and xylenes — the BTEX analytes. Contaminant levels should be compared with maximum contaminant levels (MCLs) as established by federal or state agencies.

Regulatory Guidelines

Many regulatory agencies, including the Colorado Department of Public Health and Environment, have established cleanup *guidelines* rather than absolute standards. The effort recognizes that cleanup requirements and risks can vary from site to site; requirements need not be as stringent everywhere as, for example, within a quarter mile of a municipal drinking water well.

Figure VIII–2. (Facing page) Site Assessment, Planning, and Evaluation. Figure VIII–1 noted that the overall process of site assessment and evaluation must begin with a planning phase to set priorities and protocols. The actual process is a complex mix of interactions with regulatory agencies, consultants, laboratory personnel, design engineers, and construction experts. All of the information generated must lead to the selection of the appropriate technology given site–specific conditions, available budget, and regulatory guidelines. The crucial steps are data integration and technology screening since cost estimates are based on these features. Sampling is important to generate data, but data must be integrated into information. At each point it is important to maintain contact with regulatory personnel. Final technology selection should be straightforward if the preceding steps have been followed properly. Evaluation parameters must be established to allow comparison with preestablished goals. Continuous monitoring is necessary to be sure the system works within design parameters.

Integrated Site Planning

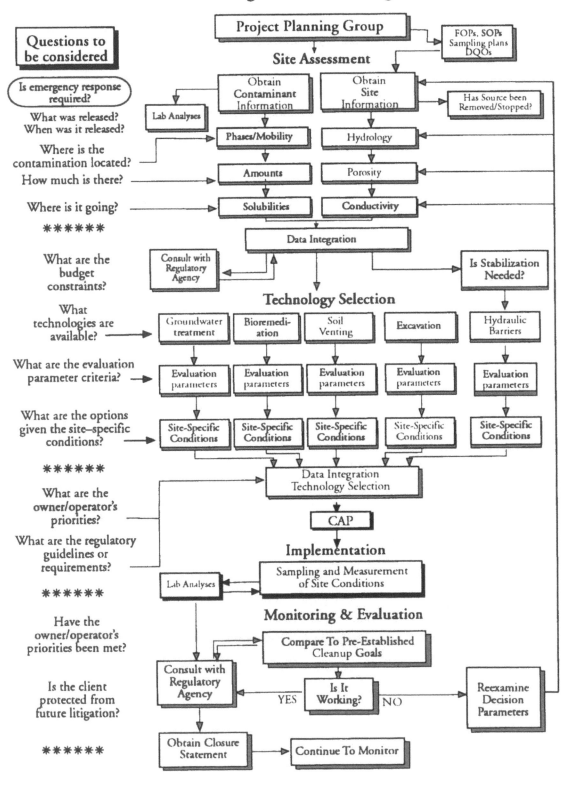

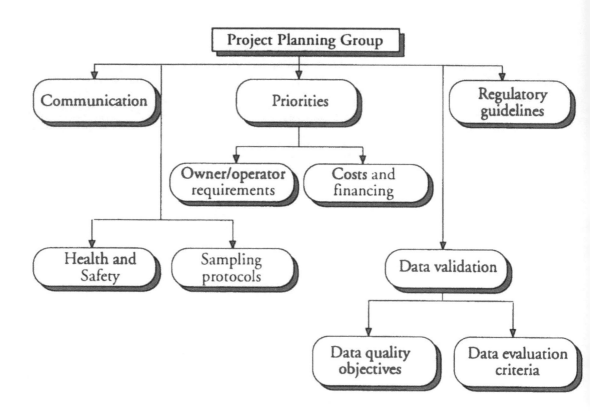

Figure VIII–3. The Planning Phase. Planning is an essential step even for relatively simple sites. The planning phase is the time to choose data requirements, priorities, protocols, and begin to identify potential problems. Potential site safety and health problems must be evaluated at this point in order to plan appropriate procedures to ensure worker safety. Input from the implementing agency is essential. This is the point where budget development and priorities will guide the project for months into the future. For a simple site, planning is straightforward; for a complex site, planning may take months.

Figure VIII–4. (Facing page) Initial Site Assessment. Site assessment begins with gathering information on contaminant levels, physical characteristics of both site and contaminated zones, and site use. Regulatory guidelines must be factored in at an early stage. Human health and environmental risk assessment must be performed to protect workers and the environment. Appropriate document searches yield valuable information on prior site use, ownership, possible secondary contaminants, adjacent land use, and proximity to drinking water wells. Hydrogeology information may be available from the state geological survey or equivalent, or the information will need to be developed onsite.

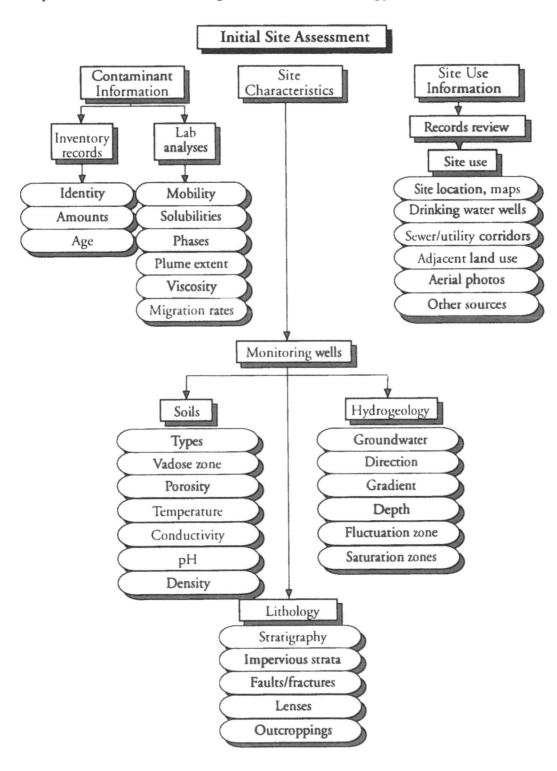

Guidelines — as opposed to absolute standards — give environmental consultants and regulators a certain amount of flexibility regarding cleanup levels. Guidelines are more flexible with respect to soil cleanup levels, as opposed to groundwater cleanup levels, which are usually much tighter. The use of guidelines rather than strict standards allows each site to be evaluated based on site–specific characteristics. Presumably that translates as reasonableness and a professional working relationship between consultant and regulator. Owner/operators who want the cheapest possible cleanup should be in favor of guidelines. Guidelines are much less acceptable to regulators for whom avoidance of responsibility is of paramount importance.

Guidelines and negotiated cleanup levels are a much more serious problem for the purchaser of "clean" property that has residual contamination. This includes problems for environmental professionals doing Phase I Environmental Assessments, who must decide whether residual contamination rates a "flag." The bottom line for UST sites is that closure means "legally clean" within a regulatory jurisdiction. The normal result is a document that reads, "No further action required at this time." It is hard for the owner to take that to the bank.

In many cases guidelines will be adequate to establish cleanup criteria and risk levels. If guidelines are not acceptable or a site is complex, it may be necessary to evaluate potential exposure routes for humans, wildlife, and/or the environment on a site–specific basis.

Dose/response assessment describes the quantitative relationships between exposure and extent of injury to humans, wildlife, or land use values. For straightforward petroleum contamination, uncomplicated by other wastes, dose/response assessment is well established. Dose/response and exposure pathways for common petroleum products are listed in **Table 3–5**, in Chapter 3.

Exposure assessment determines human, wildlife, or environmental receptors that could potentially be exposed to the contaminants. The assessment considers the following pathways: contact, airborne, groundwater, surface water, and crop uptake. Pathways must be evaluated for both in situ exposure and exposure through non–in situ remediation methods.

Risk characterization determines the degree of potential risk to human and wildlife health, the environment, and land use values as a result of the predicted exposure levels assuming various cleanup criteria. For groundwater that is used as a source for public or private drinking water, cleanup standards are quite strict. For soils, cleanup standards are more flexible.

Risk management considers potential risks, remediation feasibility, costs and regulatory guidelines to select the most appropriate residual contaminant levels.

Data Acquisition

The second stage is the data acquisition stage in which relevant physical and chemical parameters are obtained. Physical parameters include data for both contaminant hydrocarbons, the vadose zone, and groundwater. The goals at this point are as follows:

- Acquire site–specific physical parameters;
- Acquire chemical analyses of contaminants.

The results of even a preliminary site investigation should include the physical and chemical nature of the contaminants released. It then becomes possible to estimate the following parameters:

- How the contaminant may partition in the subsurface;
- What phases are likely to be encountered;
- How readily the plume will migrate away from the site;

  and

- Whether the contamination is likely to degrade (or has degraded) significantly over time.

For recent gasoline releases, most of the compounds (excluding methanol and phenol that are highly soluble in water) will still be in the bulk liquid or vapor phase, with only relatively small amounts in the dissolved and adsorbed phases.

Table 8–1. Contaminant Phases and Mobility. Some General Rules for Determining Contamination Phase. Distinguishing between free product and adsorbed or dissolved hydrocarbons is important in choosing a remedial technology. The factors listed in this table are indicators of contaminant phases (adapted from UST Guide).

- **Evidence of a bulk liquid hydrocarbon phase**
  High concentrations of contaminants in several soil samples

  TPH values above 5,000 ppm;

  Soil gas analyses are above 10,000 parts per million.

- **Evidence of volatile hydrocarbon vapors**

  Presence of residual liquid contaminants;

  Significant concentrations on several soil gas samples.

- **Evidence of a dissolved phase in groundwater or soil water**

  Significant concentrations of contaminants in several analyses of pore water or groundwater;

  Presence of residual liquid contaminants *and* a significant soil moisture content.

Technology Selection

The first step in making a decision on an appropriate technology to treat soils is to decide between onsite versus offsite technology; that is, to decide whether or not excavation and transportation are justified or cost effective (**Figure VIII–5**). If excavation is not justified, the choice is an in situ method. In situ methods affect remediation in the ground without the necessity of removing contaminated soils. If the choice is for excavation, the next decision is between onsite and offsite treatment. Some technologies require onsite treatment (**Figure VIII–6**). For example, landfarming or land treatment of groundwater is often carried out onsite. Requirements for other technologies vary. Onsite thermal treatment usually requires a mobile unit, otherwise the soil must be shipped offsite.

Therefore, the planning process should examine those factors that mitigate for or against excavation. One set of circumstances that must be considered is the timetable and necessity for speed. If, for example, the contamination is threatening a municipal water source, or is migrating at an unacceptably rapid rate, excavation can remove the contamination quickly, even if a final treatment method has not been chosen. In situ methods, on the other hand, often require remediation times on the order of one to several years. Excavation is usually not an option at very large sites where contamination is spread over a large volume of soil.

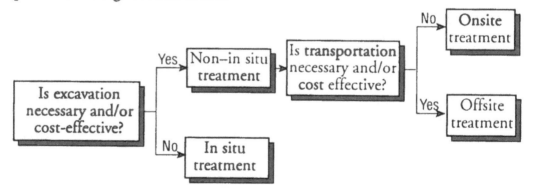

Figure VIII–5. On/Offsite Decision Tree. The initial planning process must examine the factors for and against excavation and transportation. If excavation is required or cost effective, the excavated soils can be treated on or offsite. If transportation of contaminated soils is justified, an offsite treatment method can be chosen. If excavation and transportation are not justified, the choice must be an in situ method.

Performance Monitoring

After a technology has been selected and installed it is important to track performance and effectiveness during cleanup activities, and to match performance with preestablished goals. Questions to be addressed include the following:

- What is "effective"?
- What indicators are significant?
- How frequently should the site be monitored?
- Is the sampling representative of the site?
- Are the preestablished goals being met?
- Is the project within budget?

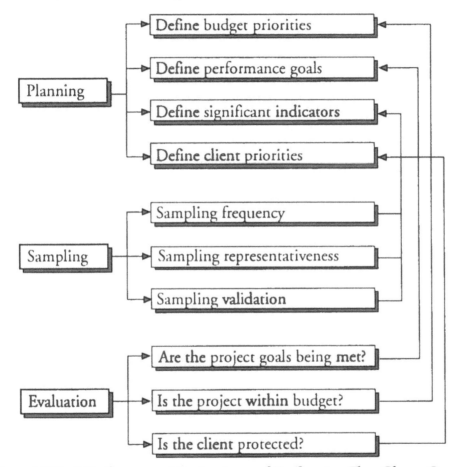

Figure VIII–6. Performance Monitoring and Evaluation Flow Chart. Once the technology has been selected and the remediation design has been implemented, it must be monitored on a continuing basis. Criteria are selected that will accurately reflect whether or not project goals are being met within budget. Monitoring normally requires sampling on an appropriate periodic basis.

Monitoring and follow-up are essential because of the uncertainty surrounding subsurface conditions. Complete definition of the areal extent of contamination is essential. Subsurface monitoring must be at well–defined, appropriate intervals and, if performance is to be evaluated, each sampling set must be representative of the site as a whole. Monitoring data are validated by selecting suitable QA/QC criteria, which are discussed elsewhere in this chapter and in Chapter 7.

Figure VIII–7. Technology Decision Tree. Matching the appropriate technology to site–specific conditions is critical to eventual success of the project. A plume that is an imminent threat to public health or that is rapidly migrating requires aggressive action. Containment, diversion, or isolation must be considered first. If hazardous wastes are present, the entire site may fall under Subtitle C (RCRA) regulations. At some point a decision to excavate must be adopted or rejected. Once the choice has been narrowed to in situ or non–in situ, evaluation parameters (decision criteria) are matched to site–specific conditions, and a preliminary selection made for remedial technology. Only proven methods are listed in the figure. As noted in Chapters 9 and 10 remedial technologies are not mutually exclusive. Bioventing, partial excavation, air sparging, etc., are discussed in more detail in later chapters. Final technology selection is made after cost comparisons.

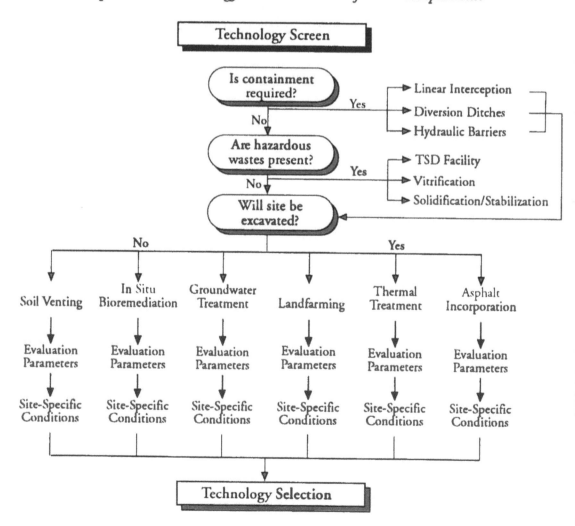

Performance monitoring means that the project must remain within budget and the client's priorities must be observed as far as possible. If the technology performance is poor, the selection will need to be reevaluated or the design will need to be modified. If, for example, unsuspected subsurface features, such as lenses or other strata will affect flow rates and extraction efficiencies. Along the way, regulatory agencies must be kept informed of progress toward established cleanup levels.

Evaluation Parameters for In Situ Technologies

This section, from here to the end of the chapter, provides information on in situ remedial technologies. These methods are carried out without excavation, or with minimal excavation. The terminology "onsite" and "offsite" and "in situ" and "non–in situ" have been retained to coincide with usage of **On–site Treatment of Hydrocarbon Contaminated Soils** (ESI, 1990). The usage departs somewhat from **Remedial Technologies for Leaking Underground Storage Tanks** (EPRI, 1988). For large sites such as petroleum refineries onsite treatment is often feasible. For a service station or bulk storage plant typically there is no room for onsite treatment. Thus, the choice of onsite or offsite becomes inevitable, specified by the physical location of the site. There is, however, a significant need to make a choice between in situ and non–in situ treatments.

The figures that follow in this section, called **Evaluation Fact Sheets**, give a summary of the most significant evaluation parameters for each technology discussed in the next two chapters. The Fact Sheets that are presented in this section are to be used in connection with **Figure VIII–2**, *Integrated Project Planning*, and **Figure VIII–6**, *Technology Decision Tree*. A separate version of evaluation parameters called *Considerations and Recommendations* is given at the end of each section in Chapters 9 and 10.

Each figure begins with financial or operational considerations. Without financing remediation remains at a standstill; it is an essential first step. In spite of the availability of state LUST funds not all sites qualify for reimbursement. Orphan tanks may not be included or the UST system may not have been in compliance at the time the release occurred. Operational factors include, for example, the necessity of keeping a business open during remediation or the high cost of transporting excavated soils.

The figures usually include chemical and physical properties of the contaminants and site–specific information. Factors vary depending on the methodology, but usually the chemical composition of the contamination is important — gasoline or other; hydrocarbons or mixed waste. Soil composition and porosity are important to most in situ methods.

Similarly, the figures conclude with client and legal considerations. The goal of the remedial effort is certainly to clean the environment. However, the following are important:

• To limit as far as possible the client's future liability exposure;

• To leave the property with a clean bill of health for future use;

- To obtain a closure statement from the appropriate regulatory agency;
- To remain within budget.

Volatilization Fact Sheet

Technology. In situ volatilization or soil venting is the removal of volatile organic compounds from subsurface soils by applying a vacuum to the interstitial spaces of the soil. It is the most common means of removing free gasoline product from contaminated soils. Volatilization is a relatively low cost, proven method applicable to small to large areas where excavation is impractical or undesirable.

Application: This technique is useful for the more volatile hydrocarbons, primarily gasoline contamination. With horizontal drilling techniques extraction lines may be extended beneath buildings, roadways, and other on– or offsite permanent facilities.

Combinations. Volatilization can be operated along with groundwater treatment as air sparging techniques. It can be combined with bioremediation to decrease the time required for remediation. Pneumatic soil fracturing can be used to increase efficiency of volatilization from clays.

Equipment. The following equipment is required:
- Extraction wells;
- Air pumps and vapor treatment units;
- Injection wells;
- Sampling points;
- Vapor or groundwater monitoring wells.

Only the first two are essential in all installations. Sampling wells are essentially a universal requirement. Air treatment units are necessary in areas where the extracted air cannot be vented to the atmosphere — which is most urban areas in the U.S. — and where hazardous VOCs are mixed with the gasoline.

Limitations. The method treats only volatile hydrocarbons and is really suitable only for gasoline. It works best in loose, porous soils that allow vapors to migrate freely. Soil venting alone does not work below or near the water table. Cold soils are less suitable since volatility is a function of temperature. However, this method is widely used in Alaska where the soils are largely loose glacial tills.

Permitting. Extracted air can be recycled or discharged after treatment. An air quality permit is normally required for discharge. Discharged air must be cleaned by catalytic converters, carbon filtration, or scrubbing. Exhaust air should be sampled on a routine basis.

Figure VIII–8. (Facing Page) Evaluation Parameters for Soil Venting. Factors for and against soil venting are summarized. The method is a standard, low cost, proven technology, accepted by all regulatory authorities. It is used to extract gasoline vapors from contaminated soils. Remediation is usually a matter of years to completion.

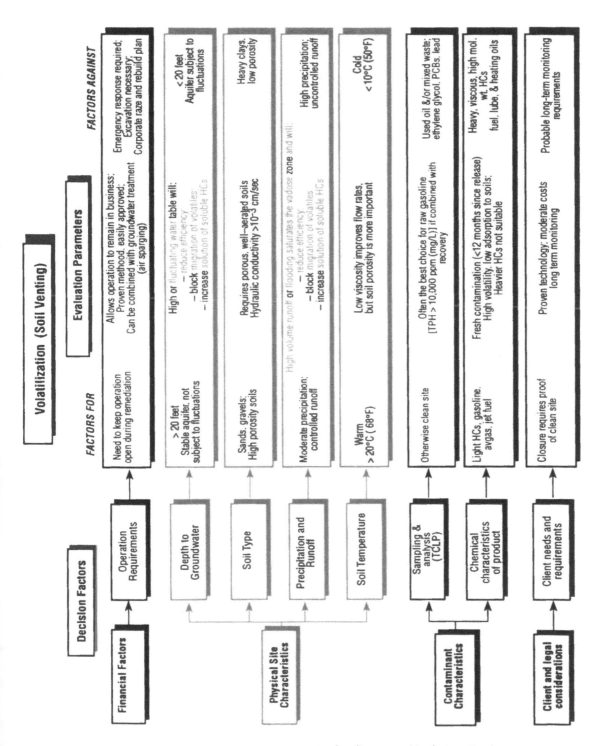

Evaluation Parameters For Volatilization (Soil Venting)

In Situ Bioremediation Fact Sheet

Technology. Bioremediation is the process of supplying nutrients to encourage bacterial degradation of hydrocarbon compounds in soils. It is more difficult to control nutrient parameters in situ than through non–in situ bioremediation. Because in situ bioremediation is not a "standard" technology, it may be necessary to justify the method to regulators. On the other hand, in feasible situations in situ bioremediation allows remediation to continue with minimal disruption of onsite operations and with relatively low costs. The method works best in warm, loose, porous soils.

Application: In situ bioremediation is often the only method available to treat very large sites such as refineries, bulk distribution facilities, airports, and railroad fueling yards. This method can also affect remediation of a mixed waste site containing low levels of volatile organic solvents.

Combinations. Bioremediation has been combined with soil venting to improve removal rates. Groundwater can be extracted, treated, mixed with nutrients, and reinjected. Groundwater treatment effectiveness can be improved by treating contaminated water in an external vessel or by combining groundwater extraction with land treatment.

Equipment. Major equipment requirements include the following:
- Injection wells, infiltration galleries, or sprinklers to provide nutrients to subsurface microbial populations;
- Extraction wells for groundwater treatment;
- Mixing vessels for nutrients;
- Treatment units for groundwater;
- Diversion ditches, dams, or ground covers to control surface runoff.

Limitations. The major difficulties are the following:
- A viable population of indigenous soil organisms;
- Supplying nutrients to the microbial population;
- The length of time required to clean up soils may allow contaminants to migrate offsite;
- The presence of mixed wastes including heavy metals, arsenic, mercury, chlorinated solvents, pesticides, herbicides, and low pH (<6).

Permitting. Bioremediation requires a permit to operate. Permits can include a water quality permit, NPDES, a special regulatory agency permit, or all of the above. The moderate cost associated with in situ bioremediation can be offset by the increased costs of more frequent sampling to satisfy permitting requirements.

Figure VIII–9. (Facing Page) Evaluation Parameters for In Situ Bioremediation. Factors for and against in situ bioremediation are summarized. The method can be useful and cost–effective in the proper circumstances, but requires careful management and control of nutrient parameters.

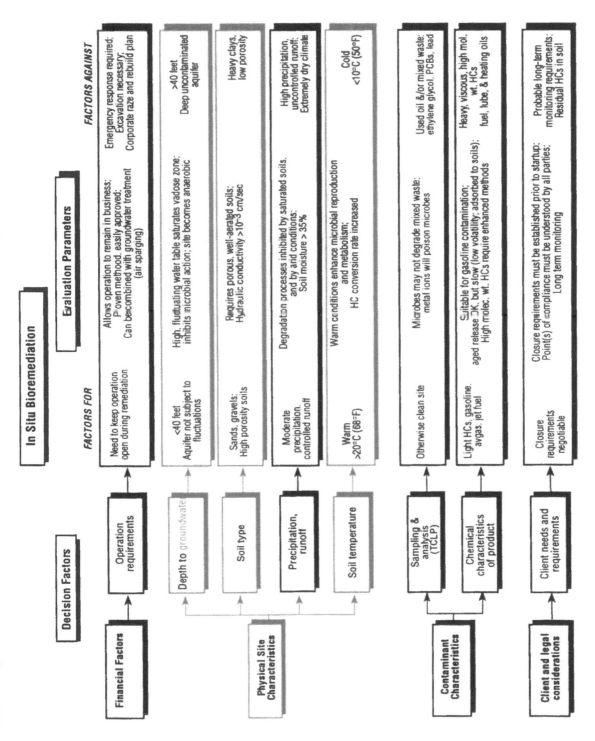

Evaluation Parameters For In Situ Bioremediation

In Situ Passive Bioremediation Fact Sheet

Technology. Passive remediation is simply bioremediation without any efforts to encourage bacterial growth and degradation rates. Passive remediation goes on at all times anywhere soil bacteria exist. Soil bacteria consume hydrocarbons at a rate determined by the natural bacterial population and the available nutrient supply. Under optimum conditions the conversion rate is relatively high, in poor conditions conversion rates are negligible. The limiting nutrient in the subsurface is always oxygen. Soils are increasingly anaerobic as depth increases.

Application. Surface gasoline contamination, as for example a spill or release from an AST, has the best chance for passive remediation because the soil bacteria population is usually higher near the organic soil horizon. Conversely, deeper subsurface contamination, even in the vadose zone, may persist for years without appreciable degradation simply because the bacterial population is too low to be effective. Hydrocarbons heavier than gasoline have little chance of being degraded naturally if the contamination is very deep.

Combinations. Volatilization and leaching also contribute to passive removal of hydrocarbons.

Equipment. There are no equipment needs in passive bioremediation other than the usual sampling well requirements.

Limitations. The major difficulties are the following:

- A viable population of indigenous soil organisms large enough to sustain hydrocarbon conversion;
- The length of time required to clean up soils may allow contaminants to migrate offsite;
- The presence of mixed wastes including heavy metals, arsenic, mercury, chlorinated solvents, pesticides, herbicides, and low pH (<6). These inhibit biological processes and may require faster action than passive remediation allows;

- Uncertainty in future land use (property could be flagged in subsequent Phase I assessment).

Permitting. Passive remediation is rarely approved by regulatory authorities, but may be suitable under very special circumstances. If approved, passive bioremediation normally requires one or more permits to operate. Permits can include a water quality permit, air quality permit (due to volatilization), NPDES, a special regulatory agency permit, or all of the above. Periodic monitoring is required. The moderate cost associated with passive bioremediation can be partially offset by the increased costs of more frequent sampling to satisfy permitting requirements.

Figure VIII–10. (Facing Page) Evaluation Parameters for Passive Remediation. Factors for and against in situ passive bioremediation are summarized. Passive remediation is seldom approved by regulatory agencies as part of a CAP. However, it can be justified under some circumstances.

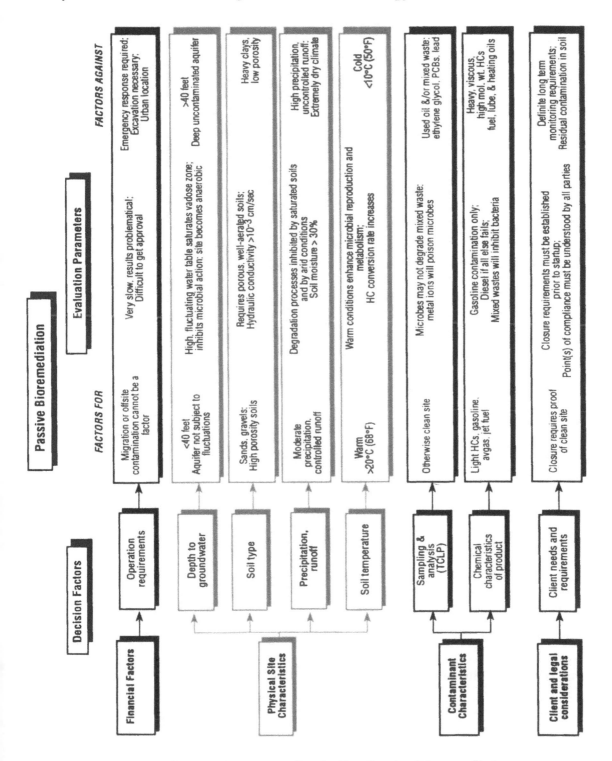

Evaluation Parameters For In Situ Passive Bioremediation

In Situ Leaching and Chemical Extraction Fact Sheet

Technology. In situ leaching, or soil washing, is exactly what the name implies, washing contaminated soil with clean water to gradually remove hydrocarbons. If fresh water is injected into or mixed with contaminated soils, more soluble hydrocarbons will be removed. Hydrocarbons are not very soluble, but with a continuous supply of clean water, a slow extraction can be carried out in which the more soluble hydrocarbons are removed preferentially. Extracted contaminated water can be treated by external treatment (oil/water separation), filtration, or land treatment.

Application. Soil washing extracts the lighter aliphatic and aromatic petroleum hydrocarbons. Care must be exercised to determine the type of contamination present. Water will extract other contaminants preferentially, for example, metals, lead, arsenic, cadmium, pesticides, VOCs, and phenols.

Combinations. This method is difficult to justify as a stand–alone technology. There are difficulties in obtaining water quality permits (NPDES and/or injection), air quality permits, and regulatory approval. However, in combination with other methods, such as landfarming or groundwater treatment, it could be effective.

The technology is probably most effective when the contamination has extended to include groundwater, and is combined with groundwater extraction, treatment, and recirculation.

Equipment. Leaching equipment requirements are similar to those of groundwater treatment and include the following:

- Injection wells;

- Extraction wells;

- External treatment unit — oil/water separator, filtration, or ozonolysis;

- Land treatment facilities — to treat extracted water;

- Surfactant mixing tank;

- Leachate treatment or storage unit;

Limitations. As with all in situ methods, relatively long periods of time are required to achieve regulatory cleanup levels. Indeed, it is doubtful that complete cleanup could be obtained without an expensive carbon filtration polishing unit.

Permits. If process water is to be discharged into a surface stream or storm drainage, an NPDES permit will be required. If groundwater is recirculated, an injection permit is required. Other possible permits include air quality, sanitary sewer, and/or special local permits.

Figure VIII–11. (Facing Page) Evaluation Parameters for Soil Washing and Chemical Extraction. Factors for and against soil washing are summarized. The method can be effective, particularly when combined with recirculating groundwater extraction and treatment. Use of surfactants improves remediation efficiency, but introduces added complications of disposal of the oil/water emulsion.

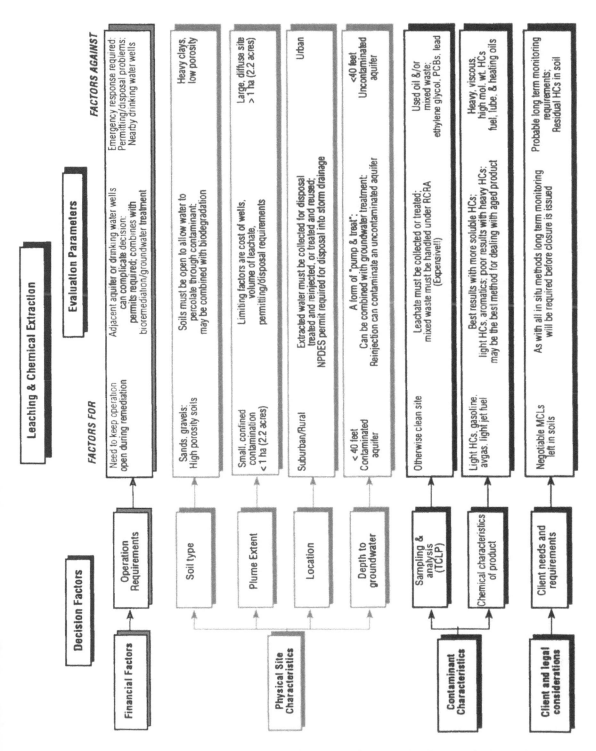

Evaluation Parameters For In Situ Leaching and Chemical Extraction

In Situ Vitrification Fact Sheet

Technology. In situ vitrification is a method that uses electrical energy to melt soil and form a glass. Typical aluminosilicate soils melt between 1,500 and 2,000°C (2,500 to 3,600°F). Organic contaminants are volatilized and pyrolyzed. Inorganic contaminants are immobilized. Soils are converted into a glass or glass–like substance that may be left in place or removed and landfilled.

Electrodes are placed in an array around the contaminated area and a current passed through the soil. As the temperature rises soils melt and, because most soils are ionic, become more conductive. As the melt spreads, volatile organic compounds are vaporized and migrate to the surface. At the surface in the presence of high temperature oxygen, organic material is pyrolyzed. Off–gases must be collected and treated, depending on the type of contamination present.

Application. Vitrification has limited use, but in specialized applications it is the method of choice. Sites for which in situ vitrification should be considered include small mixed waste sites containing petroleum hydrocarbons plus mixed wastes that include the following: radioactive wastes, heavy metals (mine or mill tailings or plating sludges), refractory inorganic substances (nitrates, fluorides, sulfates), and sludges.

Combinations. None.

Equipment. Required equipment includes the following:
- Silicon or graphite electrodes;
- Trenching and drilling equipment;
- Electrical power source;
- Tent to cover site;
- Off–gas collection and treatment equipment.

Limitations. Vitrification cannot be used on sites with high groundwater or high soil moisture levels, excessive surface runoff, near buried utility corridors, with highly combustible contaminants, as for example, free gasoline. However, the method is suitable for contaminated clays.

Permitting. In situ vitrification is categorized as experimental (ESI, 1990). Because it is a nonstandard method, application must be justified and special regulatory approval must be obtained. Vitrification must be accompanied by off–gas collection and treatment. An air quality permit is normally required. Since the affected area is confined and limited in area, human exposure to contaminants is minimized.

Figure VIII–12. (Facing Page) Evaluation Parameters for In Situ Vitrification. Factors for and against in situ vitrification are summarized. The method must be considered experimental and applicable in special circumstances. It is an alternative in treating mixed waste sites, especially those containing radioactive materials, heavy metals, or refractory inorganic materials.

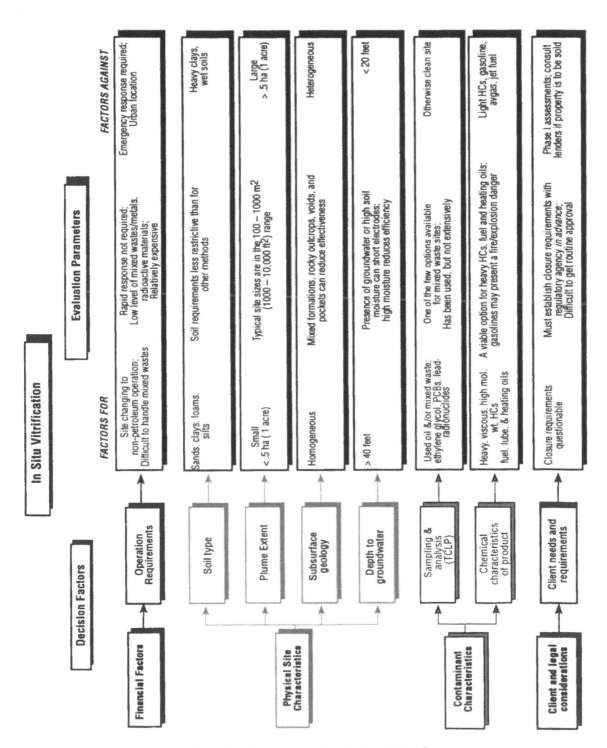

Evaluation Parameters For In Situ Vitrification

Isolation and Containment Fact Sheet

Technology. Containment is neither treatment nor remediation. This method is required in response to a plume that is migrating at unacceptable rates or in unacceptable directions. The method is also an alternative in the event of prolonged litigation to resolve percent responsibility for each PRP. Contamination is contained by means of physical barriers that prevent contamination of adjacent areas.

Containment consists of diversion ditches, berms, curtain walls, subsurface and surface barriers to trap and divert migrating contamination. Isolation consists of controlling and diverting surface runoff and groundwater flow. This technology is suitable for mixed waste sites as well as sites contaminated with petroleum. It is only a temporary remedy, particularly if groundwater is being contained. Groundwater will eventually find a way around barriers.

Containment is relatively low cost unless extensive excavation is required to install subsurface barriers. Berms and walls are typically bentonite or cement. Asphalt or clay surface covers are used to divert runoff.

Combinations. Groundwater extraction wells can productively be placed upgradient from berms or containment walls. The increased hydraulic pressure head is used to enhance groundwater extraction and treatment.

Equipment. Equipment requirements are mainly normal construction and earth moving equipment.

- Earth moving equipment for trenching operations;
- Surface covers;
- Drains and dikes.

Limitations. Containment is not a substitute for remediation. It is a temporary measure to contain groundwater. Eventual permanent remediation is required. Once in place, the containment time frame always tends to be prolonged. This fact should be considered when planning for technique and materials. Property transfer is not feasible with containment still in place.

Permitting. Normally none required. Water discharge permit may be required if runoff is discharged into surface drainage or sewers. Since no remediation is involved, containment will need justification to regulatory agencies.

Figure VIII–13. (Facing Page) Evaluation Parameters for Isolation and Containment. Factors for and against isolation and containment are summarized. Containment is an option to buy time in the event of a rapidly moving plume or prolonged litigation. It is not treatment, but is useful as a temporary or emergency measure. Containment can be considered as a low cost insurance against third–party impairment.

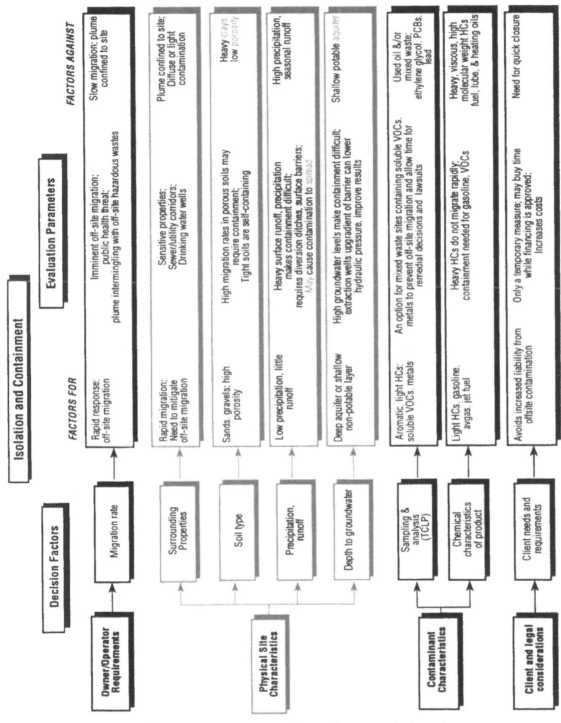

Evaluation Parameters For Isolation and Containment

Solidification and Stabilization Fact Sheet

Technology. Solidification and stabilization result in immobilization of contaminants for permanent containment, reuse, or eventual disposal. It has been used for permanent containment for heavy residues and sludge ponds in production fields and refineries. The technology is — and is not — a remediation technology. Contaminants are immobilized in concrete or fly ash, but not treated to remove the contamination. Formulation of solidification mixtures is exacting, but the technology is otherwise straightforward. Depending on local regulations the stabilized material can be disposed of in sanitary landfills.

Application. Solidification and stabilization are suitable for the following contaminants:

- Bottom sludges from USTs (including sludges from leaded gasoline storage tanks) (also see Appendix E);

- Crude petroleum sludges;

- Drilling muds and sludges;

- Waste oils (including waste oils contaminated with mixed wastes);

- Lubricating and machinery oils;

- Refinery sludges;

- Acidic wood preservatives (pentachlorophenols, creosotes);

- PCBs;

- Heavy metals;

- Radioactive wastes (tailings).

Combinations. None.

Equipment. Equipment requirements are primarily limited to material handling equipment, include earth moving equipment, mixers, pumps, etc.

Limitations. If contaminated soil is solidified in place, future land use or transactions may be difficult. Long term stability data are lacking; contaminants could leach over time. The method has been used for petroleum hydrocarbons in production fields and at refineries, but has not been widely used otherwise.

Permitting. This technology must be justified to regulatory agencies on a case–by–case basis. No special permit exists.

Figure VIII–14. (Facing Page) Evaluation Parameters for Solidification and Stabilization. Factors for and against solidification and stabilization are summarized. Solidification and stabilization is not a remediation technology in that contaminants are immobilized and prevented from leaching into the environment. Contaminants are not treated or destroyed. It is an option for heavy sludges, waste oils (including waste oil containing mixed wastes), crude sludges, drilling muds, and UST sludges.

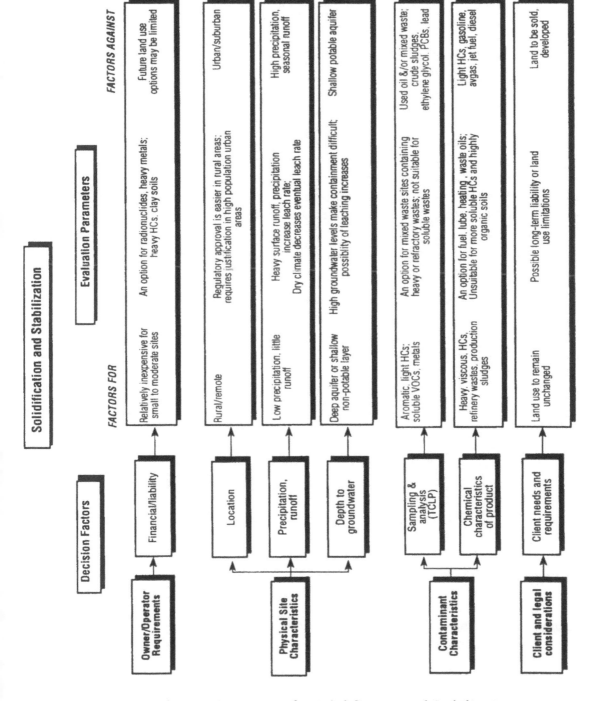

Evaluation Parameters for Solidification and Stabilization

Evaluation Parameters for Non–In Situ Technologies

Non–in situ remediation is very attractive for several reasons:
- It allows precise control of variables in the remedial process;

- Cleanup values are easily verified;

- Complete cleanup (to zero ppm) can be achieved

Non–in situ remediation must, of course, be preceded by excavation which adds to the expense. Proven non–in situ methods include landfarming or land treatment of groundwater, asphalt incorporation, and other methods of beneficial reuse (ESI, 1990), soil washing and chemical extraction, and thermal treatment. In addition to these established methods, combined methods are proving successful and cost–effective. Combined methods include biofiltration, bioventing, air sparging, groundwater reinjection, land treatment of groundwater, and waste water reuse.

Excavation and Landfilling Fact Sheet

Technology. Excavation is physical removal of all or part of the contaminated soil at a site. Excavation need not be complete removal of all contaminated soil.

Excavation is still the leading method to remediate petroleum contaminated soils, but is not as common as it once was. At one time it was common to simply excavate contaminated soils up to the property line, backfill, and leave. Now it would be surprising to slip that scenario past regulatory authorities. According to a recent EPA report, excavation and landfilling constitute 55% of the remedial activities across the country (WTN, 1993; EPA, 1993). In some areas of the country, most notably in California, landfilling contaminated soils has become rare.

Combinations. Very heavily contaminated soils (10,000 ppm TPH) act as a source for the spread of contaminants. Limited excavation of only the most heavily contaminated soil can greatly speed in situ remediation. Excavation may be followed by storage at a permitted landfill or by non–in situ treatment. Landfill storage is decreasing in many parts of the country since storage space is at a premium and transportation costs rise rapidly as the distance to a landfill increases. Other options include thermal treatment, landfarming, and asphalt incorporation.

Equipment. Only the usual earth–moving equipment is required.

Limitations. Excavation treats only contaminated soils. Heavy gasoline contamination can pose a safety hazard. Landfilling is economic only when a permitted landfill is within a reasonable distance.

Permitting. None, other than construction permits for excavation.

Figure VIII–15. (Facing Page) Evaluation Parameters for Excavation. Factors for and against excavation are summarized. Excavation is necessary prior to non–in situ remediation. Excavation is followed by land filling or by treatment onsite or offsite. Partial excavation may be combined to increase the efficiency of non–in situ methods.

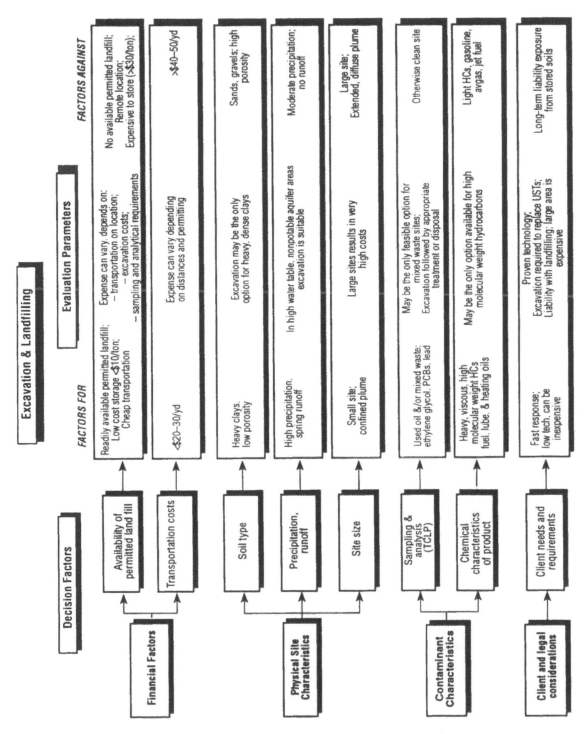

Evaluation Parameters for Excavation & Landfilling

Landfarming

Technology. Landfarming, or non–in situ bioremediation, is the process of spreading excavated soils over a large area to encourage natural remediation. Landfarming is a more efficient form of bioremediation since it allows more precise control of nutrient, moisture, and temperature parameters. However, bacterial activity is not the only form of natural remediation that occurs. Up to 40% of the actual remediation is due to volatilization. After suitable surface preparation, contaminated soils are spread in 12– to 18–inch deep rows. Periodically the soil is turned with rototiller, disk, or harrow. A sprinkler system is used to control moisture content. Land farming may be carried out onsite or offsite. In some regulatory jurisdictions offsite treatment may be hindered by restrictions on transporting contaminated soils. Groundwater treatment is generally economic only if done onsite.

Combinations. Landfarming has been combined with groundwater extraction. Contaminated groundwater is sprayed over otherwise clean soil. Indigenous or enhanced bacterial degradation cleans the water, which is collected and reinjected. Groundwater treatment effectiveness can be improved by pretreating contaminated water in an external vessel. The water is mixed with nutrients and aerated in an external mixing vessel before being sprayed over the soil.

Equipment. Major equipment requirements include the following:
- Earth moving equipment, tractors, disks, rototillers;
- Mixing tanks and sprinklers to provide nutrients to microbial populations;
- Extraction wells for groundwater treatment;
- Treatment units for groundwater;
- Diversion ditches, dams, or ground covers to control runoff.

Limitations. The major difficulties are the following:
- A large land area is required;
- The location must have restricted access to protect human health and safety;
- The length of time required to clean up soils may allow contaminants to migrate offsite;
- The presence of mixed wastes including heavy metals, arsenic, mercury, chlorinated solvents, pesticides, herbicides, and low pH (<6).

Permitting. Landfarming requires a permit to operate. Permits can include a water quality permit, NPDES, a special regulatory agency permit, or all of the above. The moderate cost associated with landfarming can be offset by the increased costs of more frequent sampling to satisfy permitting requirements.

Figure VIII–16. (Facing Page) Evaluation Parameters for Landfarming. Factors for and against in situ bioremediation are summarized. The method can be useful and cost–effective in the proper circumstances, but requires large land area. The method provides accurate control of nutrients and bacterial populations. Land farming is normally used to treat contaminated soils, but can be used to treat groundwater.

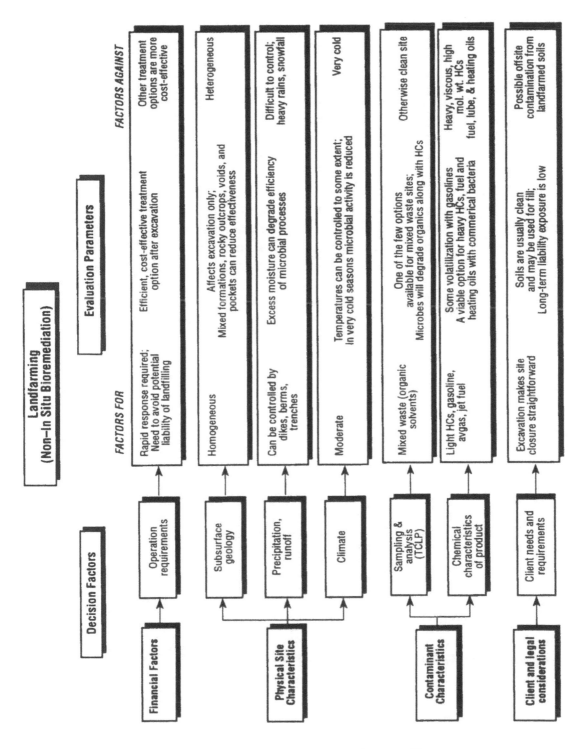

Evaluation Parameters for Landfarming

Low Temperature Thermal Stripping Fact Sheet

Technology. Low temperature thermal stripping (LTTS) is a technology in which contaminated soil is fed into a chamber and heated to between 200 to 260°C (400 to 500°F). Volatile hydrocarbons and VOCs are vaporized. Off–gases are oxidized by passing through a flame, catalytic combustion system, or burned as fuel in an engine. LTTS treated soils are normally clean enough to be returned to the excavation as backfill.

The technology is classed as developmental (ESI, 1990), although mobile units have been used on the east and west coasts for several years. It is suitable for petroleum hydrocarbons through diesel fuels.

Applications. LTTS competes with in situ volatilization. Some advantages of LTTS are that it vaporizes a wider range of hydrocarbons than in situ volatilization. LTTS can treat up through diesel fuels that are difficult to recover through in situ volatilization. In situ volatilization requires treatment times on the order of years, whereas LTTS requires much shorter periods of time. The trade–off is in higher operating costs for LTTS systems.

Combinations. None. LTTS is similar to asphalt incorporation.

Equipment. The following equipment may be mobile or fixed:

- Soil handling equipment;

- Screens to remove large debris;

- Heating/volatilization chamber;

- Emissions control/monitoring system;

- Air flow equipment.

Limitations. LTTS is suitable for volatile hydrocarbons and other organics only. Saturated soils cannot be put through an LTTS system due to the risk of fire or explosion. It is unsuitable for mixed wastes containing halogenated hydrocarbons, pesticides, or inorganics. The process of combustion of halogenated hydrocarbons produces hydrochloric acid, HCl. Pesticides tend to be resistant to lower temperature combustion. Metals cause problems with catalytic converters and emission control systems. Acidic or alkaline soils can cause corrosion in equipment. The method is unsuitable for large sites, such as airports and refineries, where jet fuels and diesels are common in soils. The constraint is usually cost of excavating large sites and processing large amounts of soil in a batch processing system. Clays and other tight soils are usually unsuitable. Soils with a high organic or soil moisture content are not suited to LTTS.

Permitting. LTTS systems must be permitted to treat mixed wastes that include hazardous wastes. Most jurisdictions require an air quality permit and a NPDES if there is waste water discharge.

Figure VIII–17. (Facing Page) Evaluation Parameters for Low Temperature Thermal Stripping. Factors for and against low temperature thermal stripping, LTTS. Contaminants are vaporized from soils and burned. Mobile units are available for onsite treatment.

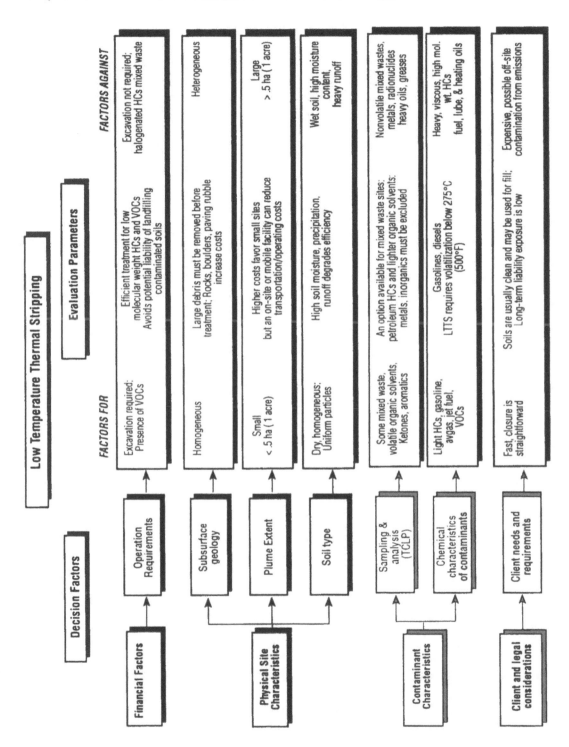

Evaluation Parameters for Low Temperature Thermal Stripping

High Temperature Thermal Treatment Fact Sheet

Technology. High temperature thermal treatment (HTTT) is a technology in which contaminated soil is fed into a chamber and burned at temperatures between 800 to 1100°C (1500 to 2200°F). A typical treatment facility is a modified asphalt plant. Temperatures are kept as low as possible to prevent vitrification. Even so moisture content in treated soils is below 3%. Treated soils are not suitable for reuse, but the soils are clean. The destruction and removal efficiency (DRE) typically exceeds 99.99% (ESI, 1990).

Applications. Contaminated soils that must be cleaned completely. Mixed wastes should be treated only with permission. HTTT is one of the few means of treating contaminated clays. Soils with a high organic or soil moisture content are not suited to HTTT.

Combinations. None

Equipment. The following equipment may be mobile or fixed, but is generally fixed. Converted asphalt batch plants are used for HTTT.

- Earth moving equipment;

- Soil handling equipment;

- Screens to remove large debris;

- Combustion chamber;

- Emissions control/monitoring systems;

- Sampling access and equipment.

Limitations. HTTT is suitable for volatile and nonvolatile hydrocarbons and other organics. It is unsuitable for mixed wastes containing halogenated hydrocarbons, pesticides, or inorganics unless modified exhaust scrubbers are used. Pesticides, PCBs, and chlorinated hydrocarbons require a waste treatment permit. Metals cause problems with catalytic converters and emission control systems, and contaminate residues. Exceptionally acidic or alkaline soils can cause corrosion in equipment. The method is suitable for large sites, such as airports and refineries, where jet fuels and diesels are common in soils. The constraint is usually cost of excavating large sites and processing large amounts of soil in a batch processing system.

Permitting. HTTT systems must be permitted to treat mixed wastes that include hazardous wastes. Most jurisdictions require an air quality permit and a NPDES if there is waste water discharge.

Figure VIII–18. (Facing Page) Evaluation Parameters for High Temperature Thermal Treatment. Factors for and against high temperature thermal treatment are summarized. HTTT is suitable for a wide variety of soil type and contaminants. The disadvantage is the requirement for excavation and transportation beforehand. Treated soils can be disposed of in a sanitary landfill without special permitting.

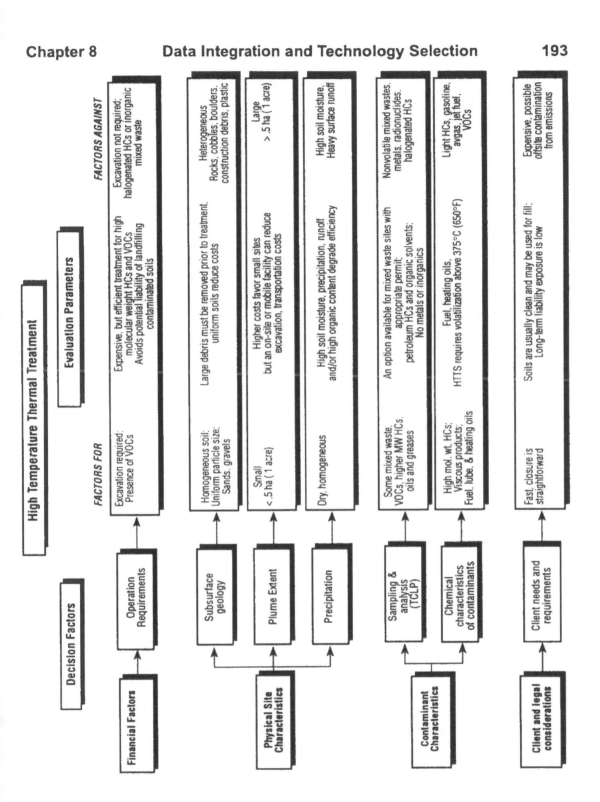

Evaluation Parameters for High Temperature Thermal Treatment

Asphalt Incorporation Fact Sheet

Technology. Contaminated soils are mixed with asphalt as whole or partial substitute for the customary aggregate. Subsequently the asphalt is either used in road construction or is broken up and landfilled, or crushed and used for road base. Low molecular weight petroleum hydrocarbons are volatilized and pyrolyzed; higher molecular weight hydrocarbons are incorporated into the asphalt. Lighter molecular weight hydrocarbons are incompatible with asphalt; they act as solvents and soften and degrade the asphalt. Large debris from excavating is unsuitable, as is plastic — notably Visqueen. Soils with a high organic content, including plant debris are unsuitable.

Even if the contaminated soils are unsuitable for asphalt incorporation this technique offers possibilities for disposal of contaminated soils since used asphalt is not hazardous. As such, asphalt incorporation becomes similar to solidification. The contaminated soils are immobilized in the asphalt and, as long as there is no possibility of leaching, the used asphalt can be disposed of in a sanitary landfill.

Combinations. None, other than the similarity to solidification noted above. The method is almost indistinguishable from LTTS for soils contaminated with low molecular weight hydrocarbons.

Equipment. The following equipment may be mobile or fixed, but is generally fixed. In addition to the asphalt batch plant itself, the following is required:

- Earth moving equipment;

- Soil handling equipment;

- Screens to remove large debris;

- Emissions control/monitoring systems;

- Sampling access and equipment.

Limitations. Most contaminated soils are not really suited to asphalt incorporation. Asphalt manufacture requires carefully graded aggregate (gravel) and soils seldom meet the specifications. Soils containing mixed hazardous wastes are completely unsuitable. Asphalt cannot be applied in cold weather.

Permitting. Asphalt incorporation systems must be permitted to treat mixed wastes that include petroleum contaminated soils. Most jurisdictions will require an air quality permit or an exemption, especially if the plant is considered a HTTT facility instead of an asphalt plant.

Figure VIII–19. (Facing Page) Evaluation Parameters for Asphalt Incorporation. Factors for and against asphalt incorporation are summarized. Contaminated soils are mixed with asphalt as whole or partial substitute for the customary aggregate. Typical mixing ratios call for about 5% contaminated soil. Low molecular weight hydrocarbons are vaporized and burned; higher molecular weight hydrocarbons are incorporated into asphalt and used in road construction. The limiting factor is suitability of the soil. Relatively few soils are actually suited to manufacture of asphalt.

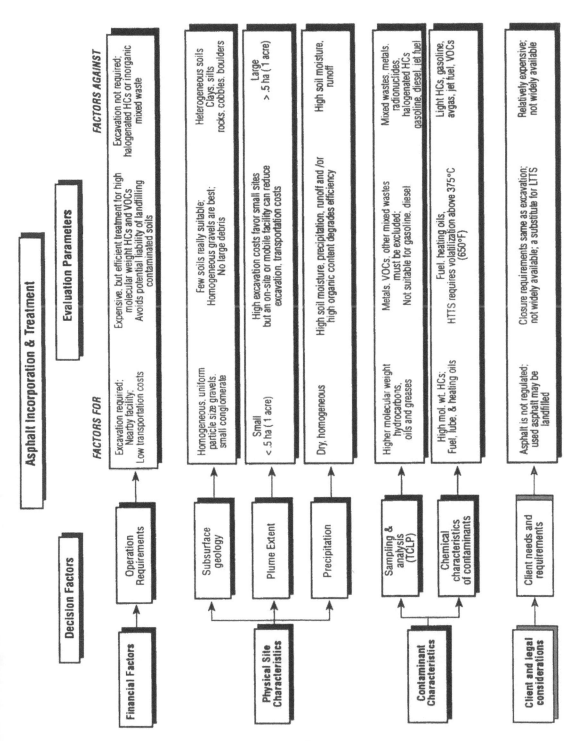

Evaluation Parameters for Asphalt Incorporation & Treatment

9 In Situ Remediation Technologies

In Situ vs. Non–In Situ

The choice of remediation strategy is the result of a frequently complex calculation based on a number of variables, many of which were covered in the preceding chapter. The purpose of all the data and information collected up to this point is to remove as much contamination from the site as is feasible. The current trend appears to be away from excavating and landfilling, and in the direction of "benevolent reuse"; that is, cleaning and reusing contaminated soil instead of merely adding to the pollution burden by storing soil at a permitted landfill.

Currently there are only four methods for remediating petroleum contaminated soil in widespread use:

- Excavation and removal, followed by either landfilling or further treatment;

- Volatilization or soil venting, removes volatile hydrocarbons from the vadose zone; usually used to treat raw gasoline contamination;

- Groundwater treatment to treat contaminated groundwater;

  and

- Bioremediation, used to treat contaminated soil both in situ and non–in situ.

Each method has its pros and cons and will be discussed in turn in the following sections.

Excavation means removing all or nearly all of the contaminated soil from the ground. Excavation is followed by one of the following:

- storage at a secure landfill; or

- some form of further treatment to remove the contamination.

The latter may be carried out on site or at a remote location. The principal disadvantage of landfilling is that there is no remediation involved other than whatever passive remediation or volatilization occurs as the soil is stored. Non–in situ technologies that require excavation are discussed in the second half of this chapter following in situ technologies.

In situ technologies approach soil remediation at the site of contamination. Remediation is carried out without excavating or otherwise disturbing the subsurface or, at least, with minimal disturbance. The main advantage of this approach is that the commercial operation can usually be left in service while the remediation continues. The main disadvantage is the length of time required for remediation; excavation is measured in terms of weeks while in situ methods may take years, accompanied of course by periodic monitoring. Both approaches to remediation can be equally efficient and cost effective. (Ludvigsen et al., 1992; ESI, 1990)

Volatilization

Volatilization is a proven technology that has been in use for a number of years. It has been used to remediate both gasoline– and chemically–contaminated soil, usually volatile organic compounds (VOAs). Other terms used to describe the same technology include soil venting, soil vacuuming, vapor extraction, or vapor extraction system (VES) (ESI, 1990). It is a treatment for contaminated soil, although a modification of the technology, air sparging, is used to treat groundwater.

Volatilization means removal of volatile hydrocarbon vapors from subsurface soil, specifically from the vadose zone, more specifically from the interstitial spaces within the vadose zone. Hydrocarbon vapors are entrained in a flow of extracted air and removed from the pore spaces of contaminated soil. In porous soil, extraction of volatiles can be rapid; in heavy clays the method is useless.

Successful removal of product from contaminated soil depends on a number of site–specific factors. Chief among these are the following (Johnson et al., 1993; Noonan and Curtis, 1990):

- Depth to groundwater;

- Soil moisture content;

- Volatility of the product;

- Kinematic viscosity;

- Porosity of the soil;

  and

- Soil temperature.

Each of these factors affects the flow rate of volatile contaminants to the wellhead. The first two, depth to groundwater and soil moisture content, are inversely proportional to flow rates. The more saturated the soil the slower the migration rate and extraction becomes that much less efficient.

The area from which a given well is able to extract volatiles is called the *radius of influence* or, since the area is seldom a sphere, a preferable term is *capture zone* or *zone of influence*. Vapor flow rates are a function of the pressure gradient between ambient pressure and the negative pressure at the wellhead.

Volatilization is efficiently combined with various other methods to enhance rates. It is combined with bioremediation as *bioventing*, and with groundwater treatment as *air sparging*. Extracted vapors usually cannot be vented directly into the air. Filtration, scrubbing, or combustion is used to clean extracted air before exhausting.

Technical basis: The basis for this technology relies on Henry's law, which is an empirical description of the equilibrium distribution of hydrocarbons between air and dilute aqueous solutions; that is, the volatility of hydrocarbons in water. Henry's law is written as follows:

$$p_a = K_H C_a$$

where p_a is the vapor pressure of a dissolved hydrocarbon, C_a is the concentration of the hydrocarbon in water, and K_H is a proportionality constant usually called the Henry's law constant for a dissolved gas.

In effect K_H is a partitioning coefficient or a "stripping coefficient" (Noonan and Curtis, 1989). By continuously replacing contaminated air with fresh air, the volatile hydrocarbon components are stripped out of the interstitial spaces. Fresh air flows in either by natural flow or through injection wells.

The presence of vapor phase hydrocarbons can be determined by sampling; either by a variety of soil gas techniques or by monitoring wells. As a general rule, benzene levels (as indicated by BTEX analyses) in the neighborhood of 200 ppm or greater indicate free gasoline. If the release has not aged significantly, free gasoline is accompanied by a vapor phase.

Soil gases may be monitored by sampling wells or by using stainless steel vapor probes driven into the ground. Hand–driven probes can be used up to about 14 feet, but are usually limited to a maximum depth of 5 to 8 feet. Canister techniques such as Draeger® tubes and Petrex™, are more problematical and less easily interpreted. All require porous soil.

Comments: At a minimum, volatilization requires installation of extraction wells, and frequently requires injection wells for optimum results (**Figure IX–1**). In colder climates the injected air may need to be heated to achieve good recovery rates. Soil temperatures are less important than soil porosity.

The more volatile components of gasoline, aviation gasoline, and jet fuel are removed easily and rapidly from open, porous soil; heavier components, including diesel, are removed reluctantly; heating oils and up are not removed at all. The method does not work in tight, nonporous clays or for heavy hydrocarbons with low volatility. Experimental results from both lab and field studies indicate that gasoline can be 100%

removed within about 100 days from reasonably porous soil; however, fuel oil contaminants were still at high levels after 120 days (EPRI, 1988).

Mixed waste sites can be treated if the additional contaminants are also VOCs, including chlorinated hydrocarbons. Alcohols, including methanol, ethanol, and propanol, and 2–propanol, ketones, for example, MEK, and aromatic solvents are adsorbed onto charcoal or can be burned as fuel. Halogenated compounds are more complicated and are best handled by pretreatment in biofilters.

Health and Safety: Workers and inspectors routinely exposed to hydrocarbon vapors should be monitored for adverse health effects. The more volatile components are also the more toxic components, and occur largely in gasolines. These components include n–pentane, n–hexane, n–octane, benzene and alkylbenzenes. Potential adverse health effects result from inhalation and adsorption through the skin. More volatile components are more likely to cause damage to lung tissue and be absorbed into the bloodstream from the lungs.

Volatile hydrocarbons in heavily contaminated soil, in recovery containers, trenches, or in diversion ponds are potential fire or explosion hazards. Hydrocarbon vapors are uniformly heavier than air (except for methane and ethane) and accumulate in the lowest points of excavations, ditches, and ponds. Precautions must be taken to avoid smoking materials and other sources of ignition, which can include field instruments, battery chargers, and steel sampling equipment. These precautions should be built into the site safety plan.

In many jurisdictions, extracted air must be treated to remove essentially all hydrocarbons before being exhausted into the air. At a minimum the more toxic components, such as aromatics, should be removed. Hydrocarbons in the atmosphere are converted fairly rapidly to CO_2, but aggravate CO and O_3 problems.

Requirements: Volatilization extracts the gasoline components present in the vapor phase within the vadose zone. Therefore, lighter more volatile contaminants are extracted more readily than heavier contaminants. Lighter contaminants include butanes, pentanes, and hexanes, the aromatics, benzene, toluene, xylenes, ethylbenzene, and other alkylbenzenes. Diesel, heating oils, used or waste oil and fuel oil are poorly suited to vapor extraction.

Vapor extraction can be combined with recovery of free product (gasoline) for more efficient operation. If free gasoline is present, recovery wells to pump free gasoline are set in the region of highest concentration and allowed to run. Vapor recovery wells are sited in areas of lower concentration to recover migrating vapor–phase hydrocarbons.

Figure IX–1. Soil Vapor Extraction System. A complete soil venting system consists of an injection and extraction manifold, sampling ports, vapor treatment and associated equipment. In many applications the injection side is not installed, but, in some cases, can improve efficiency. The injection air heating unit is necessary only during operations in cold climates. Carbon treatment of extracted vapors is expensive and is often used as a final polishing step prior to release to the atmosphere. Sampling ports on the extraction manifold are necessary to estimate hydrocarbon recovery rates (Adapted from EPRI, 1988).

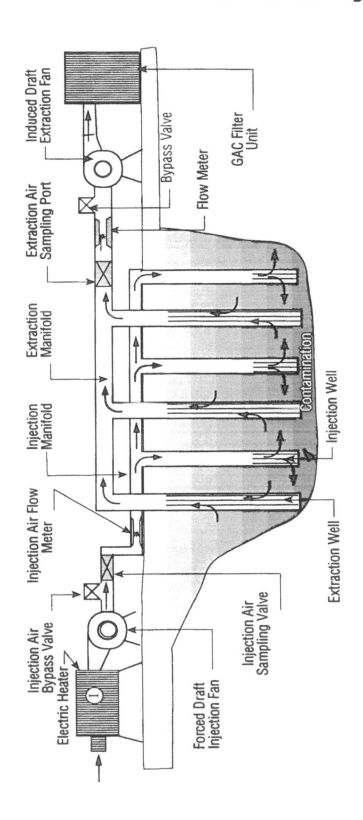

Vapor recovery from contaminated soil rarely results in completely clean soil. Typical cleanup levels are 100 to 1,000 ppm TPH. Liability is a factor in choosing this method if the property is to be sold after remediation. A residual TPH level of say, 500 ppm, may be adequate for the regulatory agency, and "no further action is required." But it may not be adequate for a potential buyer or lending institution.

Extraction wells and possibly injection wells are required for efficient operation. The zone of influence of each well varies from 20 to 50 feet depending on soil type and pump size. If the radius of influence is smaller than about 40 feet, volatilization may not be a suitable method for remediation. Typical pumping rates are 20 to 30 cfm (cubic feet per minute). Care must be exercised to avoid exceeding the capacity of the pumps during operations. Preheated air improves recovery, particularly in colder weather but of course adds to the cost of operation. Steam injection has been used with success in field tests, but can also simply diffuse and spread the contamination.

In most areas an air permit or variance or exemption must be obtained and exhaust air must be treated before release into the atmosphere. Common techniques for treating hydrocarbon contaminated air include:

- Oil/water separation in which contaminated vapors flow into and through a large water reservoir. Hydrocarbons go preferentially into the aqueous phase. As the tank is aerated, oil separates and floats to the top. A floating skimmer continuously removes the oil from the surface (Testa and Winegardner, 1991).

- Granulated activated charcoal filters require approximately 100 lbs carbon for each 15 lbs of hydrocarbons filtered. GAC filters will clean the exhaust air, but are too expensive to be the primary hydrocarbon removal method. Weekly maintenance is required since saturated filters are useless.

- Burning as fuel or a fuel supplement in an internal combustion engine. In this configuration extraction rates of up to 50 pounds of hydrocarbons per hour can be achieved.

- Catalytic converters similar to the devices used in automobile exhaust systems are available. These devices remove hydrocarbons by oxidizing them to CO_2 and H_2O. They are expensive to operate mostly due to high maintenance costs.

- Ozonolysis has been used to oxidize mainly aromatic hydrocarbons. Air that has been passed under ultraviolet lamps is injected into separation tanks and circulated.

- Biofilters have been used recently to remove hydrocarbons efficiently (Diks and Ottengraf, 1991). A filter material such as sphagnum moss is packed into a column. The packing material is charged with bacterial cultures and nutrients. Contaminated vapors are passed through the column. Contaminants are adsorbed onto the high surface area of the filter material and degraded by bacterial action.

Vapor phase contaminants are entrained and removed rapidly in reasonably porous soil by drawing vapor through the pore spaces; however, adsorbed hydrocarbons are more difficult to remove. In the long term it is possible to extract adsorbed components due to equilibria shifts of the following type:

Vapor   ⇆   Liquid   ⇆   Solid

In somewhat more detail, the equilibria are complex (**Figure IX–2**).

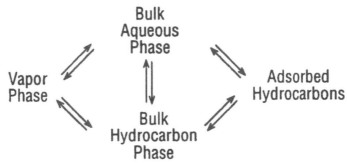

A. Phase equilibria in the unsaturated zone of contaminated soils

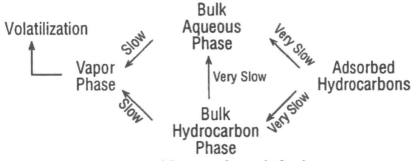

B. Equilibrium phase shifts during soil venting operations

Figure IX–2. Subsurface Phase Equilibria. (A) In aged, stable contaminated soil petroleum hydrocarbons are found in an equilibrium distribution across multiple phase boundaries. Hydrocarbon vapor, bulk hydrocarbon liquid, and adsorbed (solid) phases are normally in equilibrium. If the hydrocarbons are perched on the water table, a dissolved aqueous phase (hydrocarbons dissolved in water) is also present. If groundwater fluctuates seasonally, a bulk hydrocarbon phase becomes trapped in the aqueous layer. Solubilization, adsorption, and ganglia formation are increased. Equilibrium is reached among the various phases in relatively short time periods. (B) Hydrocarbon vapors are removed rapidly in relatively porous soil; however, complete removal of all or most of the hydrocarbons requires time for equilibria redistribution. As the equilibria shift to restore equilibrium, bulk liquid and adsorbed hydrocarbons migrate into the vapor phase. Dissolution of dissolved hydrocarbons also occurs.

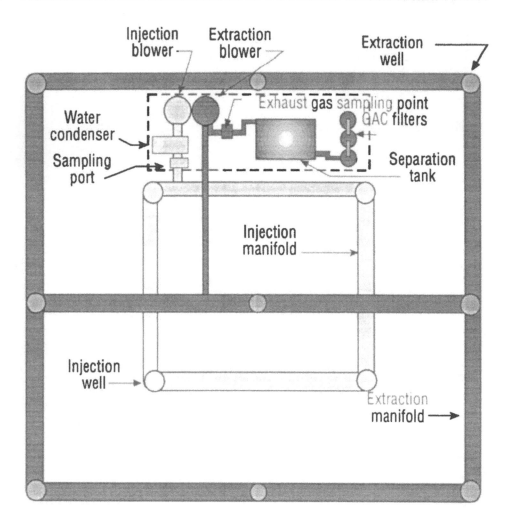

Figure IX–3. Soil Vapor Recovery Manifolds. Schematic diagram of a typical soil venting system using extraction and injection wells. Extraction wells are located according to calculated capture zones. Injection wells are placed to enhance subsurface vapor flow. Extracted petroleum vapors are recovered using granulated activated charcoal (GAC) filters after treatment in a bioreactor (biofiltration). (Adapted from EPRI, 1988).

To a large degree the equilibrium distribution of phases depends on the following:

- The extent to which the hydrocarbons are adsorbed onto soil particles;

  Hydrocarbons are most strongly attracted to clays, less strongly attracted to sands and gravels.

- The extent to which the contamination has aged;

Fresh leaks in which the gasoline is still largely in the bulk phase are more readily remediated than older, aged hydrocarbons containing higher percentages of acids and heavier hydrocarbons and less volatile compounds.

• The extent to which the liquid hydrocarbon phase has been "smeared" by fluctuating groundwater.

Table 9–1. Considerations and Recommendations: Volatilization. Vapor extraction, volatilization, or soil venting is a proven technology that is widely used for both petroleum contaminated and hazardous waste contaminated soil, but is effective over a limited range of hydrocarbons. The method treats contaminated soil in the vadose zone and is suitable for gasoline–range contaminants, especially for removal of free product. Soil venting requires porous soil for efficient operation. Extraction wells are required; injection wells may be necessary for efficient operation.

Volatilization (Soil Venting, Vapor Extraction)

Soil type:	Gravel	Silty Sand	Clay
	Yes	Maybe	No
Contaminant type:	Gasoline	Diesel	Fuel Oil
	Yes	No[a]	No[a]
Soil temperature:	20°C (70°F)	10°C (50°F)	5°C (40°F)
	Yes	Yes [b]	No
Contaminant extent:	Large	Moderate	Small
	Yes	Yes	Yes
Contaminant phase:	Free Product	Aged Gasoline	Dissolved
	Yes	Maybe	No [c]
Depth to groundwater:	50 ft [d]	30 ft	10 ft
	Yes	Maybe	No
Subsurface zone:	Vadose	Fluctuation	Groundwater
	Yes	Maybe	No [c]

a Diesel and heavier hydrocarbons are not volatile enough.
b Requires porous soils or heated air injection.
c Groundwater treatment can be accomplished with air sparging.
d Specific site and groundwater conditions can vary these numbers.

In areas where oxygenated fuels are mandated, MTBE may be the primary product recovered in the initial phases of soil venting. MTBE, which is an ether, is quite volatile (see Chapter 3). In many cases MTBE recovery can be used as a measure of efficiency of initial recovery rates.

The volatility of the xylenes forms an approximate boundary between easily recovered and less easily recovered hydrocarbons. Hydrocarbons more volatile (lighter) than m– or p–xylene (molecular weight = 106) are readily removed from soil; hydrocarbons less volatile (heavier) than m– or p–xylene are removed less readily. Diesel fuels and above (molecular weights greater than 170) are not candidates for volatilization.

Volatilization rates are generally greater from dryer soil than wetter soil. The method works slowly, or not at all, if any of the following conditions are present (Friesen, 1990; Turkall, et al., 1992):

- If the contamination is mainly in groundwater;

- If groundwater is near the surface;

  or

- If groundwater levels fluctuate over short time periods, as for example, tidal or seasonal fluctuations.

A recent variation that combines volatilization with groundwater treatment is called air sparging. In this technique, air is bubbled through contaminated groundwater. Dissolved hydrocarbons must compete with air (N_2 and O_2) of solubility. The bubbles form air–water interfaces, at which solubility is reduced. The net effect is to force dissolved hydrocarbons out of solution. As the vapors leave solution, they are picked up by the soil venting system and removed.

Bioremediation

Biodegradation, bioremediation, biorestoration, landfarming, and passive remediation are all related, just different approaches to the same basic technology. "Letting nature take its course" is an appealing concept that unfortunately is either seldom borne out in practice, or it is achieved but over a very long time span. Soil bacteria have evolved which exist on plant detritus present in the organic layer in soil. Because biological molecules and plant residues are similar to petroleum hydrocarbons they can be metabolized by native soil bacteria.

The detailed chemistry of hydrocarbon degradation is complex. Lighter hydrocarbons are more easily degraded by indigenous bacteria. A complete degradation of even low molecular weight hydrocarbons requires multiple strains of indigenous bacteria.

Technical Basis: Bioremediation is a process of aerobic microbial oxidation (oxidative degradation) of hydrocarbon contaminants to CO_2 and H_2O using naturally–occurring soil bacteria. The overall process may be "enhanced" or accelerated by providing nutrients to the bacteria, by providing laboratory cultured bacteria, or both.

Comments: The apparent low cost and simplicity have made this method appear very attractive to owners of contaminated sites. However, success is highly site–specific. Many of the restrictions noted above for volatilization are also restrictions on biaremediation. It requires relatively porous soil, and high molecular weight hydrocarbons are normally not suitable candidates for biodegradation.

This method may be carried out in situ or non–in situ. The non–in situ version is often called landfarming. Non–in situ land farming is generally more efficient and gives better results, but requires excavation. In situ technology is less expensive, but requires long periods of time for remediation during which the site must be continuously monitored.

Landfarming is complicated by the fact that up to 60% of the actual remediation is due to volatilization of petroleum hydrocarbons rather than degradation by microbes. If carried out on an extensive basis, the method could pose an air quality hazard. As a result landfarming of petroleum contaminated soil is carefully regulated. In some jurisdictions an air emissions permit is required.

Careful analysis and consideration must be given when deciding to adopt this technique in situ *versus* excavating and landfarming. From an engineering viewpoint, the deciding factor is usually the extent of contamination, transportation costs, and the cost of excavating. In real life the deciding factor is whether or not the state regulatory agency will approve landfarming as an option.

Despite the cost, bioremediation may be the only technology available to deal with the following special cases:

- The contamination extends deep or extensive enough to preclude excavation;
- Very large sites with extensive petroleum contamination;
- Diesel or jet fuel contamination.

Example of sites where each of these conditions exist are petroleum refineries, railroad fueling operations, and airports.

Health and Safety: Workers should be protected from exposure to hydrocarbon vapors, especially during landfarming operations. Excavating, drilling and/or handling petroleum contaminated soil may be hazardous if the exposure extends over a protracted period of time and/or is carried out on a routine day–to–day basis. Workers should be aware of normal construction safety standards.

Requirements: Most microbial activity is carried out in an aqueous layer that exists on the surface of soil particles. This is the reason that soil moisture content is necessary information to be collected during sampling operations. It is also the reason for controlling the moisture to 40 to 60%. The rate of conversion of hydrocarbons (HCs) to carbon dioxide, CO_2, and water, H_2O, is most strongly dependent on the following factors:

- Soil porosity;
- O_2 content in the pore (interstitial) spaces of the soil;
- Moisture content in the interstitial spaces of the soil;
- Macronutrient content, phosphorous, P and nitrogen, N;
- Soil temperature; and
- pH.

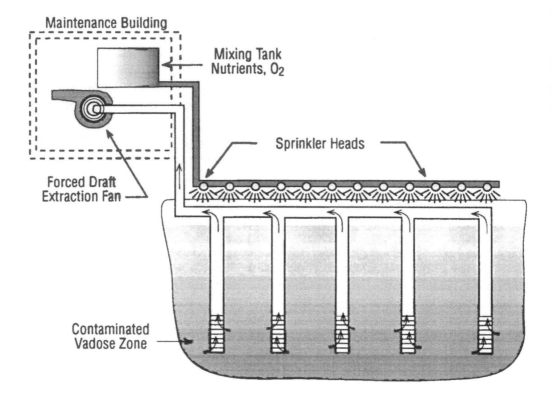

Figure IX–4. Extraction Wells and Nutrient System for In Situ Bioremediation.
Schematic diagram of typical extraction wells to provide an optimum environment
for microbial activity. The gradient produced by the vapor extraction system
improves the flow of fresh O_2 through the vadose zone. The infiltrated solution
includes nutrients, usually a dilute, well aerated fertilizer solution or hydrogen
peroxide solution.

The overall stepwise process may be represented as follows (Lehman, 1992):

$$(HC)_x \ + \ (N,P,S) \ + \ O_2$$
Contaminants Nutrients Oxygen

$$\rightarrow \ \{Acid\ Intermediates\}$$

$$\rightarrow \ CO_2 \uparrow \ + \ H_2O \ + \ NO_3^- \ + \ SO_4^{2-}$$

$$+ \ \boxed{Energy \ + \ Bacterial\ Cells}$$

Up to 60% of hydrocarbons:
Used by bacteria for reproduction
and cell materials

In this schematic outline $(HC)_x$ represents generic petroleum hydrocarbons and (N,P,S) represents the nutrients; nitrogen, N, phosphorous, P, and sulfur, S, which are required for normal metabolism and population growth. The hydrocarbons are broken down in a stepwise manner by different strains of bacteria. Degradation usually occurs two carbons at a time. For example, a given strain of bacteria will degrade an octane (8 carbons) into a hexane (6 carbons), another strain will degrade the hexane to a butane (4 carbons) and so forth. The actual degradation products are acids. The vertical arrow indicates CO_2 is a gas, and thus percolates through soil as it is produced. Since the presence of subsurface microbial activity is reflected in elevated CO_2 levels at the surface, CO_2 mapping has been used as a soil gas survey to estimate contamination plumes. Control of the degradation process can be complicated due to the large number of variables involved.

1. *Macronutrient control:* Macronutrients are elements that bacteria need in relatively large quantities to support metabolism. For most bacteria, macronutrient requirements include carbon, nitrogen, and phosphorus. The optimum C:N:P ratio is 25:1:0.5. Carbon is supplied by the hydrocarbons themselves; nitrogen and phosphorous are supplied by fertilizers or manure. Soil treated with inorganic fertilizers gives higher conversion rates of hydrocarbons to CO_2.

2. *Micronutrient control:* Micronutrients are nutrients required in small quantities. These include nitrogen, phosphorous, potassium, and sulfur. Trace elements are elements that are required in very small to trace quantities. Trace element requirements are not drastically different from similar requirements in human nutrition. Bacteria need iron, manganese, zinc, and copper just as people do. Many commercial fertilizers contain adequate trace elements, but many native soils do not. Therefore, the process of remediation is accelerated by an ample nutritional supplement; i.e., bacteria thrive on a balanced diet.

3. *pH control:* The optimum pH range is 6.0 to 8.0 with best results as close to 7.0 as possible. Due to production of intermediate organic acids and CO_2 the pH will decrease (become more acid) during bioremediation. If the pH slips much below 6.0, bacterial activity slows dramatically. The change in pH can be rapid in unbuffered soil such as sands and some clays. Soil may be treated with limestone solutions to adjust pH. pH control is much easier in landfarming than in situ.

4. *Oxygen control:* Since the major degradation pathway is aerobic, bacterial activity proceeds more rapidly in well–aerated soil than in poorly aerated soil. Delivery of O_2 to the site of microbial activity is facilitated by porous soil. Exact reaction of octane with oxygen requires 3.5 pounds of O_2 per pound of octane. In practice each pound of hydrocarbons converted requires 3–5 pounds of available O_2. Increasing O_2 availability by treating

the soil with dilute hydrogen peroxide, H_2O_2, has been tried with mixed success. Hydrogen peroxide decomposes as follows:

$$2\,H_2O_2 \longrightarrow O_2 + 2\,H_2O$$

The idea is to inject hydrogen peroxide sufficiently deep into the soil to maximize availability to the bacteria. One problem that has been encountered is that H_2O_2 will react with *any* reduced species present; it is not selective. Consequently it reacts with plant detritus, reduced iron, and ammonia produced as a result of the degradation process itself. If H_2O_2 is the only source of available O_2 each pound of hydrocarbon requires approximately 6–10 pounds of hydrogen peroxide. An alternative method is to sprinkle aerated water on the soil through infiltration galleries and injecting aerated water directly into the contaminated volume through injection wells.

5. *Temperature control:* This is one area where the in situ process has an advantage over landfarming. The temperature of soil at a depth of six feet is nearly constant and approximately equal to the mean annual temperature of the region. Representative means are :

Phoenix	72°F	29° North latitude
Denver	48°F	40° North latitude
Juneau	40°F	60° North latitude

   Temperatures can be controlled somewhat by adding nutrients since increased biological activity will increase the soil temperatures. Heated air injection is an alternative, but also increases costs.

6. *Species control:* Mixtures of several bacterial species are required for conversion: Typical bacterial species include, in descending order of occurrence, *Pseudomonas, Arthobacter, Alcaligenes, Corynebacterium, Flavobacterium, Achromobacter, Acinetobacter, Micrococcus, Nocardia,* and *Mycobacterium*. The best mixture of species depends on the contaminant. Normal soil populations degrade gasoline and lighter hydrocarbons but not heavier hydrocarbons. However, degradation of heavier hydrocarbons can be enhanced by adding commercial preparations of bacteria, sometimes called "bag of bugs" or "bug 'n a bag." Survivability of commercial species is improved in landfarming relative to in situ methods.

7. *Moisture control:* Soil moisture levels can be crucial to success of in situ bioremediation. A soil moisture content of 50% is ideal; overall soil moisture content should be maintained between 30 and 80%. Since saturated soils rapidly lose O_2 and become anaerobic, stormwater and runoff should be controlled during remediation by means of dikes, berms, or trenches. Diversion ditches are described in the section on **Isolation and Containment** in the next chapter. Landfarmed soils are frequently covered with black plastic to control runoff and provide a source of heat.

Not surprisingly, the best results have been obtained using large bioreactors as, for example, 100,000 L fixed reaction vessels, in which the various parameters can be controlled with precision (C&EN, 1992). Unfortunately, these vessels are very expensive and practical only at large scale, Superfund remediation sites. Next best results are obtained with landfarming techniques (ex situ bioremediation), where again the goal is to control the nutrient and metabolic parameters as closely as possible (Hinchee and Olfenbuttel, 1991a). Good results in situ are more problematic and are heavily dependent on the nature of the subsurface.

Table 9–2. Considerations and Recommendations: In Situ Bioremediation. In situ bioremediation is an established technology that works well under the right conditions, but can prove frustrating under less optimum conditions. With a viable bacteria population and well–aerated, porous soil, the method is inexpensive and results in soil that can meet regulatory cleanup guidelines. Under less ideal conditions costs escalate and management of operating parameters becomes more difficult.

In Situ Bioremediation (Biorestoration)

Soil type:	Gravel	Silty Sand	Clay
	Yes	Maybe	No
Contaminant type:	Gasoline	Diesel	Fuel Oil
	Yes	Yes[a]	Yes[a]
Oxygen content:	Aerobic	Moderate	Anaerobic
	Yes	Maybe [b]	No
Contaminant extent:	Large	Moderate	Small
	Yes	Yes	Yes
Contaminant phase:	Free Product	Aged Gasoline	Groundwater
	Yes	Maybe	No
Soil temperature:	20°C (70°F)	10°C (50°F)	5°C (40°F)
	Yes	Maybe	No
Depth to groundwater:	50 ft [c]	30 ft	10 ft
	Yes	Maybe	No

a   Using commerical bacterial preparations.
b   Poor soil aeration can be improved with bioventing or O_2 injection.
c   Specific site and groundwater conditions can vary these numbers.

The half–life of petroleum hydrocarbons (as measured by TPH) has been observed to be as low as 23 days in landfarmed soil, but may be on the order of months to years for in situ applications (EPRI, 1988). In general, lighter, more volatile hydrocarbons, including aromatics, are degraded more rapidly; straight chain and heavier hydrocarbons are degraded more slowly, although results can vary if commercial bacterial mixtures are used. Therefore, bioremediation is a viable option for gasoline contaminants, with the understanding that much of the remediation is due to volatilization.

Bioremediation is a less clear–cut option for diesel contamination. If the diesel fuel can be landfarmed using commercial bacterial mixtures, remediation results are greatly improved. On the other hand in situ bioremediation is often the only choice to remediate very large sites with extensive contamination, particularly mixed contamination. Examples are railroad fueling yards, wood treatment plants, and refineries. In one study involving a railroad yard heavily contaminated with diesel fuel, bioremediation was found to be a viable, but expensive, option. Diesel degradation required not only nutrient solutions, but inclusion of surfactants as well (Battelle, 1993; Updegraff, 1993).

Table 9–2 summarizes the considerations and recommendations for deciding on in situ bioremediation.

In Situ Passive Bioremediation

Bacterial activity will occur in soil and landfills regardless of the wishes of state regulatory agencies. Therefore, a unique feature of petroleum contamination is that if you do not do anything at all it will sometimes go away. Unlike chemical contamination, particularly heavy metals and radioactivity, petroleum contamination is self–correcting to a degree. Degradation of hydrocarbons by naturally–occurring soil bacterial is called passive biodegradation (Battelle, 1993).

Technical Basis: Natural aerobic microbial oxidation of hydrocarbons to CO_2 and H_2O. This is the same as bioremediation discussed above except that the process proceeds naturally without enhancement or interference. As with all bioremediation, actual remediation includes at least some degree of volatilization and leaching. Estimates of the contribution of volatilization to the overall remediation run as high as 60% (EPRI, 1988). When contaminated soil is excavated and hauled to a landfill, this is the process by which restoration occurs.

Comments: This is the least expensive, and slowest of all the available remedial technologies. It is seldom encouraged as an option for remediation. Few state agencies approve correction action plans which include passive remediation. But passive remediation occurs continually in nature, and therefore, is a corrective action choice by default. Sites that are subject to prolonged litigation are slowly being remediated by passive biodegradation.

As a legitimate option in a CAP it is probably suitable in certain specific cases, such as remote locations where the source of the release has been contained or eliminated, and there is no chance of third–party or groundwater contamination.

Health and Safety: Workers should be protected from exposure to hydrocarbon vapors, especially during landfarming operations. Excavating, drilling and/or handling petroleum contaminated soil may be hazardous if the exposure extends over a protracted period of time or is carried out on a routine day-to-day basis. Workers should be aware of normal construction safety standards.

Requirements: Successful passive restoration of a contaminated site is a complex interaction of bioremediation, contaminant volatilization, and leaching. The requirements are essentially the same as for bioremediation, except that nutrients, oxygen, and pH are only those that are available naturally in the soil. A contaminated vadose zone can be treated by passive remediation, but conditions must be appropriate. In practice, the method must be justified to regulatory agencies on a case–by–case basis.

Table 9–3 summarizes the considerations and recommendations for deciding on passive remediation.

Table 9–3. Considerations and Recommendations: Passive Bioremediation. The factors most commonly used in evaluating the use of passive remediation are summarized. Passive remediation proceeds at all times providing that sufficient indigenous soil bacteria and nutrients are present. In most cases the degradation rates are too slow to be acceptable. In the right circumstances passive remediation can be an option, but must be carefully justified to regulatory personnel.

Passive Bioremediation (In Situ Passive Remediation)

Location:	Remote Rural	Suburban	Urban
	Yes	No[a]	No[a]
Soil type:	Gravel	Silty Sand	Clay
	Yes	Maybe	No
Contaminant type:	Gasoline	Diesel	Fuel Oil
	Yes	No[b]	No[b]
Soil temperature:	20°C (70°F)	10°C (50°F)	5°C (40°F)
	Yes	Maybe	No
Contaminant extent:	Large	Moderate	Small
	Yes	Yes	Yes
Oxygen content:	Aerobic	Moderate	Anaerobic
	Yes	Maybe	No
Depth to groundwater:	50 ft [c]	30 ft	10 ft
	Yes	Maybe	No

a   Despite the fact that passive remediation proceeds continuously, it is too slow to be used as the method of choice in populated areas.

b   Indigenous, unenhanced soil bacteria degrade diesel and higher hydrocarbons with difficulty.

c   Specific site and groundwater conditions can vary these numbers.

Soil Leaching

Soil leaching, also known as soil washing or soil flushing, is a process of gradually cleaning soil by flowing fresh water through the soil particles and extracting hydrocarbons. The idea is that while petroleum hydrocarbons are not very soluble, they are not completely insoluble. With each pass of fresh water a small portion of hydrocarbon contaminants will dissolve. The process is analogous to trying to clean a greasy spoon by washing it in water only; it will eventually become relatively clean, it just takes awhile.

Fresh water can be introduced in any of several methods, including injection wells, surface spraying, or infiltration galleries. A more stringent requirement is collection of the contaminated effluent.

Technical Basis: The principle of this method is similar to volatilization. A continuous flow of fresh water will take up hydrocarbons from regions of higher concentrations. Hydrocarbons are not very soluble in pure water, but they are soluble enough that, in time, petroleum contaminants can be flushed from contaminated soil. The controlling factors are:

- the degree to which contaminants are retained on the soil particles;

- the soil–water partitioning coefficient, Henry's law constant;

- soil porosity;

- temperature of the eluent, the injected water.

Petroleum recovery from soil is enhanced if a surfactant (detergent) is added to the flushing solution. Cost–effectiveness is improved if biodegradable surfactants are used. Surfactants are a general class of chemical compounds to which detergents belong. They are characterized by having a hydrophobic (water–hating, oil–loving) end, which is attracted to hydrocarbons, and a hydrophilic (water–loving, oil hating) end, which is attracted to water. The result is an emulsion.

Since the emulsions are usually quite stable it is necessary to collect the effluent emulsion and treat it in some manner. Land farming is an attractive option. Clean soil is spread as described below under **Landfarming**. The aqueous emulsion is sprayed over the soil and allowed to degrade by microbial activity.

Comments: New stormwater regulations are placing more severe constraints on this technology since the extracted water must meet stormwater standards and have a NPDES permit before it can be discharged into storm sewers, if it can be discharged at all. Groundwater plus the emulsified hydrocarbons, the leachate solution, must be treated to remove petroleum hydrocarbons or be otherwise disposed of properly. After treatment the water may be recycled through the system. However, the U.S. EPA is considering classifying reinjection wells as Class IV injection wells, which are banned in the United States (Testa and Winegardner, 1991).

Health and Safety: Workers must be protected from exposure to hydrocarbon vapors and contaminated soil. In a closed system contact with hydrocarbon vapors should be minimal. If the treated leachate solution is to be discharged into storm sewers or surface waterways, an NPDES permit is required.

Requirements: This method is more expensive since it requires installation of both injection and extraction wells. A treatment facility to treat the effluent water or a recycling system is required or the effluent must be collected. Discharge into storm sewers requires an NPDES permit. If a surfactant is added to the leaching solution, an oil–water emulsion is formed. The emulsion must be collected and/or treated. The water–surfactant can be separated by filtration. Alternatively, the emulsion can be treated (usually onsite) by landfarming methods (see Chapter 10). These factors must be considered in evaluating the cost–effectiveness of the method.

Table 9–4. Considerations and Recommendations: Leaching and Chemical Extraction. The factors to be considered in evaluating leaching and chemical extraction are summarized. The technology is useful when other established technologies are ruled out because of adverse circumstances. Aqueous leaching is relatively slow. Detergent extraction requires collection and disposal or collection and treatment of the leachate emulsion.

Leaching and Chemical Extraction (Soil Washing)

Soil type:	Gravel	Silty Sand	Clay
	Yes	Maybe	No
Contaminant type:	Gasoline	Diesel	Fuel Oil
	Yes	Maybe[a]	Maybe[a]
Contaminant extent:	Large	Moderate	Small
	No	Yes	Yes
Contaminant phase:	Free Product	Aged Gasoline	Groundwater
	No	Yes	Maybe [b]
Subsurface zone:	Vadose	Fluctuation	Groundwater
	Yes	Maybe	Maybe [b]
Depth to groundwater:	50 ft [c]	30 ft	10 ft
	Yes	Maybe	No

a　Diesel and heavier hydrocarbons require detergent extraction for efficient removal.
b　Leaching and extraction should not be used if it will contaminate otherwise
　　clean groundwater. Contaminated groundwater can be recycled after treatment.
c　Specific site and groundwater conditions can vary these numbers.

Chemical Extraction

Technical basis: This technique is a modification of methods used for several years for secondary recovery in production oil fields. It is based on the mobilization and extraction of hydrocarbons with chemical additives, usually detergents (surfactants), or with solvents, often water or steam, or with CO_2. When used in situ the idea is to reduce the viscosity of the hydrocarbons by emulsification. As the viscosity is reduced, flow rates are improved.

The method is more commonly done in situ, but can be either in situ or non–in situ. As an in situ method, extraction is most suitable for aged contamination with considerable adsorption. After excavation soil can be processed to remove residual contamination. Non–in situ extraction is an experimental method.

Comments: This can be an effective means for removal of hydrocarbons from soil. However, the costs are relatively high. Costs are lower when done on excavated soil (non–in situ). There is comparatively little experience with petroleum contaminated soil.

Requirements: The extraction solution is added to excavated soil and thoroughly mixed. The emulsion is separated from the soil by filtration. The soil may need to be washed to completely remove all the surfactant. The emulsion must be treated to remove all hydrocarbons, or disposed of as a regulated waste.

In situ treatment consists of drilling an array of injection wells and several extraction wells. The extraction solution is pumped through the contaminated area. The resulting emulsion is collected and treated or disposed of as a regulated waste.

Table 9–4 summarizes the considerations and recommendations for deciding on leaching or chemical extraction.

In Situ Vitrification

In situ vitrification is an interesting technology that has seen limited application since its introduction in 1980. It is a useful addition to the arsenal of methods to treat soil; in some situations it is the ideal choice. As a relatively expensive technology, it is not widely used, especially for petroleum contaminated soil. For mixed wastes containing both petroleum contamination and metals or radioactive wastes, it can be an ideal choice.

Technical Basis: Vitrification is the process of melting soil by application of high voltage, low current electrical power through trenches filled with conductive material. The amount of heat generated can produce temperatures in the neighborhood of 1,500°C (3,000°F); typical silica and alumina soil melt around 1000 to 1,500°C (2,000 to 3,000°F). In the process organic contaminants are volatilized and pyrolyzed on contact with high temperature, oxygen–rich air.

Comments: In situ vitrification is a relatively new technology, first tested in 1980, and developed by Pacific Northwest Labs (DOE). Electrodes are placed around the contaminated site and a voltage applied. As the soil melts, organic contaminants are vaporized and the gases combust at the surface as they come in contact with air. The resulting mass of vitrified soil resembles obsidian.

Health and Safety: During the vitrification process precautions must be maintained to exclude people from the remediation area, and to protect against exposure to off–gas vapors.

Requirements: This method requires considerable technical expertise. Electrodes are placed 3 to 5 meters apart in a square pattern. Trenches are excavated between electrodes and filled with a conducting material such as sodium bentonite and gypsum, $CaSO_4 \cdot 6H_2O$, to facilitate current flow. Large scale applications of this method require voltages on the order of 3,200 kW. As the treatment proceeds, vapors off–gassing must be collected and treated, either by filters or scrubbers.

The presence of volatile metals, such as cadmium, mercury, or arsenic, may require trapping off–gases. Vitrification leaves a vitrified glass that isolates and contains nonvolatile residues in a nonleachable form as indicated by TCLP results (ESI, 1990). The residual material can be left in place or broken up and removed.

Table 9–5 summarizes the considerations and recommendations for deciding on in situ vitrification.

Table 9–5. Considerations and Recommendations: In Situ Vitrification. The most important considerations for a decision on in situ vitrification are summarized. The method is useful in specific conditions. It is rarely used for petroleum contamination alone, but can be the method of choice when metals, other inorganics, or radioactive wastes are present.

In Situ Vitrification

Contaminant phase:	Free Product	Adsorbed	Dissolved
	No	Yes	No
Contaminant type:	Gasoline	Diesel	Fuel Oil
	No [a]	Maybe	Yes
Contaminant extent:	Large	Moderate	Small
	No	Maybe	Yes
Soil type:	Gravel	Silty Sand	Clay
	Yes	Yes	Maybe
Subsurface zone:	Vadose	Fluctuation	Groundwater
	Yes	No [b]	No [b]
Depth to groundwater:	50 ft [c]	30 ft	10 ft
	Yes	Maybe	No

[a] Gasoline contaminated soils are not suitable due to the fire/explosion hazard.

[b] Vitrification is not recommended in wet soils or if groundwater is near to the contaminated area.

[c] Specific site and groundwater conditions can vary these numbers.

Linear Interception

In some situations, for example, emergency response or to prevent off–site migration, it is feasible to install a system of trenches and drains or fractured strata to collect or divert migrating product. The idea is to use the natural gradient and hydraulic flow together with a path of decreased resistance to direct product flow into a recovery area. An ideal situation would be free product (gasoline) floating on a shallow water table, and moving rapidly through porous soil.

Technical Basis: Linear interception is intermediate between isolation and containment and in situ remediation. On the one hand it is a form of isolation in that the flow is redirected and contained. On the other hand it is also a form of remediation where product is collected and recycled or disposed of.

Health and Safety: Workers must be protected from, or monitored for, exposure to hydrocarbon vapors and contaminated soil or water. Since all hydrocarbons are flammable to a greater or lesser degree, and since interception may collect product in open trenches or reservoirs, precautions against fire or explosion must be observed. If the product being recovered is gasoline the health and safety risk is magnified.

Normal OSHA safety practices at construction sites and around heavy equipment must be observed.

Comments: Interception methods may be active or passive; that is, a simple ditch may suffice, or pumps and wells may be added. In other applications explosives or hydraulic pressure have been used to fracture a subsurface strata to provide a path of least resistance. Linear interception is a consideration in circumstances where groundwater depth is less than 25 feet, air quality concerns are minimal, the area is accessible by excavating equipment, and there are no underground utilities to obstruct excavation.

Requirements: Passive systems consist of open trenches set perpendicular to the downgradient flow of contamination. The trench system should drain into a collection pond or lagoon. Restrictions on applicability include the following:

- Low permeability soil that transport contaminants slowly;

- Product with low kinematic viscosity;

- Rapid or seasonal fluctuation in the water table;

- Heavy surface runoff;

- Possibility of fire or explosive hazard from open reservoirs of product.

Active systems include recovery/extraction well installation at a point designed to interdict the flow. Extraction wells will be discussed more fully later in this chapter.

Table 9–6 summarizes the considerations and recommendations for deciding on linear interception.

Table 9–6. Considerations and Recommendations: Linear Interception. The most important considerations for a decision on linear interception are summarized. The method is useful in specific conditions, especially in the event of an emergency response or imminent threat of offsite contamination. It is rarely used for petroleum contamination alone, but can be the method of choice when other hazardous VOCs are present.

Linear Interception [d]

Contaminant phase:	Free Product	Adsorbed	Dissolved
	Yes [a]	No	Maybe [b]
Contaminant type:	Gasoline	Diesel	Fuel Oil
	Yes	Maybe	No
Contaminant extent:	Large	Moderate	Small
	No	Yes	Yes
Soil type:	Gravel	Silty Sand	Clay
	Yes	Maybe	No
Subsurface zone:	Vadose	Fluctuation	Groundwater
	Yes	Maybe	Maybe [b]
Depth to groundwater:	50 ft [c]	30 ft	10 ft
	Yes	Maybe	Maybe [b]

a   Appropriate safety precautions must be taken to protect workers from exposure to gasoline vapors and from possible fire hazard.
b   Contaminated groundwater is much more difficult to contain than a migrating hydrocarbon plume. Flow rates will determine the feasibility of this method.
c   Specific site and groundwater conditions can vary these numbers.
d   One of the most important considerations is the necessity of emergency response or imminent off–site damage.

Isolation and Containment

Large complicated remediation sites are often tied up in legal skirmishes, possibly for years. Meanwhile it may not be possible to ignore the contamination and its potential spread. Therefore it is essential to have in hand techniques for temporary contaminant to prevent the released product from spreading and causing third party injury or increasing the cost of the eventual remediation.

Comments: This method has been used extensively to contain chemically contaminated areas. It has also been used to contain petroleum contamination and is particularly appropriate as an emergency measure to contain large–scale spills such as tanker accidents. Isolation and containment are not remedial methods, but rather short–term temporary measures to allow time to develop more specific remediation strategies.

Technical Basis: This process is designed to physically contain an area of contamination, and isolate it from the rest of the environment. Although there is no remediation, it is an effective means of isolating petroleum contamination until remediation can be accomplished. It should be considered when the danger of contamination of groundwater, drinking water wells, sanitary or storm sewers, or buildings is imminent.

Isolation techniques will adequately contain contamination, but do not constitute a permanent solution. First, there is no remediation involved. Second, the containment is not permanent; the plume will frequently find a path around the containment.

Health and Safety: There are no unusual health and safety requirements beyond normal construction practices.

Requirements: There are two general methods that have been developed to contain the spread of hydrocarbons:

- Subsurface barriers placed in the path of the plume;

- Surface covers or caps to reduce surface infiltration and runoff.

Subsurface barriers may be designed to divert or contain groundwater flow or the contaminant plume, and usually consist of curtain (or containment) walls. Containment walls should be constructed of materials compatible with the native soil. Typical materials are slurries (cement and/or bentonite and water), grout (portland cement or sodium silicate and water), or sheet pilings (rigid sheets of steel or concrete). If the curtain does not reach bedrock, it is often referred to as a "hanging wall." Typically if groundwater is near the surface, one or more recovery wells are drilled close to the barrier on the upgradient side to relieve hydraulic pressure and to facilitate product recovery.

Surface covers are synthetic membranes (geotextile cloth), soil/bentonite mixtures, asphalt, or concrete. The primary effect of surface covers is to redirect and channelize the flow of surface runoff away from the area of contaminated soil to prevent spread of contamination.

Barriers and surface covers may also be used in combination with berms, dikes, and diversion channels to divert surface water or groundwater around spills and contamination. Monitoring wells will be required downslope of the barriers to monitor the effectiveness of the containment.

Table 9–7 summarizes the considerations and recommendations for deciding on isolation and containment.

Table 9–7. (Facing Page) Considerations and Recommendations: Isolation and Containment. The most important considerations for a decision on isolation and containment are summarized. The method is useful in specific conditions. The most important consideration is in the case of emergency response or drawn out litigation. Containment is neither treatment nor remediation, but can buy time and prevent offsite migration. Containment may also be required in the event of prolonged litigation or CAP review.

Isolation and Containment

Contaminant phase:	Free Product	Adsorbed	Dissolved
	Yes	No	Maybe [a]
Contaminant type:	Gasoline	Diesel	Fuel Oil
	Yes	Maybe [b]	No [b]
Contaminant extent:	Large	Moderate	Small
	No	Yes	Yes
Soil type:	Gravel	Silty Sand	Clay
	Yes	Maybe	No
Subsurface zone:	Vadose	Fluctuation	Groundwater
	Yes	Maybe	Maybe
Depth to groundwater:	50 ft [c]	30 ft	10 ft
	Yes	Maybe	No

[a] Contaminated groundwater is much more difficult to contain than a migrating hydrocarbons plume.

[b] Diesel and heavier hydrocarbons require containment only in extreme situations due to the slow rate of migration.

[c] Specific site and groundwater conditions can vary these numbers.

Solidification and Stabilization

Technical Basis: In this method fly ash or cement is incorporated into contaminated soil to form an immobile mass. The basis for the remediation is not destruction of contaminants, but immobilization. Thus, it occupies a category similar to containment.

Comments: This technique has been used more extensively to immobilize chemically–contaminated soil than petroleum contaminated soil. Incorporation can be accomplished either in situ or non–in situ, but is usually in situ. Typical applications have been impoundment ponds containing oily sludges from production facilities.

Health and Safety: Workers should be protected from exposure to contaminated soil and hydrocarbon vapors, and from exposure to fly ash dust. Typical fly ash contains metals at relatively high levels.

Requirements: The exact mix of cement or fly ash is highly technical and requires exact specifications, but the process is straightforward. If required the contaminated soil is excavated and a liner placed in the excavation. The soil is replaced. The solidification agent, which can be fly ash, cement, or a mixture, is spread on the contaminated area. A rotary mixer is used until the mixture is homogeneous. Alternatively, a trackhoe–mounted injector is used to inject and mix fly ash into the contaminated area. Petroleum contaminated soils are suitable for fly ash incorporation; cement is more suitable for inorganic wastes

Table 9–8 summarizes the considerations and recommendations for deciding on solidification and stabilization.

Table 9–8. Considerations and Recommendations: Solidification and Stabilization. The most important considerations for a decision on solidification and stabilization are summarized. The method is useful in specific conditions, and has been used to stabilize waste ponds at refineries and production fields. Solidification confines a contaminant in a nonleachable matrix, but does not treat or remediate contamination.

Solidification and Stabilization

Contaminant phase:	Free Product	Adsorbed	Dissolved
	No [a]	Yes	No
Contaminant type:	Gasoline	Diesel	Fuel Oil
	No [a]	Yes[b]	Yes[b]
Contaminant extent:	Large	Moderate	Small
	Maybe [c]	Yes	Yes
Soil type:	Gravel	Silty Sand	Clay
	No [d]	Yes	Yes
Subsurface zone:	Vadose	Fluctuation	Groundwater
	Yes	Maybe	No
Depth to groundwater:	50 ft [c]	30 ft	10 ft
	Yes	Maybe	No

[a] Gasoline is not the best choice of contaminant for this process.
Light HCs destabilize the solidification agent.
[b] Diesel and heavier hydrocarbons are good candidates for stabilization due to their low viscosity and migration rates.
[c] Specific site and groundwater conditions can vary these numbers.
[d] Other remedial methods are more suitable in porous soils.

Groundwater Extraction and Treatment

If petroleum hydrocarbons have contaminated groundwater, and if the groundwater is flowing off–site, then remediation of the aquifer must have a high priority. In the most common treatment method, water must be removed from the ground and treated at the surface, either on–site or off–site. Options for treating groundwater efficiently and cost–effectively are few. Oil/water separation, air stripping, ozonolysis, and land treatment are among the options available. Air sparging is an efficient option. This method is discussed earlier in this chapter in connection with volatilization.

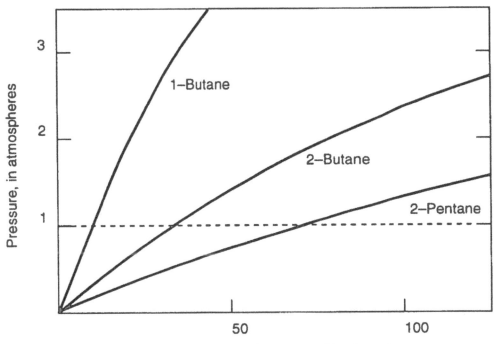

Figure IX–5. Solubility of Hydrocarbons in Water. Henry's law predicts a linear relationship for hydrocarbon solubility in water. For lighter hydrocarbons the data fit predictions quite well. The graph shows the relationship between solubility and vapor pressure for hydrocarbons dissolved in water. As the concentration of dissolved hydrocarbon increases, there is approximately a linear increase in vapor pressure of the hydrocarbon.

An excellent, more detailed reference is available in *Restoration of Petroleum Contaminated Aquifers,* by Stephen Testa and Duane Winegardner.

Groundwater extraction and treatment is a proven technology that is among the most common remediation methods for petroleum contaminated soil. Also known as "pump and treat," this method removes contaminated groundwater to the surface for treatment and reinjection.

Technical Basis: The effectiveness of this method is based on Henry's law for dissolved substances (**Figure IX–5**). Treatment of groundwater revolves around overcoming the solubility of hydrocarbons. Treatments rely on two different effects: one at the surface and one at the subsurface level. At the surface, the controlling factor is the efficiency of the hydrocarbon–water separation process. One of the following options can be assumed for the treated water:

- It can be treated and reinjected;

Reinjection requires additional wells and a permit.

- It can be released into surface drainage or storm sewers;

  This option requires an NPDES permit and/or a sewer district permit.

- It can be collected for disposal;

  This is the most expensive, least desirable option.

The other level is the degree to which contaminated soils, which are the source of dissolved hydrocarbons, are remediated. As treated water is reinjected, it acts as a soil washing agent. Adsorbed hydrocarbons are difficult to remove so this is a technology that runs on the order of months to years.

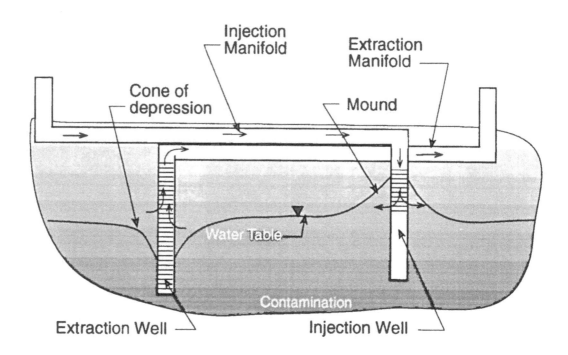

Figure IX–6. Groundwater Extraction and Injection System. Groundwater is extracted by conventional extraction wells described in following sections. A negative pressure is induced at the wellhead and water is forced up into the extraction manifold. Since there is a period of time required for the groundwater to recover, a cone of depression is created. If an injection well is used, a mound of higher groundwater is created.

Comments: The controlling factor is the efficiency of the oil–water separation. The overall rate is determined by the removal of adsorbed hydrocarbons. Petroleum recovery from soil is enhanced if a surfactant (detergent) is added to the flushing solution (see

Chemical Extraction, this chapter), but the resulting emulsion must be treated or collected for disposal.

Health and Safety: Workers exposed to contaminated soil or groundwater brought to the surface should be protected from inhalation exposure and skin contact, particularly if the contaminant is gasoline. Normally, reinjected groundwater must meet standards close to drinking water standards before it can be returned to the subsurface environment.

Requirements: This method affects removal of hydrocarbons from groundwater. Groundwater is brought to the surface and treated by one or more of the following methods, ranked in approximate order of importance:

- Oil/water separation;

- Air stripping;

- Oxidation, using ozone, O_3;

- Absorption by activated charcoal (GAC);

- Biological treatment, usually non-in situ biodegradation;

- Biofiltration.

1. Oil–water separation can be as simple as a large tank in which dissolved hydrocarbons slowly separated. This is a slow process. Separation is enhanced if the solution is aerated.

2. Air strippers are efficient, but may be contrary to air quality standards. Principles and design of air strippers are discussed in detail in Noonan and Curtis, 1990.

3. Ozone is produced in situ by irradiating an aerated flow of contaminated water with ultraviolet (uv) light. Photons at uv wavelengths convert O_2 to O_3, which is a strong oxidizing agent. Aromatic compounds and lighter aliphatic hydrocarbons are oxidized to CO_2 and water.

4. GAC filters are often used in combination with one of the other methods since GAC filtration alone is expensive. Hydrocarbons can be adsorbed at a rate of about 15 pounds of hydrocarbons per 100 pounds of carbon.

5. Land treatment means preparing a site as for landfarming except that noncontaminated soil is used. Contaminated water is sprinkled over the soil and degraded by soil bacteria. This method is discussed in more detail in the next chapter (See **Land Treatment**, Chapter 10).

6. Biofiltration is a new technology. Biofilters have been discussed in a previous section (see **Volatilization**)

7. If groundwater is reinjected, the dissolved or emulsified hydrocarbons must be disposed of properly or separated from the extracted water on–site, which is generally a difficult and expensive task.

Table 9–9 summarizes the considerations and recommendations for deciding on groundwater extraction and treatment.

Table 9–9. Considerations and Recommendations: Groundwater Extraction and Treatment. The most important considerations for a decision on groundwater extraction and treatment are summarized. The method is considered a proven technology and is widely used to treat petroleum contaminated and hazardous waste contaminated groundwater. Costs are comparable to volatilization and excavation. Air sparging can increase the efficiency of the method.

Groundwater Extraction and Treatment

Soil type:	Gravel	Silty Sand	Clay
	Yes	Maybe	No
Depth to groundwater:	10 ft	30 ft	50 ft a
	Yes	Maybe	No
Contaminant type:	Gasoline	Diesel	Fuel Oil
	Yes	Nob	Nob
Contaminant extent:	Large	Moderate	Small
	Yes	Yes	Yes
Contaminant phase:	Free Product	Adsorbed	Dissolved
	No	Maybe c	Yes
Subsurface zone:	Vadose	Fluctuation	Groundwater
	No	Maybe	Yes
Well requirements:	Injection	Recovery	Extraction
	Maybe d	Maybe	Yes

a   Specific site and groundwater conditions can vary these numbers.
b   Diesel hydrocarbons are not soluble enough to contaminate groundwater to a significant extent.
c   Aged, adsorbed gasoline can be a source of dissolved phase contamination.
d   Injection wells improve efficiency and allow extracted water to be recycled.

(Facing page) Figure IX–7. $p\epsilon$ vs. pH Diagram for Natural Degradation Pathways. Each line on the figure shows the negative log of the reduction potential, $p\epsilon$, of an electron transfer couple as a function of pH. The set of lines in the upper right represents oxidation half–reactions; the set in the lower left represents reduction half–reactions. Adding a pair of oxidation and reduction half–reactions gives a net degradation pathway. The shaded area in the center is the area of significant, naturally–occurring pathways. One of the major paths is provided by the NO_2^-/NO_3^- oxidative pathway. It is an anaerobic degradation, but O_2 is not the electron transfer agent.

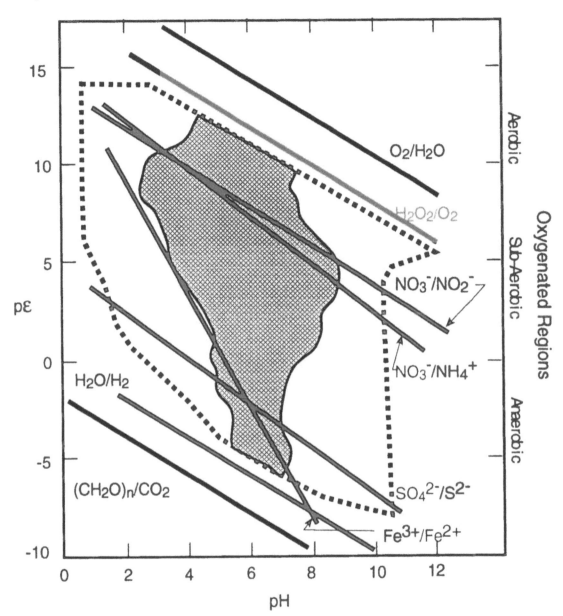

Anaerobic Biodegradation

For many years aerobic biodegradation was considered to be the only feasible means of bioremediating contaminated soil. Correspondingly, aerobic bacteria were the only microbes worth studying. In recent years, as the difficulties of carrying out aerobic degradation have become apparent, attention has turned to anaerobic degradation. In the process scientists have begun to appreciate the role of anaerobic bacteria in naturally–occurring soil remediation (Battelle, 1993; Hinchee and Olfenbuttel, 1991).

Anaerobic simply means without oxygen, O_2. In this case, the molecular process can still be properly called an oxidative degradation provided only that there is electron transport away from the organic compounds. In an anaerobic process, the electron sink is just not O_2, as it is in conventional oxidative degradation.

One of the principal oxidative pathways not utilizing O_2 uses the nitrate/nitrite couple, NO_3^-/NO_2^-, for electron transfer (**Figure IX–7**). Each line on the figure shows the reduction potential of an electron transfer couple, pε, as a function of pH. The set of lines in the upper right represents oxidation half–reactions; the set in the lower left represents reduction half–reactions. Overall reactions are obtained by matching a reduction half–reaction (lower left) with an oxidation half–reaction (upper right). The farther apart the two half–reactions, the greater the net potential. Net reactions that are viable under natural conditions fall within the shaded portion of the figure.

Other Technologies

In a 1988 review (EPRI, 1988) of remedial technologies the following were listed as emerging technologies:

- Thermal destruction with radio frequency;
- Thermal destruction with infrared radiation;
- Thermal destruction in molten salt reactors.

These are applications of known heating methods and are used to strip soil of hydrocarbons before secondary treatment.

- Supercritical leaching with CO_2 or water;
- Soil shredding;
- Soil freezing and excavation.

A the time of writing none had emerged as a useful technology. New developments have consisted of combinations of existing technologies such as air sparging and bioventing. Other new technologies include electrical discharge, a modified form of electrophoresis, and magnetic separation. All of these methods remove or treat hydrocarbons in soil; the major obstacle is simple economics; they are expensive. A more subtle but effective obstacle is the necessity of obtaining approval for a new technology from regulatory agencies. For example, several LTTS mobile units are available, but have limited approval. Regulatory approval may keep a promising technology on the sidelines, but it also keeps out the questionable methods.

Well Design

Virtually all remediation projects require the installation of wells. Wells are required for some or all of the following:

- Monitoring and sampling;

- Vapor extraction;

- Groundwater extraction;

- Air injection;

- Steam, detergent, or nutrient injection;

- Water injection or reinjection;

- Product recovery.

The design and placement of wells depend as much on experience as on engineering. Each site is different and wells should be designed for a single purpose. Trying to combine wells for multiple purposes is seldom cost–effective. Monitoring wells are rarely efficient as extraction wells, and extraction wells do not make good sampling wells.

This section details the construction of typical monitoring or injection wells. Of necessity the explanations are somewhat generic. No two regulatory agencies agree entirely on the specifics of well design. To paraphrase an old joke, three regulators will come up with four designs for well installation and development. However, there are standard features that will be discussed here. The focus is on monitoring well design. Injection and extraction wells are similar. Recovery wells are more specialized and are designed to custom specifications. As noted above, all wells should be designed for a specific purpose.

The discussion deals only with vertical wells. However, recent developments in horizontal drilling techniques are promising. Horizontal wells allow sampling and vapor extraction from beneath streets and buildings. Some experience suggests that horizontal wells through the most concentrated volume of soil contamination are more efficient than vertical wells.

The location of monitoring wells involves not only siting (lateral placement), but also depth (vertical placement). Initial location of monitoring wells is based on an estimate of the plume dimensions, which in turn relies on one or more of the following:

- Soil vapor surveys using hand– or hydraulic–driven probes;

- Known or approximate groundwater flow direction;

- Indirect evidence (vapors, complaints, seeps, distressed vegetation);

- Educated guess.

Placement of monitoring wells should follow plume boundaries rather than property boundaries. Although this may be difficult in practice, it is a requirement of federal corrective action regulations (see Appendix B). Permission to place wells in off–site locations is a necessary step. In some cases where a property owner is recalcitrant, it may be necessary to work through the implementing agency.

A thorough site characterization must be performed prior to placement of a substantial number of wells. That is, it is common to develop a small number of wells initially to gather data on the extent of the plume. As this information is processed, additional wells are located to refine and complete characterization of the site. Access to the site may be an influence in locating wells, but generally wells should be located both downgradient and upgradient to establish contaminant levels and a representative background, respectively.

The placement of wells relative to each other depends on the "capture zone," sometimes incorrectly referred to as "sphere of influence." The capture zone is simply the volume through which groundwater or vapors are drawn to the wellhead at sustainable rates, and is typically horseshoe–shaped, but depends heavily on the details of the subsurface strata. Numerous computer programs are available to estimate the capture zone.

The placement of wells relative to each other depends on the "capture zone," sometimes incorrectly referred to as "sphere of influence." The capture zone is simply the volume through which groundwater or vapors are drawn to the wellhead at sustainable rates, and is typically horseshoe–shaped, but depends heavily on the details of the subsurface strata. Numerous computer programs are available to estimate the capture zone.

Once the location is established, drilling should be carried out with a minimum of disturbance of the native subsurface and groundwater. Hollow–stem augers are popular and widely used, but are not always feasible. Drilling fluids and additives should be avoided at all costs to prevent contamination and disturbance of volatile contaminants.

Figure IX–8. (Facing page) Groundwater Sampling Well. Schematic diagram of a typical monitoring well installed to monitor groundwater. Details vary widely depending on state regulations. The design shown includes features most commonly required. Bores are typically 8– to 10–inches in diameter for a 4–inch well. Common well casing material is 4–inch i.d. PVC. For 2–inch well casings, the bore is 4– to 6– inches in diameter. 2–inch diameter wells are suitable for sampling only. The surface apron is concrete that is spread out to form a surface shield. In traffic areas the wellhead cover may be set flush, but surrounding paving should slope up to the access cover to prevent surface water intrusion. Bentonite, grout, or cement plugs are required to exclude surface runoff intrusion and to ensure that samples are of indigenous groundwater only. Barriers are often required to be installed from the bottom up with a tremie pipe to insure proper packing and a good seal. A layer of bentonite pellets may be required between the seal and the sand/gravel filter pack. Clean sand or fine gravel is used to backfill around the casing. Threaded joints are normally required rather than glued joints. For groundwater sampling, the screened interval must be well above the highest permanent groundwater level to allow floating contaminants access to the well. For vapor sampling, the entire screened interval should be well above the highest permanent groundwater level.

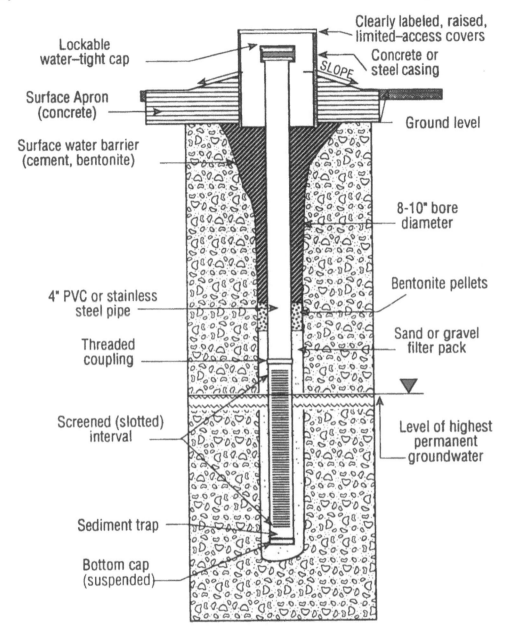

The usual construction material for wells of all types is polyvinyl chloride, PVC, pipe. Teflon™, fiberglass and stainless steel pipes are available and may be required for specialized conditions. For example, stainless steel is frequently required at mixed waste sites where contaminated areas contain solvents that can dissolve PVC. These solvents include methyl ethyl ketone and acetone. Prolonged exposure to high levels of benzene may cause some degradation in PVC. High levels of MTBE may also attack PVC over a period of time. Stainless steel is also required if it is necessary to sample for vinyl chloride since PVC can degas enough monomer to contaminate samples. Teflon and fiberglass are used for more specialized applications . In the diagrams that follow, PVC casing is assumed.

A uniform point of disagreement is the size of well casings. Some regulations permit 2–inch wells, some specify 4–inch or larger wells only. There is a definite cost advantage in installing 2–inch wells for monitoring and sampling. According to one set of regulations (CDPHE, 1991) problems with 2–inch wells that make the cost advantage "questionable" are the following:

- Proper well development is slow and costly;
- Poor down–hole performance (high turbidity);
- Down–hole restrictions (doglegs) common with deep wells;
- Size limitations for down–hole equipment;
- High cost per sample difficult to get a true formation sample;
- Aquifer drawdown tests are difficult;
- Cannot be converted into a recovery well.

Some environmental companies have a policy of only using 4–inch wells for sampling and injection and extraction. Some use 8–inch wells for recovery to accommodate pumps and equipment, others use 6–inch wells. Combination wells are used in specialized situations (see **Figure IX–9**). In the discussion below a 4–inch well is taken to be standard.

In the diagrams that follow, a 4–inch design is the default, despite the fact that 2–inch monitoring wells are less expensive to install. Since wells are — or should be — designed individually for specific purposes, conversion of wells from one purpose to another is not considered. Check with the appropriate regulatory agency for approved designs.

Materials used to construct the filter pack should be chemically inert (see **Figure IX–7**). This is particularly relevant in groundwater sampling wells. Water withdrawn from the well should sample the native soil, not the backfill. Small pea gravel or sand can be used. Cement or bentonite is used above the screened interval as a sanitary seal, usually to a depth of about 10 feet. A layer of bentonite beads is sometimes placed in between the sanitary seal and the backfill. The concrete cap should be drawn out into an apron to exclude surface runoff. Accepted practice places the well head 1–2 feet above ground level, although flush well covers are acceptable in areas subject to traffic. Wellheads should be very clearly marked and locked.

Figure IX–9. (Facing page) Product Recovery and Groundwater Extraction Well. This schematic illustrates a 12-inch diameter well designed to remove free product and groundwater simultaneously. As groundwater is removed, a cone of depression is created. The depression in the water table allows floating product to accumulate and be removed more efficiently. Two–inch PVC wells are set inside 12–inch diameter casing. The bentonite seal is brought to the top of groundwater or 10 feet whichever is higher. A flexible hose, peristaltic skimmer floats on the water surface to remove product. The borehole is backfilled with clean sand. Drawing is not to scale and some elevations are approximate.

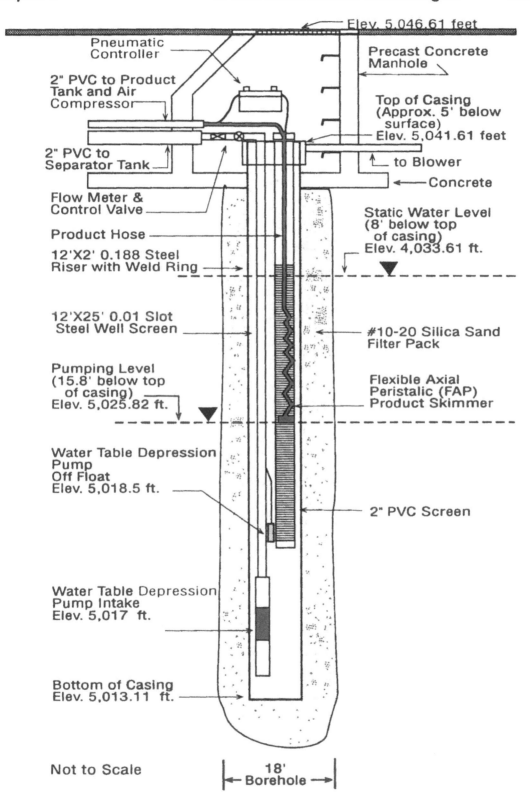

Elev. 5,046.61 feet

Pneumatic Controller

2" PVC to Product Tank and Air Compressor

Precast Concrete Manhole

Top of Casing (Approx. 5' below surface)
Elev. 5,041.61 feet

2" PVC to Separator Tank

to Blower

Flow Meter & Control Valve

Concrete

Product Hose

Static Water Level (8' below top of casing)
Elev. 4,033.61 ft.

12'X2' 0.188 Steel Riser with Weld Ring

12'X25' 0.01 Slot Steel Well Screen

#10-20 Silica Sand Filter Pack

Pumping Level (15.8' below top of casing)
Elev. 5,025.82 ft.

Flexible Axial Peristalic (FAP) Product Skimmer

Water Table Depression Pump Off Float
Elev. 5,018.5 ft.

2" PVC Screen

Water Table Depression Pump Intake
Elev. 5,017 ft.

Bottom of Casing
Elev. 5,013.11 ft.

Not to Scale

18'
Borehole

Figure IX–10. Product Recovery System. Schematic diagram of a remediation and monitoring well system at a service station location. Note that groundwater and vapor monitoring wells are separate. Groundwater and vapor extraction wells are also separate. This type of installation is more expensive, but necessary to ensure the integrity of sampling and efficiency of recovery.

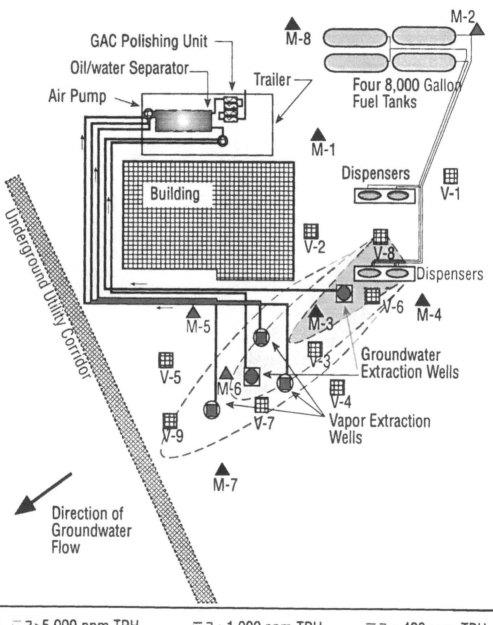

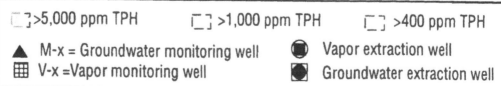

Each well must be carefully documented with respect to siting and lithologic development. Well location should be specified both on a site grid and with respect to local survey standards. As the well is being drilled successive soil characteristics should be documented, particularly if using a hollow–stem auger. PIDs can be used to measure contaminant levels as the core is removed from the auger. Groundwater levels in each well should be measured. At least one well should be designated as the reference standard and measured relative to sea level. Other well levels can be reported relative to the reference well.

Whatever the drilling method, lithologic conditions must be documented every foot. One indication of subsurface structures is to report blow counts as the sampling device is driven into the ground. Caution must be exercised when drilling through soil saturated with gasoline since a serious fire and/or explosive hazard exists. Care must also be exercised when drilling through soils contaminated with hazardous wastes. The drill cuttings must disposed of properly, and there is a potential health risk.

In designing a well the screened interval must be designed to allow sufficient flow of groundwater into the well for sampling and to minimize the passage of formation materials (silts and sediments) to prevent contamination. Materials used for construction must ensure sufficient structural integrity to prevent the collapse of the screened interval.

Figure IX–10 shows a complete vapor and groundwater recovery system at a service station location. The product release was severe and traced to a faulty fitting in the piping network. Approximately 5,000 gallons of unleaded gasoline were released before the leak was detected. Preliminary soil gas surveys placed the plume in the location shown by the contours. Maximum contaminant levels are for soil contamination. Groundwater was contaminated to a moderate extent. Vapor recovery wells were installed and operated. A groundwater extraction system was installed to interdict the groundwater flow (approximately 1 foot per day) from the utility corridor.

The following table (Table 9–10) gives a basic checklist for well documentation. It is not necessarily complete for all regulatory agencies, but it does provides the essential information that must be included.

Table 9–10. Well Documentation Checklist. A well documentation checklist should be included as part of the Standard Operating Procedure, SOP, for a site. The checklist includes the items most commonly required in well installation and development. (Modified from CDPHE, 1991).

- ☐ Date and time of construction
- ☐ Drilling method and drilling fluid
- ☐ Well location (±0.5 ft.)
- ☐ Elevation of surface level (±0.1 ft.)
- ☐ Bore hole and well casing diameters
- ☐ Well depth (±0.1 ft.)

- ❏ Drilling and lithologic logs
- ❏ Casing materials
- ❏ Screen materials and design
- ❏ Casing and screen joint type
- ❏ Screen slot size and length
- ❏ Filter pack material, size, and grain analysis
- ❏ Filter pack volume calculations
- ❏ Filter pack placement method
- ❏ Sealant materials (percent bentonite)
- ❏ Sealant volume (lbs/gal cement)
- ❏ Sealant placement method
- ❏ Surface seal design and construction
- ❏ Well development procedure
- ❏ Elevation of groundwater (±0.1 ft.)
- ❏ Type of protective well cap
- ❏ Ground surface elevation (±0.1 ft.)
- ❏ Surveyor's pin elevation to concrete apron (±0.1 ft.)
- ❏ Elevation to top of monitoring well casing (±0.1 ft.)
- ❏ Elevation to top of protective steel casing (±0.1 ft.)
- ❏ Detailed drawing of well with dimensions

10 Non-In Situ Soil Treatment Technologies

In Situ vs. Non–in Situ

All of the techniques discussed in this section require the contaminated soil to be excavated. After excavation contaminated soils are then treated onsite or offsite, or hauled elsewhere for storage. Because petroleum contaminated soils are being brought to the surface, handled, and transported, all of these methods entail worker exposure to petroleum hydrocarbon vapors and possibly contaminated dusts. Under such conditions workers must be carefully monitored, particularly when working with gasoline contaminated soils.

Although excavation of contaminated soils is certainly used on a widespread basis, excavation is neither a remediation nor treatment method. Excavation may leave a clean site, but not necessarily clean soil. Landfilling the soil simply transfers the contamination from one site to another, although it is someone else's site. The non–in situ technologies discussed in this chapter technologies are potentially efficient and cost effective since both chemical and engineering parameters can be monitored and controlled. However, the overall cost must include costs of excavation and transportation, which can be quite high. Also the cost of landfilling must be considered. For non–in situ treatment, permitting costs must be taken into account.

Decisions to adopt one of these methods are frequently based on factors other than technical or engineering considerations. For example, excavation, by definition, requires the facility to be taken out of operation — often for an extended period of time with a concomitant loss of revenue. Or since excavation is an aggressive approach to a release, it may be good public relations to rapidly excavate a site, haul the soil away, and rebuild a pristine new operation. On the other hand excavation–based remediation frequently extends onto adjacent properties which may dramatically increase costs and legal complications. In general, excavation is more suitable for a confined, smaller site with a concentrated plume than for a large site with a more diffuse plume.

Excavation and Landfilling

Excavation and landfilling, also known as "muck and truck" or "scoop and run," is the traditional method to remediate all kinds of contamination from barnyard manure to radioactive sites. It has the advantage of being technically simple, proven technology and it is relatively quick compared with in situ methods. Non–in situ treatments, preceded by excavation, are among the best options for complete remediation of contaminated soils. That is, after treatment using some of the options discussed below, soils can reach nearly nondetect in contaminant levels.

Landfill storage is the least desirable option. The main advantages are it is an inexpensive option in some areas and it requires no special technical knowledge. However, it is merely moving the contamination from one site to another and should be the method of last resort, particularly for large to very large sites. Once soils have been excavated and stored, there is undoubtedly an increase in passive remediation, so in that sense landfilling can be a treatment, albeit a long one.

Technical Basis: Excavation is the physical removal of all or almost all contaminated soil from a site. The contaminated soil is then stored at a permitted storage facility. Permitting requirements at the storage facility are intermediate between a sanitary landfill (a local landfill) and a treatment, storage, and disposal, (TSD) facility that handles hazardous wastes. Excavation itself requires no special technology beyond normal earth moving techniques. Excavation is usually limited to the depth to which a trackhoe arm can reach, which is typically about 22 feet. Beyond this depth more specialized earth removal techniques must be used.

Comments: Excavation is the basic step required before any of the methods discussed in this chapter can be used, and, with or without landfill storage, is an option under the following conditions:

- A prompt response to a release or spill is required;

  Excavation provides the most rapid means of responding to a spill or release that has contaminated soil. (It is not an option for contaminated ground-water). However, caution is required when working around soil that is heavily contaminated with gasoline, particularly a fresh release.

- If the contaminated area is entirely on the original site that is owned by or controlled by the owner/operator;

  Excavation is much more feasible if the contamination is confined onsite (contiguous property) and has not migrated offsite. As soon as the plume contaminates third–party property, technical and legal complications increase rapidly. If the contamination has moved offsite, excavation is usually not an option.

- A small to moderate sized facility, such as a service station being taken out of operation or completely renovated;

  One difficulty with excavation is that excavation itself must be filled, and backfill increases costs. A large site is less suitable both because of the cost of the excavation and the cost of backfill. For the same reasons excavation is an option for small sites in which the contamination is localized. Under these conditions excavation of the entire contaminated site can be cost–effective.

- The site is open and free of surface obstacles, paving, buildings, and other structures;

  Excavation costs increase rapidly when surface obstacles are encountered or shoring is required. In some confined locations it can be physically difficult to get equipment in place or to get trucks in and out.

- Petroleum contaminated clays which cannot be treated in any other manner can be excavated.

  There are few options for treatment of contaminated clays. One of those is excavation and landfilling.

- The contaminant(s) is a high molecular hydrocarbon. There are few options for treating fuel and heating oils, lubricants, waste oil, and other heavy petroleum products. These products have sufficiently low mobility that landfilling is a viable option.

Health and Safety: Excavation is usually carried out with heavy equipment such as trackhoes, front loaders, etc., so normal safety considerations for working around heavy equipment must be observed. Heavy equipment operators are at particular risk from exposure to hydrocarbon vapors while excavating heavily contaminated soils. A benchmark value of 5,000 ppm TPH (as measured by a PID or other suitable field instrument) can be considered a warning level for heavily contaminated soils.

Since contaminated soils are being brought to the surface, workers must be protected from exposure to hydrocarbon vapors, and, in the case of gasoline contaminated soils, from fire or explosive hazards. Protection is required for exposure to excavated soils. When handling disturbed gasoline contaminated soils the potential exposure routes include inhalation, ingestion, and skin absorption.

There is potential exposure to vapors that have accumulated in the excavation itself. All petroleum vapors from propane up are heavier than air and tend to accumulate in the bottom of an excavation. Since hydrocarbon vapors displace oxygen, O_2 deficiency can result. Therefore, prudent operations require that an O_2 meter be used in the vicinity of heavily contaminated soils.

Requirements: Assuming adequate equipment and site conditions, excavation itself should be straightforward. Equipment requirements include the following:

- Heavy equipment such as backhoes, trackhoes, front end loaders, and end dumps;
- Shoring;
- Pumps to remove groundwater during excavation;
- Protective equipment including barriers.

Landfill storage normally requires permitting. Soil analysis, possibly including TCLP, may be required to assure the landfill operator that the soil being delivered is contaminated only with regulated substances, not hazardous substances. Transportation of heavily contaminated soils should be carried out safely with regard for potential fire or explosive hazards. State regulations may require a permit to transport contaminated soils.

Table 10–1. Considerations and Recommendations: Excavation and Landfilling. The most common factors to be considered in deciding on excavation are listed. Excavation requires the facility to be shut down for a period of time, resulting in economic loss for the owner/operator. Landfilling entails extended exposure to liability in the event the landfill site proves to be contaminated, operated improperly, or the regulatory requirements change.

Excavation and Landfilling [e]

Soil type:	Clay	Silty Sand	Gravel
	Yes	Yes	Maybe [a]
Contaminant type:	Gasoline	Diesel	Fuel Oil
	Yes[b]	Yes	Yes
Contaminant extent:	Confined	Moderate	Dispersed
	Yes	Maybe [c]	No
Contaminant phase:	Free Gasoline	Adsorbed	Dissolved
	Yes[b]	Yes	No
Subsurface features:	Homogenous	Mixed	Rocky [d]
	Yes	Maybe	No
Plume extent:	Small	Large	Very Large
	Yes	Maybe	No
Depth to groundwater:	50 ft [c]	30 ft	10 ft
	Yes	Maybe	No

[a] In situ methods are probably more suitable for porous soils.

[b] Caution is required when working with soils heavily contaminated with gasoline. There is a fire/explosion hazard as well as a health risk

[c] Specific site conditions, transportation, labor, and equipment costs strongly influence this decision.

[d] Bedrock close to the surface, boulders, and outcroppings discourage excavation.

[e] Excavation alone can be followed by a non–in situ treatment method. The principal argument against storing contaminated soils in a permitted landfill is the potential for long–term liability.

Landfarming

Non–in situ bioremediation is also known as landfarming, land treatment, ex situ bioreclamation, surface biodegradation, and biopile. This method is technically very similar to in situ bioremediation except that contaminated soils are excavated and treated above ground. It allows potentially very precise monitoring of parameters that control the conversion rate of hydrocarbons to carbon dioxide and water.

Technical Basis: Contaminated soils are excavated and spread over a large area to allow natural or enhanced aerobic (and possibly anaerobic) microbial oxidation of hydrocarbons to CO_2 and H_2O. Landfarming is a combination of biodegradation and soil venting. Microbial oxidation is typically accompanied by moderate to substantial volatilization depending on the chemical nature of the contaminants. For example, a substantial portion of gasoline contamination volatilizes when spread over a large area.

The biochemical basis is the same as that outlined in the preceding chapter under in situ bioremediation; namely, a stepwise, microbial oxidation of hydrocarbons ultimately to CO_2, cell components, energy, and H_2O. The aerobic oxidation of hydrocarbons, plus the inevitable nitrogen and sulfur compounds that are always present can be represented as follows:

$$\underset{\text{(Hydrocarbons)}}{(CH)_x} + \underset{\text{(Oxygen)}}{O_2} + \text{Nutrients} \longrightarrow \text{Acid intermediates}$$

$$\underset{\substack{\text{(Carbon} \\ \text{dioxide)}}}{CO_2} + \underset{\text{(Water)}}{H_2O} + \underset{\substack{\text{(60\% of all} \\ \text{organic carbon)}}}{\text{Cell components}}$$

$$+ \text{Energy} + \underset{\text{(Ammonium)}}{NH_4^+} + \underset{\text{(Sulfite)}}{SO_3^{-2}}$$

$$\underset{\text{(Nitrates)}}{NO_3^-} + \underset{\text{(Sulfates)}}{SO_4^{-2}}$$

Anaerobic microbial oxidation of hydrocarbons has received considerable attention recently. The increased attention has resulted from the realization that anaerobic processes contribute significantly to natural, in situ degradation of hydrocarbons. Anaerobic simply means without oxygen, it does not necessarily mean reduction. The oxidizing agent (the electron sink) is something other than molecular oxygen. The overall process can be represented schematically as follows:

$$(CH)_x \; + \; \text{Nutrients} \; + \; NO_3^- \; \longrightarrow \; \text{Intermediates}$$

(Hydrocarbons)   (An electron
                                transfer agent)

$$H_2O \; + \; \text{Cell components} \; + \; \text{Energy}$$

(Water)

$$+ \quad CO_2 \quad + \quad NO_3^- \quad + \quad SO_4^{2-}$$

(Carbon   (Nitrate)   (Sulfate)
dioxide)

The site is prepared by grading to control runoff. Ditches and/or berms may be necessary, along with surface liners. Specific site requirements vary with the requirements of local regulatory authorities. At a minimum the site must be protected from spreading contamination to uncontaminated areas. Groundwater at the site must be protected from contaminated runoff. Contaminated soil must be turned periodically to provide a uniform supply of oxygen. Soils or water contaminated with mixed wastes are unsuitable for landfarming without a special permit.

Comments: The advantage of this technique is that, given available land, the method is relatively inexpensive, low tech, non–labor intensive, and results in clean soils after relatively short periods of time. The disadvantage is that it requires large amounts of land, potentially spreads petroleum contamination to otherwise clean land, and can be accompanied by release of hydrocarbons into the air. The latter is the principal objection raised by many regulatory agencies; in some states landfarming must be accompanied by air emissions permits, and in California it is regulated under the state's implementation plan (SIP) under the Clean Air Act of 1990.

Health and Safety: Because landfarming is accompanied by some volatilization, workers should be protected from exposure to hydrocarbon vapors. Since contaminated water has been brought to the surface and must be handled, care must be taken to protect workers from the effects of contact with contaminated water spread over a large area. Effluent water from the site is contaminated and must be contained to prevent the spread of hydrocarbons offsite.

Requirements: Equipment requirements include earth moving and soil–handling equipment, rototillers, tractors, disc harrows, sprinklers, mixing tanks, and monitoring facilities. The site must be graded to control surface runoff; berms or ditches may be required if the runoff is heavy. Effluent from the contaminated soils or groundwater must be collected to prevent spread of contamination. Care must be taken to be sure that soils are contaminated only with petroleum and not hazardous wastes.

Contaminated soils are spread in uniform layers, usually about 12 to 18 inches deep. The area is rototilled or harrowed to mix the soils and provide aeration. Watering may necessary to maintain 40–60% of saturation. Contaminated soils that are to be spread should contain sufficient soil organisms for desired conversion rates or commercial bacterial preparations should be added.

Highly acid or alkaline soil will inhibit bacterial activity. The pH of the soil should be adjusted with lime to about 7.0 and the pH should be monitored to maintain a value close to 7. Agricultural fertilizer is added as needed to provide nutrients and maintain adequate bacterial populations. Although microbial oxidation produces considerable amounts of heat, landfarming may not be feasible in cold weather unless supplemental heat is available.

Table 10–2. Considerations and Recommendations: Landfarming. The more important and common considerations for landfarming or non–in situ bioremediation are summarized. Landfarming allows greater control over nutrients and bacterial populations than in situ bioremediation. A significant fraction of the remediation is due to volatilization. The method is suitable for gasoline and, with commercial bacterial preparations, heavier hydrocarbons. It is not suitable for hazardous wastes, especially heavy metals and other inorganics.

Landfarming (Non–In Situ Bioremediation)

Contaminant type:	Gasoline	Diesel	Fuel Oil
	Yes[a]	Yes [b]	Yes [b]
Contaminant extent:	Confined	Moderate	Dispersed
	Yes	Maybe	Maybe [c]
Contaminant phase:	Free Gasoline	Adsorbed	Dissolved
	Yes [a]	Yes	Maybe [d]
Precipitation:	Heavy	Moderate	Low
	Maybe [e]	Yes	Yes
Soil pH:	Acidic	Neutral	Basic
	No [b]	Yes	No
Climate:	Warm	Mild	Cold
	Yes	Maybe	No [g]
Plume extent:	Small	Large	Very Large
	Yes	Maybe	No

a Caution is required when working with soils heavily contaminated with gasoline. There is a fire/explosion hazard as well as the health risk

b Higher molecular weight hydrocarbons require commerical bacterial preparations.

c The decision is based on excavation and transportation costs, not on feasibility of biodegradation. For large sites onsite landfarming is definitely an option.

d Contaminated groundwater can be treated by landfarming methods. (See the next section).

g Heavy precipitation or surface runoff must be controlled to prevent spread of contamination.

f If the soil pH is outside of the range of 6-8, it must be adjusted.

g The oxidation process generates heat. With proper site preparation, landfarming can be carried out even in the winter in many areas.

Pesticides, inorganic salts, heavy metals, and halogenated hydrocarbons will inhibit microbe activity, especially at low pH. Because of the metal content, used motor oil may not be suitable for this technique. Treatment effectiveness can be monitored by periodic soil sampling. Treatment times range from 45 to 90 days.

Surface bioremediation has also been carried out in bioreactors (ESI, 1990, C&EN, 1992). These are large scale, fixed reactor vessels that process contaminated soils in batches. The method has been used more for hazardous wastes than petroleum contaminated soils. Contaminated groundwater or effluent from chemical extraction (see Chapter 9) can be treated by land treatment methods (see the next Section).

Composting is a proven method of disposing of organic material. In this method contaminated or waste material is shredded, distributed over the surface and allowed to degrade. It relies on microbial oxidation virtually identical to degradation of petroleum. Composting may be carried out similarly to landfarming or in specialized reactor vessels. The advantage of a composting facility is that petroleum contaminated soil can be combined with other organic wastes, although petroleum hydrocarbons may be more resistant to degradation than other organic wastes.

Land Treatment for Contaminated Groundwater

This method is surface bioremediation except that contaminated groundwater and leachate are treated. If the required land is available, it is a cost–effective alternative to conventional water disposal methods. Like landfarming, it allows very precise monitoring of parameters that control the conversion rate of hydrocarbons to carbon dioxide and water.

Technical Basis: Contaminated water is extracted, collected, and spread over a large area to allow natural or enhanced aerobic (and possibly anaerobic) microbial oxidation of hydrocarbons to CO_2 and H_2O. Landfarming is a combination of biodegradation and soil venting. Microbial oxidation is typically accompanied by moderate to substantial volatilization depending on the chemical nature of the contaminants. For example, substantial portion of gasoline contamination volatilizes when spread over a large area.

The biochemical basis is the same as that outlined in the preceding section under landfarming; namely, a stepwise, microbial oxidation of hydrocarbons ultimately to CO_2, cell components, energy, and H_2O. The aerobic oxidation of hydrocarbons requires oxygen, nutrient and water to proceed efficiently. In the treatment of groundwater, nutrients must be supplied from an external source. Oxygen is derived from air.

The site is prepared by grading to control runoff. Ditches and/or berms may be necessary along with surface liners. Specific site requirements vary with the requirements of local regulatory authorities. Groundwater at the site must be protected from runoff. Treated soil must be turned periodically to provide a uniform supply of oxygen. Water contaminated with mixed wastes is unsuitable for landfarming without a special permit.

Comments: The advantage of this technique is that, given available land, the method is relatively inexpensive, low tech, non–labor intensive, and results in clean water after relatively short periods of time. The disadvantage is that it requires large amounts of land, potentially spreads petroleum contamination to otherwise clean land, and can be accompanied by release of hydrocarbons into the air. The latter is the principal objection raised by many regulatory agencies; in some states land treatment must be accompanied by air emissions permits.

Health and Safety: Because land treatment is accompanied by some volatilization, workers should be protected from exposure to hydrocarbon vapors. Since contaminated soil has been excavated and must be handled, care must be taken to protect workers from the effects of contact with contaminated soil spread over a large area. Effluent water from the site is contaminated and must be contained to prevent the spread of hydrocarbons offsite.

Table 10–3. Considerations and Recommendations: Land Treatment of Contaminated Groundwater. The more important considerations for land treatment of contaminated groundwater are summarized. Treatment or disposal of contaminated water by conventional means is expensive. Land treatment is simply bioremediation using clean soil as a substrate. The method is suitable for leachate from chemical extraction, particularly if biodegradable detergents are used. It is not suitable for hazardous wastes, especially heavy metals and other inorganics. Lightly contaminated water can be used for irrigation purposes.

Land Treatment for Contaminated Groundwater

Depth to groundwater:	Shallow	Moderate	Deep
	Yes	Maybe	No [a]
Water pH:	Acidic	Neutral	Basic
	No [b]	Maybe	No
Contaminant type:	Gasoline	Diesel	Fuel Oil
	Yes[c]	Yes	Yes[d]
Contaminant chemistry:	Groundwater	+ Detergent	+ Haz wastes
	Yes	Maybe [e]	No
Climate:	Warm	Mild	Cold
	Yes	Maybe	No [f]

a Depth is a problem because it is more expensive to pump deep groundwater. The increased cost must be balanced against the efficiency of land treatment.
b Lime can be added to control acidic pH.
c Caution is required when working with gasoline contaminated soil. There is a fire/explosion hazard as well as the health risk
d High molecular weight hydrocarbons require commerical bacterial preparations.
e The detergent must be easily degraded by bacterial action.
f The oxidation process generates heat. With proper site preparation, land treatment can be carried out even in the winter in many areas.

Requirements: Equipment requirements for land treatment include earth moving and soil handling equipment, rototillers, tractors, disc harrows, sprinklers, mixing tanks, and monitoring facilities. The site must be graded to control surface runoff; berms or ditches may be required if the runoff is heavy. Effluent from the contaminated soil or groundwater must be collected to prevent spread of contamination. Care must be taken to be sure that extracted water is contaminated only with petroleum and not hazardous wastes.

The base soil is spread in uniform layers, usually about 12 to 18 inches deep. Contaminated water is spread over the area. The area is rototilled or harrowed to mix the soil and provide aeration. Additional water may necessary to maintain 40–60% of saturation. Treated soil that is being used should contain sufficient soil organisms for desired conversion rates or commercial bacterial preparations should be added.

Pesticides, inorganic salts, heavy metals, and halogenated hydrocarbons will inhibit microbe activity, especially at low pH. Because of the metal content used motor oil may not be suitable for this technique. Treatment effectiveness can be monitored by periodic soil sampling. Treatment times range from 45 to 90 days.

The method could be carried out in bioreactors (ESI, 1990, C&EN, 1992) provided the site was large enough to justify the cost. Alternatively small transportable bioreactors are feasible. Provided a permit is available and the regulatory agency is agreeable, the method is advantageous in combination with chemical extraction since the leachate is expensive to handle and treat by conventional disposal methods.

Highly acid or alkaline soil will inhibit bacterial activity. The pH of the base soil should be adjusted with lime to about 7.0 and the pH should be monitored to maintain a value close to 7. Agricultural fertilizer is added as needed to provide nutrients and maintain adequate bacterial populations. Although microbial oxidation produces considerable amounts of heat, land treatment may not be feasible in cold weather unless supplemental heat is available.

Thermal Treatment Techniques

Technical Basis: In the last chapter a form of in situ thermal treatment was examined, specifically in situ vitrification. In that method the native soil was melted to ensure that all organic contaminants were vaporized and pyrolyzed. If the soil is excavated prior to being treated, treatment parameters can be controlled more precisely. Temperatures can be controlled more precisely and the energy cost is lower. Non–in situ thermal treatment is divided into two broad categories: low temperature thermal stripping, LTTS, and high temperature thermal treatment, HTTT.

In high temperature thermal treatment (HTTT) methods contaminated soil is exposed to high to moderately high temperatures which results in pyrolysis (incineration) of petroleum hydrocarbons to CO_2 and H_2O. Alternatively, soil may be heated to strip the soil of hydrocarbons which are then incinerated (LTTS). LTTS units preheat the soil to volatilize organic compounds, then pyrolyze the compounds in an afterburner. As a general rule, LTTS is best suited for lighter soils and lighter petroleum fractions. The method requires intimate contact between heated air and soil particles. Clays and silty clays form clumps and are frequently too coarse to treat efficiently. Heavy petroleum fractions, which do not volatilize readily enough, are more suitable for HTTT.

Requirements: LTTS requires more precise control of temperatures, but operates at lower temperatures. Temperatures vary depending on soil residence times in the incinerator, but are typically below 800°C (1500°F). LTTS is characterized by lower fuel consumption rates.

Typical temperatures are as follows:
- Soil stripping temperatures for light hydrocarbons: 50 to 160°C (120 to 320°F);

- Soil stripping temperatures for heavy hydrocarbons: 200 to 300°C (390 to 570°F);

- Pyrolysis temperatures: 800 to 2,000°C (1,500 to 3,500°F);

- Vitrification of soils: 2,000 to 3,000°C (3,500 to 5,400°F).

The two methods are sufficiently distinct that they will be discussed separately. LTTS is discussed first, then HTTT. For all non–in situ thermal treatments there are similar limitations that are as follows:
- Corrosive soils;

  Soil pH values outside the range 5 to 9 are usually not suitable.

- Soils with a high metal content;

  No thermal treatment is suitable for metal contamination except in situ vitrification. The softer metals such as mercury, lead, silver, tin, and cadmium, vaporize and become a toxic inhalation hazard. The more refractory metals such as zinc, nickel, iron, and chromium, and inorganic anions such as sulfate, nitrate, and phosphate clog or corrode equipment.

- Soil with a high moisture content;

  Water has a high heat capacity which reduces the amount of hydrocarbon vaporized per gram of fuel. Operating costs go up. Wet soil tends to form clumps which also reduces efficiency.

- Soils with a high organic content;

  Soils that contain large amounts of organic debris such as leaves, roots, weeds, grass, that reduce operating efficiency.

- Large particle size;

  Good efficiency requires a small particle size and large surface area for maximum volatilization. Large clumps of soil reduce operating efficiency.

- Construction debris and plastic;

Soil excavated from UST sites often contains asphalt or concrete blocks and debris and pieces of construction plastic. Large pieces of debris will break equipment, Plastic melts and fouls high temperature equipment.

Low Temperature Thermal Stripping

Low temperature thermal stripping (LTTS) is a method of heating contaminated soils to a temperature high enough to volatilize the hydrocarbons but not high enough to cause ignition and combustion. Soils are treated in rotating kilns, which are adaptations of asphalt drums and in indirect heat exchangers (ESI, 1990). After volatilization, the vapors can be burned or collected for disposal.

Technical Basis: Contaminated soil is fed into a stripping chamber where high temperature gases cause hydrocarbon contaminants to volatilize. Vapors are entrained and either burned or collected. Typically the stripper chamber is rotated or agitated to provide good contact between soil particles and hot gases. Vapors from the stripper chamber are passed through a flame (an afterburner), a catalytic converter, or a carbon filter. The choice depends on whether the unit is mobile or fixed. Exhaust gases can be passed through a wet scrubber or baghouse to remove fine particulates, then vented to the atmosphere (ESI, 1990).

Comments: The advantage of this method is that it provides a relatively simple means of cleaning heavily contaminated soils. The treatment time is short and treated soils can be used as backfill. The disadvantages are that costs are comparatively high and specialized equipment that is not widely available is required. Not all soil types can be treated, and the processing is somewhat weather dependent.

The method has some popularity in areas of heavy petroleum contamination where it is economically feasible to construct fixed processing units. With mobile units, which are more expensive to operate, the method may used on site at smaller operations, thereby reducing or eliminating transportation costs.

Health and Safety: Workers handling contaminated soil that has been brought to the surface should be protected from inhalation exposure, especially for gasoline contaminated soil. Soils heavily contaminated with gasoline can be a fire or explosive hazard.

Routine safety rules for working around high temperature, moving equipment must be observed. In the process of loading, handling, and processing contaminated soil, considerable dust is generated. Appropriate precautions must be in place to prevent inhalation exposure. Ash produced from thermal incinerators may be toxic and should be handled accordingly. Local clean air regulations normally require monitoring effluent gases.

Table 10–4. Considerations and Recommendations: Low Temperature Thermal Stripping. A summary of the factors that most commonly influence the decision to use low temperature thermal stripping. LTTS is a fast means of cleaning contaminated soils with the right characteristics. It is relatively expensive and not widely available. The cost of transportation is often the deciding factor. For LTTS, mobile units are available in some areas.

Low Temperature Thermal Stripping

Soil type:	Gravel/Sand	Silty Sand	Clay
	Yes	Yes	Maybe a
Contaminant type:	Gasoline	Diesel	Fuel Oil
	Yesb	Yes	No
Contaminant extent:	Confined	Moderate	Dispersed
	Yes	Yes	Maybe c
Contaminant phase:	Free Gasoline	Adsorbed	Dissolved
	No b	Yes	No
Soil moisture content:	High	Moderate	Low
	No	Maybe	Yes
Depth to groundwater:	50 ft d	30 ft	10 ft
	Yes	Maybe	No
Transportation costs:	Low	Moderate	High
	Yes	Maybe	No

a Particle size for clays must be small enough for good volatilization.
b Lower molecular weight, more volatile hydrocarbons are better suited.
   However, there is a fire/explosion hazard with soils heavily contaminated with gasoline.
c Specific site conditions, transportation, and equipment costs strongly
   influence this decision.
d Depths are arbitrary. Site specific conditions will change these numbers.

Requirements: There are several types of LTTS units available, including mobile units. In the rotating kiln type, contaminated soil is fed through a hopper into a stripper chamber that is similar to, or modified from, an asphalt mixing drum. Vapors are passed through an afterburner, then a catalytic converter, then vented to the atmosphere. Other designs use heated screws or infrared heaters to volatilize contaminants (ESI, 1990).

Other problems that will preclude or limit the use of LTTS include the following:
* A high metal content, which can poison catalytic converters;
* High organic content, which reduces efficiency;
* Large particle size, which limits the surface area and volatilization;

- Large debris such as boulders, concrete or asphalt blocks, roots, branches, etc. which will damage equipment; and

- Plastic, which will foul the stripping chamber.

LTTS is best suited for lighter soils since it requires a large surface area and small particle size. It is also better suited for lighter petroleum fractions. Hydrocarbons through the diesel range, including avgas and jet fuel. LTTS will also handle volatile organics other than petroleum hydrocarbons provided the requisite RCRA permit is available. However, LTTS has a broader molecular weight application range than in situ volatilization and treatment times are short.

Corrosive soils with a pH outside the range of 5 to 11 are too corrosive to be put through an LTTS system. Clays and silts are generally too coarse to treat efficiently. Heavy petroleum fractions, which do not volatilize readily enough, are more suitable for HTTT. Soils with a high moisture content have high heat capacities and are not cost–effective in LTTS systems.

High Temperature Thermal Treatment

Technical basis: HTTT methods have the common feature that they incinerate contaminated soil. Under some conditions these methods could be considered non–in situ vitrification. In the rotating kiln incinerator method, contaminated soil is fed into a closed chamber and incinerated with the exhaust from a propane flame. The process resembles asphalt production. All petroleum hydrocarbons are pyrolyzed to CO_2 and H_2O. Other organic compounds are oxidized to the corresponding oxides — NO_x, SO_x, P_xO_y. Typical combustion temperatures are in the $750 - 1,100°C$ $(1,400 - 2,000°F)$ range. Fluidized bed incinerators are somewhat similar in operation, but incorporate a bed of sand, or other refractory material, for agitation and heat transfer. Typical combustion temperatures are in the $750 - 870°C$ $(1,400 - 1,600°F)$ range.

Comments: These technologies are classed as developmental, particularly as applied to petroleum contaminated soils. However, rotary kiln incineration is used almost routinely where the facilities are available. HTTT is suitable for soil that is otherwise difficult to treat. For example, petroleum sludges and waste oil that do not contain hazardous wastes can be treated.

Health and Safety: Workers handling contaminated soil that has been brought to the surface should be protected from inhalation exposure, especially for gasoline contaminated soils. Soils heavily contaminated with gasoline can be a fire or explosive hazard.

Routine safety rules for working around high temperature, moving equipment must be observed. In the process of loading, handling, and processing contaminated soil, considerable dust is generated. Appropriate precautions must be in place to prevent inhalation exposure. Ash produced from thermal incinerators may be toxic and should be handled accordingly. Local clean air regulations normally require monitoring effluent gases.

Table 10–5. Considerations and Recommendations: High Temperature Thermal Treatment. A summary of the factors that most commonly influence the decision to use high temperature thermal treatment. HTTT is a fast means of cleaning contaminated soil with the right characteristics. It provides nearly complete destruction of organic wastes. The ash generated is not suitable for backfill and may not be allowed in sanitary landfills.

High Temperature Thermal Treatment

Soil type:	Gravel/Sand	Silty Sand	Clay
	Yes	Yes	Maybe [a]
Contaminant type:	Gasoline	Diesel	Fuel Oil
	No[b]	Yes	No
Contaminant extent:	Confined	Moderate	Dispersed
	Yes	Yes	Maybe [c]
Contaminant phase:	Liquid	Adsorbed	Dissolved
	Maybe [b]	Yes	No
Soil moisture content:	High	Moderate	Low
	No	Maybe	Yes
Chemical content:	Petroleum	Used Oil	Metals [d]
	Yes	Maybe	No
Transportation costs:	Low	Moderate	High
	No	Maybe	Yes

a   Particle size for clays must be small enough for good volatilization.
b   Higher molecular weight, less volatile hydrocarbons are better suited. There is a fire/explosion hazard for gasoline contaminated soils.
c   Specific site conditions, transportation, and equipment costs strongly influence this decision.
d   Volatile metals cause problems with emission controls and may not be legal to treat.

Requirements: There are many types of incinerators available, including rotating kilns, fixed kilns, rotating cement kilns, fluidized bed incinerators, and low temperature strippers. Rotating kiln and fluidized bed incinerators are available as transportable units. The applicability of these units is heavily dependent on local regulations.

In a rotating kiln contaminated soil is fed from a hopper into an oxidation chamber where a high–temperature flame pyrolyzes all organic components. Typically the chamber is sloped to a bin where ash is drawn off. Units are normally equipped with some combination of scrubbers, cooling towers, and afterburners. Stack gas monitors are usually mandatory.

HTTT is best suited for light to moderately heavy soils. Lighter petroleum fractions, such as gasoline and jet fuel are unsuitable due to the fire or explosion hazard. Clays and silts, which are generally unsuitable to treat efficiently by any other method can be treated in rotating kilns. Heavy petroleum fractions, which do not volatilize readily enough for LTTS, are treatable by HTTT.

Some problems that will preclude or limit the use of HTTT include the following:

- A high metal content, which can poison catalytic converters;

  Volatile metals, including mercury, tin, arsenic, lead, and silver, contaminate emissions.

- High organic content, which reduces efficiency;

  Natural organic material, such as plant debris can be present; it simply reduces efficiency and increases costs.

- Large particle size, which limits the surface area and reduces volatilization necessary for combustion;

- Large debris such as boulders, concrete or asphalt blocks, roots, branches, etc. which will damage equipment;

  and

- Plastic, which will foul the combustion chamber.

Asphalt Incorporation

Asphalt incorporation is a correct descriptive title, but may be misleading as a remedial technology. Very few contaminated soils are actually suited to the manufacture of asphalt for roadway use. If the soil particle size is compatible with the asphalt specifications and if the contaminants are in the right molecular weight range, then the method is suitable as a treatment.

The method is used to incorporate contaminated soil into a form in which any remaining contamination is immobilized. Soil is incorporated as a few percent (normally not greater than about 5%) of the asphalt. Processing is accompanied by volatilization and incineration of some lighter contaminants, immobilization of heavier ones. Asphalt that is unsuited for road use can be landfilled without a permit and, presumably, without endangering the environment.

Technical basis: This method treats contaminated soil by a poorly quantified mixture of volatilization, immobilization, and partial low temperature thermal (500 to 800°F) volatilization and/or pyrolysis. Aggregate is passed through a dryer where lighter hydrocarbons are volatilized. Higher molecular weight compounds, in the diesel and up range, and soil particles are incorporated directly into an asphalt mixture as a substitute for the normal aggregate. In the mixing chamber, aggregate, contaminated soil, and hot asphalt oil are mixed together. The mixture is transferred to a truck and delivered to the site of use.

Table 10–6. Considerations and Recommendations: Asphalt Incorporation. The most common factors to be considered in deciding on asphalt incorporation are listed. Asphalt incorporation is not widely used because contaminated soils are rarely suited for use in asphalt. As a thermal treatment it is presumed to be effective, particularly for soils contaminated with high molecular weight hydrocarbons. Wet soils or soils with a high moisture content are unsuitable.

Asphalt Incorporation

Soil type:	Clay	Silty Sand	Gravel
	Maybe	Yes	Yes [a]
Contaminant type:	Fuel Oil	Diesel	Gasoline
	Yes[b]	Yes	No
Contaminant extent:	Confined	Moderate	Dispersed
	Yes	Maybe	Maybe [c]
Contaminant phase:	Free Gasoline	Aged Gasoline	Groundwater
	No [b]	Yes	No
Soil moisture content:	High	Moderate	Low
	No	Maybe	Yes
Depth to groundwater:	50 ft [d]	30 ft	10 ft
	Yes	Maybe	No
Transportation costs:	Low	Moderate	High
	Yes	Maybe	No

a Particle size for gravel and soils must be appropriate for asphalt specifications.

b There is a fire/explosion hazard for gasoline contaminated soils.
   High molecular weight hydrocarbons work well. Asphalt oil is simply a
   very high molecular weight hydrocarbon.

c Specific site conditions, transportation, and equipment costs strongly
   influence this decision.

d Depths are arbitrary. Site specific conditions will change these numbers.

Lower molecular weight hydrocarbons are incompatible with asphalt and appear to act as a solvent. Soils heavily contaminated with volatile hydrocarbons, such as gasoline, are not suitable for this method. Other soil parameters which complicate asphalt incorporation include the following:

- Organic content;

- Particle size;

- Moisture content;

- Presence of other contaminants.

Comments: Operators of asphalt facilities must be alert that incoming soils are contaminated only with regulated petroleum products and not with hazardous materials. Treating hazardous wastes without a permit carries very stiff fines.

This method is classed as developmental (ESI, 1990). The basic engineering research to fix the parameters for asphalt incorporation has not been done, at least to the extent of carefully defining the type and mix of contaminants and soils that can successfully be incorporated. Therefore, the method remains largely empirical and dependent on the experience of the facility operator. Even when the experience is there, the method may not meet approval of regulatory agencies.

Health and Safety: Workers handling contaminated soil that has been brought to the surface should be protected from inhalation exposure, especially for gasoline contaminated soils. Soils heavily contaminated with gasoline can be a fire or explosive hazard, especially around high temperature, flame–driven equipment.

Routine safety rules for working around high temperature, moving equipment must be observed. In the process of loading, handling, and processing contaminated soil, considerable dust is generated. Appropriate precautions must be in place to prevent inhalation exposure. Ash produced from thermal incinerators may be toxic and should be handled accordingly. Local clean air regulations normally require monitoring effluent gases.

Requirements: Large quantities of lighter petroleum hydrocarbons, such as gasoline kerosenes, and diesel fuels, are incompatible with asphalt. These components are volatilized in the premix dryer. To meet air quality standards, vapors must be incinerated in an afterburner, catalytic converter, or filtered. In addition volatile hydrocarbons are a fire or explosion hazard in the asphalt mixing chamber.

Soil should be incorporated at a rate of less than 5% of the final asphalt product. The soil type is a consideration: it should have neither a large organic fraction nor a large moisture content. Clays and sands do not incorporate well. Gravels and conglomerate are suitable.

Soil must be carefully screened to exclude organic material, large debris, and plastic. Some problems that will preclude or limit the use of asphalt incorporation include the following:

- A high metal content;

  Metals can poison catalytic converters and emission controls, and treatment of listed hazardous wastes is probably not legal.

  Volatile metals, including mercury, tin, arsenic, lead, and silver;

These metals contaminate emissions and emission controls.

- High organic content, which reduces efficiency;

  Natural organic material, such as plant debris can be present, it simply reduces efficiency and increases costs.

- Soil contaminated with waste or used oil;

  Waste oil usually contains hazardous wastes such as ethylene glycol, PCBs, metals, and pesticides that cannot be treated legally by asphalt incorporation.

- Large particle size, which limits the surface area and reduces volatilization necessary for combustion;

- Large debris such as boulders, concrete or asphalt blocks, roots, branches, etc. which will damage equipment;

  and

- Plastic, which will foul the mixing chamber.

Asphalt solidifies into a solid matrix that immobilizes remaining contaminants. Even if the resulting asphalt is not suitable for road construction, it can be ground into small particles and used for roadbase — with regulatory agency approval. An alternative to true asphalt incorporation is to use the asphalt batch plant as a high temperature thermal treatment facility during slack times.

References

In this section, books are denoted by **bold**, technical journals are in ***bold italic***, journal articles are in *italic text*, and government documents are in ***bold italics***. "FR" is the Federal Register and "CFR" is Code of Federal Regulations. The first section contains a list of general references that refer to specific subjects covered in this book. These references often contain more detailed or specific information.

General References

1. (ESI, 1990) **Onsite Treatment of Hydrocarbon Contaminated Soils**, prepared by Environmental Solutions, Inc., Irvine, CA, under contract by Western States Petroleum Association. (No copyright date) [ca. 1990].

2. (Testa and Winegardner, 1991) **Restoration of Petroleum Contaminated Aquifers**, S.M. Testa and D.L. Winegardner, Lewis Publishers, Chelsea, MI. 1991.

3. (EPRI, 1988) **Remedial Technologies for Leaking Underground Storage Tanks**, EPRI, Lewis Publishers, Inc., Chelsea, MI. 1988.

4. (Noonan and Curtis, 1990)**Groundwater Remediation and Petroleum, A Guide for Underground Storage Tanks**, D.C. Noonan and J.T. Curtis, Lewis Publishers, Chelsea, MI. 1990.

5. (Kostecki and Calabrese, 1991a) **Hydrocarbon Contaminated Soils and Groundwater**, P.T. Kostecki, and E.J. Calabrese, Eds., Volume 1. Lewis Publishers, Inc., Chelsea, MI. 1991.

6. (Kostecki and Calabrese, 1992a) **Hydrocarbon Contaminated Soils and Groundwater,** P.T. Kostecki, and E.J. Calabrese, Eds., Volume 2. Lewis Publishers, Inc., Chelsea, MI. 1992.

7. (Kostecki and Calabrese, 1993a) **Hydrocarbon Contaminated Soils and Groundwater,** P.T. Kostecki, and E.J. Calabrese, Eds., Volume 3. Lewis Publishers, Inc., Chelsea, MI. 1993.

8. (Kostecki and Calabrese, 1991b) **Hydrocarbon Contaminated Soils,** P.T. Kostecki, and E.J. Calabrese, Editors, Volume 1. Lewis Publishers, Inc. Chelsea, MI. 1991.

9. (Kostecki and Calabrese, 1992b) **Hydrocarbon Contaminated Soils,** P.T. Kostecki, and E.J. Calabrese, Editors, Volume 2. Lewis Publishers, Inc. Chelsea, MI. 1992.

10. (Kostecki and Calabrese, 1993a) **Hydrocarbon Contaminated Soils,** P.T. Kostecki, and E.J. Calabrese, Editors, Volume 3. Lewis Publishers, Inc. Chelsea, MI. 1993.

11. (Kostecki and Calabrese, 1989a) **Petroleum Contaminated Soils,** P.T. Kostecki and E.J. Calabrese, Eds., Volume 1. Lewis Publishers, Inc. Chelsea, MI. 1989.

12. (Kostecki and Calabrese, 1989b) **Petroleum Contaminated Soils,** P.T. Kostecki and E.J. Calabrese, Eds., Volume 2. Lewis Publishers, Inc. Chelsea, MI. 1990.

13. (Kostecki and Calabrese, 1990c) **Petroleum Contaminated Soils,** P.T. Kostecki and E.J. Calabrese, Eds., Volume 3. Lewis Publishers, Inc. Chelsea, MI. 1991.

14. (Hinchee and Olfenbuttel, 1991a)   **On Site Bioreclamation: Processes for Xenobiotic Hydrocarbon Treatment,** R.E. Hinchee and R.F. Olfenbuttel, Eds. Butterworth–Heinemann, Boston, 1991.

15. (Hinchee and Olfenbuttel, 1991b)   **In Situ Bioreclamation: Application and Investigations for Hydrocarbon Contaminated Site Remediation,** R.E. Hinchee and R.F. Olfenbuttel, Eds., Butterworth–Heinemann, Boston, 1991.

16. (UST Guide) **Underground Storage Tank Guide,** Thompson Publishing Group, Washington, DC 20006.

17. (UST Bulletin) *Underground Storage Tank Bulletin,* Thompson Publishing Group, Washington, DC 20006.

18. (Mueller, et al.) *Recent Developments in Cleanup Technologies,* J.G. Mueller, J.E. Lin, S.E. Lantz, and P.H. Prichard, *Remed.* 3 (3) 1993. p. 369.

19. (Cole, 1991) **Underground Storage Tank Installation and Management,** G. M. Cole, Lewis Publishers, Inc., Chelsea, MI 48118. 1992.

Chapter 1 Introduction

1. (EPA, 1988) *Technical Standards For Operation of Underground Storage Tank Systems,* 52 FR, September 23, 1988.

2. (EPA, 1993) Cleaning Up the Nation's Hazardous Waste Site: Markets and Technology Trends.U.S. EPA Document No. PB93-140762, National Technical Information Service, Pueblo, CO.

3. (WTN, 1993a) Waste Tech News, July, 1993.

4. (WTN, 1993b) Waste Tech News, August 23, 1993.

5. (UST Guide) **Underground Storage Tank Guide,** Thompson Publishing Group, Washington, DC. May, June, 1993.

6. (Cole, 1992) **Underground Storage Tank Installation and Management,** G.M. Cole, Lewis Publishers, Inc., Chelsea, MI, 1992.

Chapter 2 Environmental Legislation

1. (WEST, 1991) **Environmental Law Statutes,** West Publishing Co. 1991.

2. (Findley and Farber, 1992) **Environmental Law,** R.W. Findley and D.A. Farber, 3rd Edition. West Publishing Co., 1992.

3. (FR, 1988) 53 FR 37082, September 23, 1988.

4. (FR, 1990) 55 FR 17753, April 27, 1990.

5. (Krendl and Gibson, 1992) *The Impact of The New EPA Regulation on Lender Liability,* C. S. Krendl and T. J. Gibson, *The Colorado Lawyer,* Nov. 1992. p. 233.

6. (Tolman, 1993) *RCRA Regulatory Developments*, A.J. Tolman, Indust. Wastewater, vol. 1, June/July, 1993. p.30.

7. (Dunn, 1990) *The Black Hole of Lender Liability*, D.J. Dunn, *Environ. Liab. in Comm. Trans.*, Volume 1 (1) Nov. 1990.

8. (Env. Liab., 1991) *Proposed Lender Liability Rule*, [56 FR 28,798 (June 28, 1991)], in *Environ. Liab. in Comm. Trans.*, Volume 1 (10) Aug. 1991.

Chapter 3 Petroleum Hydrocarbons

1. (Lents and Kelly, 1993) *Clearing Up The Air In Los Angeles*, J.M. Lents and W.J. Kelly, Scientific American, October, 1993. p. 32.

2. (ET, 1990) **Encyclopedia of Technology**, McGraw–Hill Book Company, 1990.

3. (EST, 1985) **Encyclopedia of Science and Technology**, Kirk and Othner, Eds., John Wiley, New York, 1985.

4. (Riddick, et al., 1986) **Organic Solvents Physical Properties and Methods of Purification**, J.A. Riddick, W.B. Bunger, and T.K. Sakano, 4th Edition, John Wiley & Sons, New York. 1986.

5. (NIOSH/OSHA, 1987) **Pocket Guide to Chemical Hazards**, NIOSH/OSHA DHHS Publication #85–114, February, 1987.

6. (ACGIH) **Threshold Limit Values and Biological Exposure Indices for 1990–1991**, The American Conference of Governmental Industrial Hygienists, ACGIH, 6500 Glenway Ave., Bldg D-7, Cincinnati, OH.

7. Memos from the Dept. of Labor regarding the applicability of 29 CFR 1910.120 to UST remedial operations have circulated privately since 1988.

Chapter 4 Soils and Subsurface Characteristics

1. (Hoffman and Bell) *Application of In–Situ Bioremediation to a Gasoline Spill in Fractured Bedrock*, A.H. Hoffman and R.A. Bell, *Remediation*, 3, (3) 1993. p. 353.

2. (Mackay and Cherry,1985) *Transport of Organic Contaminants in Ground-water*, D.M. Mackay, P.V. Roberts, and J.A. Cherry, *Environ. Sci. Technol.* v. 19(5): 384-392 (1985).

3. (Mackay and Cherry,1989) *Groundwater Contamination: Pump and Treat Remediation*, D.M. Mackay, and J.A. Cherry, *Environ. Sci. Technol.* v. 23(6): 630-636 (1989).

4. (McKee, et al., 1972) *Gasoline in Groundwater*, J.E. McKee, F.B. Laverty, and R.M. Hertzel, *J. Water Pollut. Control Fed.*, 44 (2) 1972. p. 293.

5. (Dragun,1988) **The Soil Chemistry of Hazardous Materials**, J. Dragun, Hazardous Materials Control Institute, Silver Springs, MD, 1988.

6. (Friesen, 1990) *Chemical Changes of Biodegraded Gasoline in a Laboratory Simulation of the Vadose Zone*, K.A. Friesen, Thesis, Colorado School of Mines, Golden, 1990.

7. (Turkall, et al., 1990) *The Effect of Soil Type on Absorption of Toluene and Its Bioavailability*, R.M. Turkall, G. A. Skowronski, and M.S. Abdel-Rahman, in **Petroleum Contaminated Soils**, P.T. Kostecki, and E.J. Calabrese, Eds., Volume 3, Lewis Publishers, Inc. 1992. p. 382.

Chapter 5 Environmental Assessments

1. (Krendl and Gibson, 1992) *The Impact of The New EPA Regulation on Lender Liability*, C. S. Krendl and T. J. Gibson, *The Colorado Lawyer*, Nov. 1992. p. 233.

2. (Sweeney, 1993) *Lender Liability and CERCLA*, S. Sweeney, *ABA Journal*, Feb. 1993, p. 68.

3. (Kulla and Bellomo, 1991) *Sharing Responsibilities: A New Approach to Groundwater Remediation Involving Multiple Potentially Responsible Parties*, J.B. Kulla and A.J. Bellomo, in **Hydrocarbon Contaminated Soils and Groundwater**, P.T. Kostecki and E.J. Calabrese, Eds., volume 1, Lewis Publishers, Chelsea, MI, 1991. p. 91.

4. (ASTM, 1993a) *Standard Practice for Environmental Site Assessments: Transaction Screen Process, Standard E. 50.02.1.* ASTM, Philadelphia, PA, June, 1993

5. (ASTM, 1993b) *Standard Practice for Environmental Site Assessments: Phase I Environmental Site Assessment Process, Standard E. 50.02.2.* ASTM, Philadelphia, PA, June, 1993.

6. (ASTM, 1993b) *Standard Practice for Environmental Site Assessments: Phase I Environmental Site Assessment Process, Standard E. 50.02.2.* ASTM, Philadilphia, PA, June, 1993. ¶1.1.1.

7. (FR, 1992) 57 FR 18344, 1992; and 40 CFR §300.1100.

8. (Bulletin, 1992) Petroleum Marketeers Association testimony to Senate subcommittee.

9. (Clark, 1991) *Lender Liability and Its Effect on UST Owners and Operators,* A.M. Clark, in Thompsons Underground Storage TankGuide, ¶194, p. 263, 1991.

10. (Jain, et al., 1993) **Environmental Assessment,** R.K. Jain, L.V. Urban, G.S. Stacey, and H.E. Balbach, McGraw–Hill, Inc. New York. 1993.

Chapter 6 Site Assessments

1. (ASTM, 1993a) *Standard Practice For Environmental Site Assessments: Transaction Screen Process, Standard E. 50.02.1.* ASTM, Philadelphia, PA, June, 1993.

2. (ASTM, 1993b) *Standard Practice for Environmental Site Assessments: Phase I Environmental Site Assessment Process, Standard E. 50.02.2.* ASTM, Philadelphia, PA, June, 1993.

3. (CDPHE, 1991) *Guidance Documents for Remediation of Petroleum– Contaminated Sites,* Colorado Department of Public Health and Environment, Denver, CO. 1991.

4. (LUFT, 1989) *Leaking Underground Fuel Tank — Field Manual,* (LUFT), California State Water Board, Sacramento, CA. 1989.

5. (Selby, 1991) A Critical Review of Site Assessment Methodologies, D.A. Selby, in **Hydrocarbon Contaminated Soils and Groundwater,** P.T. Kostecki and E.J. Calabrese, eds. volume 1, Lewis Publishers, Chelsea, MI, 1991. p. 149.

6. (Johnson and Haeberer, 1991) *Importance of Quality for Collection of Environmental Samples; Planning, Implementing, and Assessing Field Sampling Quality at CERCLA Sites,* G.L. Johnson and A.E. Haeberer, Eighteenth Annual National Energy Division Conference, 1991.

7. (Johnson, et al., 1990) *Estimates for Hydrocarbon Vapor Emission Resulting from Service Station Remediations and Buried Gasoline–Contaminated Soils,* P.C. Johnson, M.B. Hertz, and D.L. Byers, in **Petroleum Contaminated Soils,** P.T. Kostecki, and E.J. Calabrese, Eds, Volume 3. Lewis Publishers, Inc., Chelsea, MI. 1993. p. 295.

Chapter 7 Environmental Sampling and Laboratory Analysis

1. (EPA, 1983) *Methods for Chemical Analysis of Water and Wastes,* U.S. Environmental Protection Agency, EP–600/4–79–020 (Environmental Monitoring and Support Laboratory, Cincinnati, OH, March, 1983).

2. (EPA, 1984) *Test Methods for Evaluating Solid Waste: Physical/Chemical Methods,* 2nd Edition, U.S. Environmental Protection Agency, EPA SW–846. (Office of Solid Waste and Emergency Response, Washington, D.C., 1984.)

3. (EPA, 1986) *Test Methods for Evaluating Solid Waste: Physical/Chemical Methods,* 3rd Edition, U.S. Environmental Protection Agency, EPA SW–846. (Office of Solid Waste and Emergency Response, Washington, D.C., 1986.)

4. (USEPA) *Drinking Water Methods,* 40 CFR §141 and 142.

5. (CDPHE, 1991) *Guidance Documents for Remediation of Petroleum–Contaminated Sites,* Colorado Department of Public Health and Environment. 1991.

6. (LUFT, 1989) *Leaking Underground Fuel Tank—Field Manual,* (LUFT), California State Water Board. 1989.

7. (NJDEP, 1988) *Field Sampling Procedures Manual,* New Jersey Department of Environmental Protection, February, 1988.

8. (Bryden and Smith, 1989) *Sampling for Environmental Analysis: Part I Planning and Preparation,* G.W. Bryden and L.R. Smith, *Am. Lab.* 21 (7), 1989. p. 30.

9.   (FR, 1984) 49 FR October 26, 1984.

10. (Keith,1988) **Principles of Environmental Sampling**, L.H. Keith, American Chemical Society, Washington, D.C., 1988.

11. (Thompson, 1992) **Sampling**, S.K. Thompson, John Wiley & Sons, Inc. New York, NY, 1992.

12. (Goerlitz and Brown, 1972) *Methods for Analysis of Organic Substances in Water*, D.F. Goerlitz and E. Brown, USGS–TWRI, Book 5, Chapter A3, 1972. Available from U.S. Geological Survey, Branch of distribution, 1200 South Eads St. Arlington, VA 22202.

13. (Black, 1988) in **Principles of Environmental Sampling**, L.K. Keith, Ed. American Chemical Society, Washington, DC, 1988.

14. (Lewis, 1988) in **Principles of Environmental Sampling**, L.K. Keith, Ed. American Chemical Society, Washington, DC, 1988.

15. (Taylor, 1988) *Defining the Accuracy, Precision, and Confidence Limits of Sample Data*, J.K. Taylor in **Principles of Environmental Sampling**, L.K. Keith, Ed. American Chemical Society, Washington, DC, 1988. p. 102.

16. (Maskarinec, 1988) in **Principles of Environmental Sampling**, L.K. Keith, Ed. American Chemical Society, Washington, DC, 1988. p. 102.

17. (Glade, 1993) Personal communication, M.D. Glade, Inman, Flynn, & Crabtree, P.C. 1993.

18. (FR, 1990a) 55 FR March 29, 1990, No. 61, p. 11798.

19. (FR, 1990b) 55 FR June 29, 1990, No 62, pp. 12967.

20. (Barcelona, 1988) in **Principles of Environmental Sampling**, L.K. Keith, Ed. American Chemical Society, Washington, DC, 1988.

21. (Biedry and R. Martin, 1991) *Understanding the Toxicity Characteristic Rule*, J. Biedry and R. Martin, *Hazmat World*, June, 1991. p. 50.

22. (Clarke, et al., 1990) *Environmental Project Definition with the Analytical Laboratory*, A.N. Clarke, D.R. Davis, and J.H. Clarke, *Amer. Environ. Lab.*, 10, 1990, p. 9.

23. (Guide, 1993) **Underground Storage Tank Guide,** (5) #7, April, 1993. Thompson Publishing Group, Washington, DC.

24. (FR, 1993) 58 FR February 12, 1993, No 65, p. 8504.

25. (FR, 1992) 57 FR 21524, 1992.

26. (FR, 1991) 56 FR 48001, 1991.

27. (Blacker and Neptune, 1991) *EPA's Quality Assurance Planning Process and the Observational Approach — Strengths and Weaknesses,* S.M. Blacker and M. D. Neptune, Eighteenth Annual National Energy Division Conference, American Society for Quality Control. 1991.

28. (Harris, et al., 1991) *Comparison of Data Quality Objectives and Observational Method: Application of the Observational Method to an Operable Unit Feasibility Study — A Case Study Update,* J. Harris, K. Look, and L. Otis, Eighteenth Annual National Energy Division Conference, American Society for Quality Control. 1991.

Chapter 8 Data Integration and Technology Selection: The Corrective Action Plan

1. (Wong, et al., 1991) *Looking Past Soil Cleanup Numbers,* J.J. Wong, G.M. Schum, E.G. Butler, and R.A. Becker, **Hydrocarbon Contaminated Soils and Groundwater,** P.T. Kostecki and E.J. Calabrese, Eds. Lewis Publishers, Chelsea, MI, 1991. volume 1, p. 1.

2. (Daugherty, 1991) *Regulatory Approaches to Hydrocarbon Contamination from Underground Storage Tanks,* S.J. Daugherty, **Hydrocarbon Contaminated Soils and Groundwater,** P.T. Kostecki and E.J. Calabrese, Eds. Lewis Publishers, Chelsea, MI, 1991. Volume 1, p. 23.

3. (UST Bulletin) *Underground Storage Tank Bulletin,* Thompson Publishing Group, Washington, DC 20006.

4. (Michaud, et al., 1991) *Human Health Risks Associated with Contaminated Sites: Chemical Factors in the Exposure Assessment,* J.M. Michaud, A.H. Parsons, S.R. Ripple, and D.M. Paustenbach, **Hydrocarbon Contaminated Soils and Groundwater,** P.T. Kostecki and E.J. Calabrese, Eds. Lewis Publishers, Chelsea, MI, 1991. Volume 1, p. 283.

5. (Edmisten–Watkin, et al, 1991) *Health Risks Associated with the Remediation of Contaminated Soils,* G. Edmisten–Watkin, E.J. Calabrese, and R.H. Harris, in **Hydrocarbon Contaminated Soils and Groundwater,** P.T. Kostecki and E.J. Calabrese, Eds. Lewis Publishers, Chelsea, MI, 1991. Volume 1, p. 293.

6. (Lazzaretto, 1991) *Challenges Encountered in Hydrocarbon Contaminated Soil Cleanup,* A.C. Lazzaretto, in **Hydrocarbon Contaminated Soils and Groundwater,** P.T. Kostecki and E.J. Calabrese, Eds. Lewis Publishers, 1991. Chelsea, MI, Volume 1, p. 99.

7. (Mackay and Cherry,1985) *Transport of Organic Contaminants in Groundwater,* D.M. Mackay, P.V. Roberts, and J.A. Cherry, *Environ. Sci. Technol.* Volume 19 (5): 384-392 (1985).

8. (WTN, 1993) Waste Tech News, August 23, 1993.

9. (EPA, 1993) *Cleaning Up the Nation's Hazardous Waste Sites: Markets and Technology Trends,* U.S. Environmental Protection Agency, National Technical Information Sertvice, Pueblo, Colorado, USEPA No PB93-140762. 1993.

10. (ESI, 1990) **Onsite Treatment of Hydrocarbon Contaminated Soils,** prepared by Environmental Solutions, Inc. under contract by Western States Petroleum Association.

11. (CDPHE, 1991) *Guidance Documents for Remediation of Petroleum–Contaminated Sites,* Colorado Department of Public Health and Environment. 1991.

12. (LUFT, 1989) *Leaking Underground Fuel Tank — Field Manual,* (LUFT), California State Water Board. 1989.

13. (EPA, 1988) *Technology Screening Guide for Treatment of CERCLA Soils and Sludges.* U.S. Environmental Protection Agency, EPA 540/2–88/004. 1988.

14. (EPA, 1991) *Stabilization Technologies for RCRA Corrective Actions,* U.S. Environmental Protection Agency, EPA/625/6-91/026. Washington, DC. August, 1991.

15. (McCullough and Dagdigian, 1993) *Evaluation of Remedial Options for Treatment of Heavy Metal and Petroleum–Contaminated Soils,* M.L. McCullough and J.V. Dagdigian, Remediation, (3) 3, 265 (1993).

Chapter 9 In Situ Remediation Technologies

1. (EPRI, 1988) **Remedial Technologies for Leaking Underground Storage Tanks,** L.M. Preslo, Project Manager, Electric Power Research Institute, Inc., Lewis Publishers, Chelsea, MI, 1988.

2. (ESI, 1990) **Onsite Treatment of Hydrocarbon Contaminated Soils,** prepared by Environmental Solutions, Inc. Irvine, CA, under contract by Western States Petroleum Association. (ca. 1990).

3. (Noonan and Curtis, 1990) **Groundwater Remediation and Petroleum, A Guide for Underground Storage Tanks,** D.C. Noonan and J.T. Curtis, Lewis Publishers, Chelsea, MI, 1990.

4. (Testa and Winegardner, 1991) **Restoration of Petroleum Contaminated Aquifers,** S.M. Testa and D.L. Winegardner, Lewis Publishers, Chelsea, MI, 1991.

5. (Ludvigsen, 1992) *Petroleum Release Decision Framework,* P.L. Ludvigsen, D.H. Chemin , C.C. Stanley, and D. Draney, in **Petroleum Contaminated Soils,** P.T. Kostecki, and E.J. Calabrese, Eds., Volume 3, Lewis Publishers, Chelsea, MI, 1992. p. 19.

6. (ESI, 1990) **Onsite Treatment of Hydrocarbon Contaminated Soils,** prepared by Environmental Solutions, Inc. under contract by Western States Petroleum Association. (ca 1990). Appendix A.1, p. A.1–1.

7. (Johnson, et al., 1992) *Estimates for Hydrocarbon Vapor Emissions Resulting from Service Station Remediation and Buried Gasoline–Contaminated Soils;* P.C. Johnson, M.B. Hertz, and D.L. Byers, in **Petroleum Contaminated Soils,** P.T. Kostecki, and E.J. Calabrese, Eds. Volume 3, Lewis Publishers, Chelsea, MI, 1992.

8. (Preslo, 1988) **Remedial Technologies for Leaking Underground Storage Tanks,** L.M. Preslo, Project Manager, Roy F. Weston, Inc., Lewis Publishers, 1988.

9   (Friesen, 1990) *Chemical Changes of Biodegraded Gasoline in a Laboratory Simulation of the Vadose Zone,* K.A. Friesen, Colorado School of Mines, Thesis, 1990.

10. (Hoffman and Bell) *Application of In–Situ Bioremediation to a Gasoline Spill in Fractured Bedrock,* A.H. Hoffman and R.A. Bell, *Remediation, 3,* (3) 1993. p. 353.

11. (Turkall, et al., 1992) *The Effect of Soil Type on Absorption of Toluene and Its Bioavailability*, R.M. Turkall, G. A. Skowronski, and M.S. Abdel-Rahman, in **Petroleum Contaminated Soils**, P.T. Kostecki, and E.J. Calabrese, Volume 3, Lewis Publishers, Chelsea, MI, 1992. p. 382.

12. (Lehman, 1992) **Biochemistry**, M.A. Lehman, John Wiley and Sons, New York, 1991. 2nd Edition.

13. (Hinchee and Olfenbuttel, 1991a)**On Site Bioreclamation: Processes for Xenobiotic Hydrocarbon Treatment**, R.E. Hinchee and R.F. Olfenbuttel, Eds., Butterworth–Heinemann, Boston, MA, 1991.

14. (Hinchee and Olfenbuttel, 1991b) ***In Situ* Bioreclamation: Application and Investigations for Hydrocarbon Contaminated Site Remediation**, R.E. Hinchee and R.F. Olfenbuttel, Eds., Butterworth–Heinemann, Boston, 1991.

15. (Singer and Finnerty, 1984) *Microbial Metabolism of Straight–chain and Branched Alkanes*, M.E. Singer and W.R. Finnerty, in **Petroleum Microbiology**, R.M. Atlas, Eds., Macmillan Publishing Co., New York, 1984. p. 1.

16. (Perry, 1984) *Microbial Metabolism of Cyclic Alkanes*, J.J. Perry, in **Petroleum Microbiology**, R.M. Atlas, Ed., Macmillan Publishing Co., New York, 1984. p. 61.

17. (Cerniglia, 1984) *Microbial Transformation of Aromatic Hydrocarbons*, C.E. Cerniglia, in **Petroleum Microbiology**, R.M. Atlas, Ed., Macmillan Publishing Co., New York, 1984. p. 99.

18. (Bossert and Bartha, 1984) *The Fate of Petroleum in Soil Ecosystems*, I. Bossert and R. Bartha, in **Petroleum** Microbiology, R.M. Atlas, Ed., Macmillan Publishing Co., New York, 1984. p. 435.

19. (Battelle, 1993) *In Situ* and On–Site Bioreclamation, The Second International Symposium, April 5–8, 1993, San Diego.

20. (Baker and Herson, 1994) **Bioremediation**, K.H. Baker and D.S. Herson, McGraw–Hill, Inc. New York, 1994.

21. (C&EN, 1991) *Bioremediation*, A.M. Thayer, *Chemical and Engineering News*, American Chemical Society, August 26, 1991.

22. (Cheremisinoff, 1993) *Oil/Water Separation,* P.N. Cheremisinoff, Nat. Environ. J. vol. 3, May/June, 1993. p. 32.

23. (Diks and Ottengrat, 1991) **In Situ and On–Site Bioreclamation,** The Second International Symposium, April 5–8, 1993, San Diego.

24. (CDPHE, 1991) *Ground Water Monitoring Guidance,* Solid Waste and Incident Management Section, 1991 and *Guidance for Construction of Ground Water Monitoring Wells,* Water Quality Control Division, 1987, Colorado Department of Public Health and Environment.

25. (EPA, 1991) *Stabilization Technologies for RCRA Corrective Actions,* U.S. Environmental Protection Agency, EPA/625/6-91/026. Washington, DC. August, 1991.

26. (Updegraff, 1993) D.M. Updegraff, Colorado School of Mines, personal communication.

27. (Noonan & Curtis, 1990) **Groundwater Remediation and Petroleum, A Guide for Underground Storage Tanks,** D.C. Noonan and J.T. Curtis, Lewis Publishers, Chelsea, MI, 1990. p. 18.

28. (McCullough and Dagdigian, 1993) *Evaluation of Remedial Options for Treatment of Heavy Metal and Petroleum–Contaminated Soils,* M.L. McCullough and J.V. Dagdigian, *Remediation,* (3) 3, 265 (1993).

29. (Pamukcu, et al., 1990) *Study of Possible Reuse of Stabilized Petroleum Contaminated Soils as Construction Material,* S. Pamukcu, H.M. Hijazi, and H.Y. Fang, in **Petroleum Contaminated Soils,** P.T. Kostecki, and E.J. Calabrese, Volume 3, Lewis Publishers, Chelsea, MI, 1992. p. 203.

Chapter 10 Non–In Situ Soil Treatment Technologies

1. (Preslo, 1988) **Remedial Technologies for Leaking Underground Storage Tanks,** L.M. Preslo, Project Manager, Electric Power Research Institute, Lewis Publishers, Chelsea, MI, 1988.

2. (ESI, 1990) **Onsite Treatment of Hydrocarbon Contaminated Soils,** prepared by Environmental Solutions, Inc. under contract by Western States Petroleum Association.

3. (C&EN, 1991) *Bioremediation,* A.M. Thayer, *Chemical and Engineering News,* American Chemical Society, August 26, 1991.

4. (NIOSH, 1987) *A Guide to Safety in Confined Spaces, DHHS* (NIOSH) Publication No. 87-13, Department of Health and Human Services, National Institute for Occupational Health and Safety, 4676 Columbia Parkway, Cincinnati, OH, 45226.

5. (EPA, 1986) *Handbook for Stabilization/Solidification of Hazardous Wastes,* EPA 540/2-86/001, U.S. Environmental Protection Agency, 1986.

6. (Valezquez and Nowland, 1989) *Low Temperature Stripping of Volatile Compounds in Petroleum Contaminated Soils,* Valezquez, L.A. and J.W. Nowland, in Petroleum Contaminated Soils. Remediation Techniques, Environmental Fate, and Risk Assessment, P.T. Kostecki and E.J. Calabrese, Eds., Lewis Publishers, Chelsea, MI, Volume 1, 1989.

7. (API, 1989) Assessment and Remediation of Underground Petroleum Releases, API Publication 1628, 2nd Edition. August, 1989.

8. (Chopey, 1993) Handbook of Chemical Engineering Calculations, N.P. Chopey, McGraw–Hill, Inc. New York, 1993.

Appendix A: Petroleum Products

There are thousands of commercially available petroleum products, from fuels to cosmetics. All are derived from crude petroleum. Variations from one product to another is usually a matter of selecting the appropriate distillation cut, as for example in the case of lubrication oils and heating oils. In many cases distillation fractions from the refinery are redistilled to produce a more pure product. Kerosene can be used as a charcoal starter but has an unpleasant odor and taste. Commercial charcoal starter is redistilled to produce a "narrower" cut; i.e., a more pure product with less odor and taste.

An immense number of commercial products are derived from crude petroleum. **Table A–1** lists the components of typical light and heavy Arabian crude.

By far the most important product derived from petroleum is gasoline. **Table A–2** lists the components of a gasoline blend. The mixture represents a typical "unleaded regular" formulation.

The chromatographs shown in **Table A–3** illustrate the component hydrocarbons for each product shown in the legend. A sample is injected into a gas chromatograph where the sample is vaporized. The vapor is swept along by a stream of inert gas through a separation column. The column is packed with a silica–based material (usually methyl silicone). Component hydrocarbons move through the column by diffusion. The movement is more or less rapidly through the column material depending on molecular weights, speed of diffusion, and individual chemical structure. Those hydrocarbons that are heavier move through the column more slowly; those that are lighter flow more rapidly. The result is to separate the component hydrocarbons into a flowing column of vapor approximately proportional to molecular weight (see Chapter 7).

For higher resolution, as the vapor (or liquid) flow exits the column, it can be directed into a mass spectrometer where molecules are ionized and passed through a magnetic field. The field deflects the ions along paths that are proportional to their masses (weight). A detector measures the number of ions of the same weight. The data output of the GC/MS is a spectrum of curves. The area of each curve is proportional to the amount of that hydrocarbon present in the original mixture. The abscissa (x–axis) is labeled "retention time" and by number of carbons. Thus, the x–axis is approximately proportional to molecular weight.

Table A–1. Components of Arabian Crude. The fractions typical to crude petroleum are listed. The boiling point ranges are in °C. Data are listed for both light and heavy crudes. Fractions heavier than kerosene are nearly 100% alkanes (adapted from EST, 1985).

Fraction		Range, °C	Specific Gravity	Alkanes, vol%	Cycloalkanes, vol%	Aromatic Compounds, vol%
Crude	Heavy		0.886			
	Light		0.831			
Light Naphtha	Heavy	20–100	0.669	89.6	9.5	0.9
	Light	20–100	0.677	87.4	10.7	1.9
Heavy Naphtha	Heavy	100–150	0.737	70.6	21.4	8.3
	Light	100–150	0.744	66.4	20.0	13.7
Kerosene	Heavy	150–235	0.787	58.0	23.7	18.3
	Light	150–235	0.788	58.9	20.5	20.6
Light Gas Oil	Heavy	235–343	0.846	~100	n/a	n/a
	Light	235–343	0.838	~100	n/a	n/a
Heavy Gas Oil	Heavy	343–565	0.923	~100	n/a	n/a
	Light	343–565	0.905	~100	n/a	n/a
Residual Oil	Heavy	>343	0.984	~100	n/a	n/a
	Light	>343	0.924	~100	n/a	n/a
Residual Oil	Heavy	>565	1.044	~100	n/a	n/a
	Light	>565	0.990	~100	n/a	n/a

Table A–2. (Facing and following pages) Components of Gasoline. A representative blend of gasoline is listed by hydrocarbon in order of increasing molecular weight. Percentages are in volume (adapted from EST, 1985)

Compound	Number of Carbons	Weight Percent
Straight–Chain Alkanes		
Propane	3	0.01–0.14
n–Butane	4	3.93–4.70
n–Pentane	5	5.75–10.92
n–Hexane	6	0.24–3.50
n–Heptane	7	0.31–1.96
n–Octane	8	0.36–1.43
n–Nonane	9	0.07–0.83
n–Decane	10	0.04–0.50
n–Undecane	11	0.05–0.22
n–Dodecane	12	0.04–0.09
Branched Alkanes		
Isobutane	4	0.12–0.37
2,2–Dimethylbutane	6	0.17–0.84
2,3–Dimethylbutane	6	0.59–1.55
2,2,3–Trimethylbutane	7	0.01–0.05
Neopentane	5	0.02–0.05
Isopentane	5	6.07–10.17
2–Methylpentane	6	2.91–3.85
3–Methylpentane	6	2.4
2,4–Dimethylpentane	7	2.91–3.85
2,3–Dimethylpentane	7	0.23–1.17
3,3–Dimethylpentane	7	0.32–4.17
2,2,3–Trimethylpentane	8	0.02–0.03
2,2,4–Trimethylpentane (Isooctane)	8	0.09–0.23
2,3,3–Trimethylpentane	8	0.05–2.28
2,3,4–Trimethylpentane	8	0.11–2.80
2,4–Dimethyl–3–ethyl–pentane	9	0.03–0.07
2–Methylhexane	7	0.36–1.48
3–Methylhexane	7	0.30–1.77
2,4–Dimethylhexane	8	0.34–0.82
2,5–Dimethylhexane	8	0.24–0.52
3,4–Dimethylhexane	8	0.16–0.37
3–Ethylhexane	8	0.01
2–Methyl–3–ethyl–hexane	9	0.04–0.13
2,2,4–Trimethylhexane	9	0.11–0.18
2,2,5–Trimethylhexane	9	0.17–5.89
2,3,3–Trimethylhexane	9	0.05–0.12
2,3,5–Trimethylhexane	9	0.11–1.09
2,4,4–Trimethylhexane	9	0.02–0.16

Compound	Number of Carbons	Weight Percent
2–Methylheptane	8	0.48–1.05
3–Methylheptane	8	0.63–1.54
4–Methylheptane	8	0.22–0.52
2,2–Dimethylheptane	9	0.01–0.08
2,3–Dimethylheptane	9	0.13–0.51
2,6–Dimethylheptane	9	0.07–0.23
3,3–Dimethylheptane	9	0.01–0.08
3,4–Dimethylheptane	9	0.07–0.33
3–Ethylheptane	9	0.02–0.16
2,2,4–Trimethylheptane	10	0.12–1.70
2,3,5–Trimethylheptane	10	0.02–0.06
2–Methyloctane	9	0.14–0.62
3–Methyloctane	9	0.34–0.85
4–Methyloctane	9	0.34–0.85
2,6–Dimethyloctane	10	0.06–0.12
2–Methylnonane	10	0.06–0.41
3–Methylnonane	10	0.06–0.32
4–Methylnonane	10	0.04–0.26
Cycloalkanes		
Cyclopentane	5	0.19–0.58
Methylcyclopentane	6	n/a
1–Methyl–cis–2–ethyl–cyclopentane	8	0.06–0.11
1–Methyl–trans–3–ethyl–cyclopentane	8	0.06–0.12
1–cis–2–Dimethylcyclopentane	7	0.07–0.13
1–trans–2–Dimethylcyclopentane	7	0.06–0.20
1,1,2–Trimethylcyclopentane	8	0.06–0.11
1–trans–2–cis–3–Trimethylcyclopentane	8	0.01–0.25
1–trans–2–cis–4–Trimethylcyclopentane	8	0.03–0.16
Ethylcyclopentane	7	0.14–0.21
n–Propylcyclopentane	8	0.01–0.06
Isopropylcyclopentane	8	0.01–0.02
Cyclohexane	6	1.01–3.02
Methyl cyclohexane	7	2.01–4.02
1–trans–3–Dimethylcyclohexane	8	0.05–0.12
Ethylcyclohexane	8	0.17–0.42

Compound	Number of Carbons	Weight Percent
Straight Chain Alkenes		
cis–2–Butene	4	0.13–0.17
trans–2–Butene	4	0.16–0.20
1–Pentene	5	0.33–0.45
cis–2–Pentene	5	0.43–0.67
trans–2–Pentene	5	0.52–0.90
cis–2–Hexene	6	0.15–0.24
trans–2–Hexene	6	0.18–0.36
cis–3–Hexene	6	0.11–0.13
trans–3–Hexene	6	0.12–0.15
cis–3–Heptene	7	0.14–0.17
trans–2–Heptene	7	0.06–0.10
Branched Alkenes		
2–Methyl–1–butene	5	0.22–0.66
3–Methyl–1–butene	5	0.08–0.12
3–Methyl–2–butene	5	0.96–1.28
2,3–Dimethyl–2–butene	6	0.08–0.10
2–Methyl–1–pentene	6	0.08–0.10
2,3–Dimethyl–1–pentene	7	0.01–0.02
2,4–Dimethyl–1–pentene	7	0.02–0.03
4,4–Dimethyl–1–pentene	7	0.06
2–Methyl–2–pentene	6	0.27–0.32
3–Methyl–cis–2–pentene	6	0.35–0.45
3–Methyl–trans–2–pentene	6	0.32–0.44
4–Methyl–cis–2–pentene	6	0.04–0.05
4–Methyl–trans–2–pentene	6	0.08–0.30
4,4–Dimethyl–cis–2–pentene	7	0.02
4,4–Dimethyl–trans–2–pentene	7	n/a
3–Ethyl–2–pentene	7	0.03–0.04
Cycloalkenes		
Cyclopentene	5	0.12–0.18
3–Methylcyclopentene	6	0.03–0.08
Cyclohexene	6	0.03

Compound	Number of Carbons	Weight Percent
Alkyl Benzenes		
Benzene	6	0.12–3.50
Toluene	7	2.73–21.80
o–Xylene	8	0.68–2.86
m–Xylene	8	1.77–3.87
p–Xylene	8	0.77–1.58
1–Methyl–4–ethylbenzene	9	0.18–1.00
1–Methyl–2–ethylbenzene	9	0.19–0.56
1–Methyl–3–ethylbenzene	9	0.31–2.86
1–Methyl–2–n–propylbenzene	10	0.01–0.17
1–Methyl–3–n–propylbenzene	10	0.08–0.56
1–Methyl–3–i–propylbenzene (Cymene)	10	0.01–0.12
1–Methyl–3–t–butylbenzene	11	0.03–0.11
1–Methyl–4–t–butylbenzene	11	0.04–0.13
1,2–Dimethyl–3–ethylbenzene	10	0.02–0.19
1,2–Dimethyl–4–ethylbenzene	10	0.05–0.73
1,3–Dimethyl–2–ethylbenzene	10	0.21–0.59
1,3–Dimethyl–4–ethylbenzene	10	0.03–0.44
1,3–Dimethyl–5–ethylbenzene	10	0.11–0.42
1,4–Dimethyl–2–ethylbenzene	10	0.05–0.36
1,3–Dimethyl–5–t–butylbenzene	12	0.02–0.16
1,2,3–Trimethylbenzene	9	0.21–0.48
1,2,4–Trimethylbenzene (Pseudocumene)	9	0.66–3.30
1,2,3–Trimethylbenzene (Mesitylene)	9	0.13–1.15
1,2,3,4–Tetramethylbenzene	10	0.02–0.19
1,2,3,5–Tetramethylbenzene	10	0.14–1.06
1,2,4,5–Tetramethylbenzene	10	0.05–0.67
Ethylbenzene	8	0.36–2.86
1,2–Diethylbenzene	10	0.57
1,3–Diethylbenzene	10	0.05–0.38
n–Propylbenzene	9	0.08–0.72
Isopropylbenzene (Cumene)	9	0.01–0.23
n–Butylbenzene	10	0.04–0.44

Compound	Number of Carbons	Weight Percent
Alkylbenzenes (continued)		
Isobutylbenzene	10	0.01–0.08
sec–Butylbenzene	10	0.01–0.13
t–Butylbenzene	10	0.12
t–Pentylbenzene	11	0.01–0.14
Isopentylbenzene	11	0.07–0.17
Other Aromatics		
Indan	9	0.25–0.34
1–Methylindan	10	0.04–0.17
2–Methylindan	10	0.04–0.17
4–Methylindan	10	0.01–0.16
5–Methylindan	10	0.09–0.30
Tetralin (Tetrahydronaphthalene)	10	0.01–0.14
Polynuclear Aromatic Hydrocarbons		
Naphthalene	10	0.09–0.49
Pyrene	16	n/a
Benz(a)anthracene	18	n/a
Bena(a)pyrene	20	0.19–2.8 mg/Kg
Benzo(e)pyrene	20	n/a
Benzo(g,h,i)perylene	21	n/a
Elements		
Bromine		80,000–34,500 µg/Kg
Cadmium		1–7 µg/Kg
Chlorine		80,000–30,000 µg/Kg
Lead		530–1,120 mg/Kg
Sodium		600–1,400 µg/Kg
Sulfur		0.10–0.15 (ASTM)
Vanadium		0.1–20 µg/Kg

Table A–3. Chromatograms of Representative Petroleum Products.
The following pages show chromatograms of gasoline, diesel, and other petroleum products. The chromatograms were obtained either onsite or from the mass spectroscopy laboratory at the Colorado School of Mines or from the Spectra Catalog of the Alcohol, Tobacco, and Firearms Agency, which is a tabulation of chromatograms used in arson investigations. Partially evaporated spectra are presented since they are similar to what would be obtained from aged gasoline contaminated soils.

Compound	Retention Time (Mins.)	Boiling Point (C°)	Compound	Retention Time (Mins.)	Boiling Point (C°)	Compound	Retention Time (Mins.)	Boiling Point (C°)
C_5	0.49	36.1	C_{13}	4.98	235.4	$C_{21}(a)$	8.38	356.5
C_6	0.69	68.95	C_{14}	5.49	253.7	$C_{22}(a)$	8.72	368.6
C_7	1.12	98.42	C_{15}	5.97	270.63	$C_{23}(a)$	9.03	380.2
C_8	1.79	125.66	C_{16}	6.43	287.0	$C_{24}(a)$	9.38	391.3
C_9	2.52	150.79	C_{17}	6.89	301.8	$C_{25}(a)$	9.77	401.9
C_{10}	3.21	174.1	$C_{18}(a)$	7.27	316.1			
C_{11}	3.85	195.9	$C_{19}(a)$	7.64	329.7			
C_{12}	4.44	216.3	$C_{20}(a)$	8.03	343			

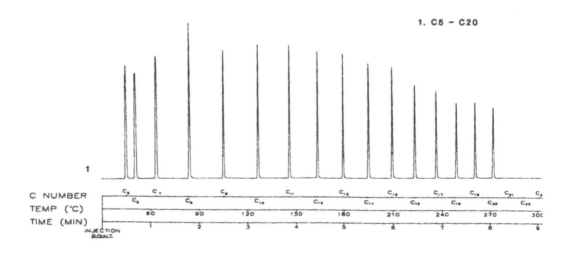

1. C5 – C20

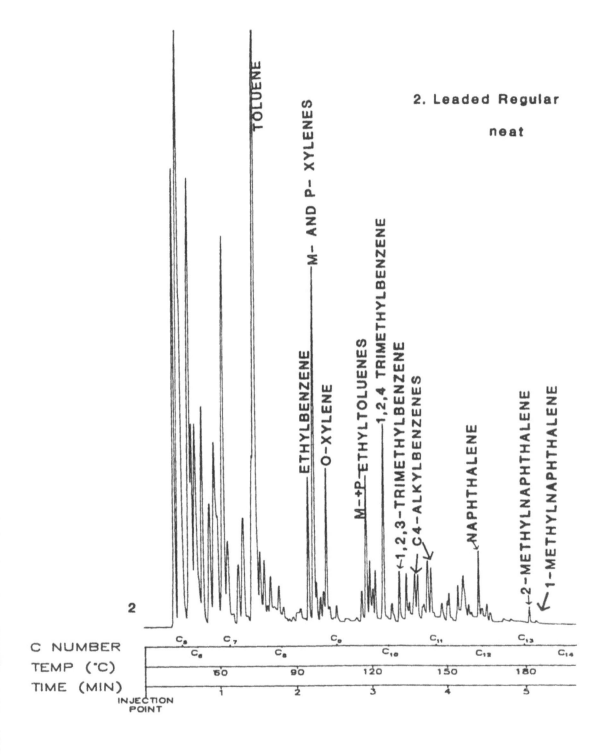

2. Leaded Regular

neat

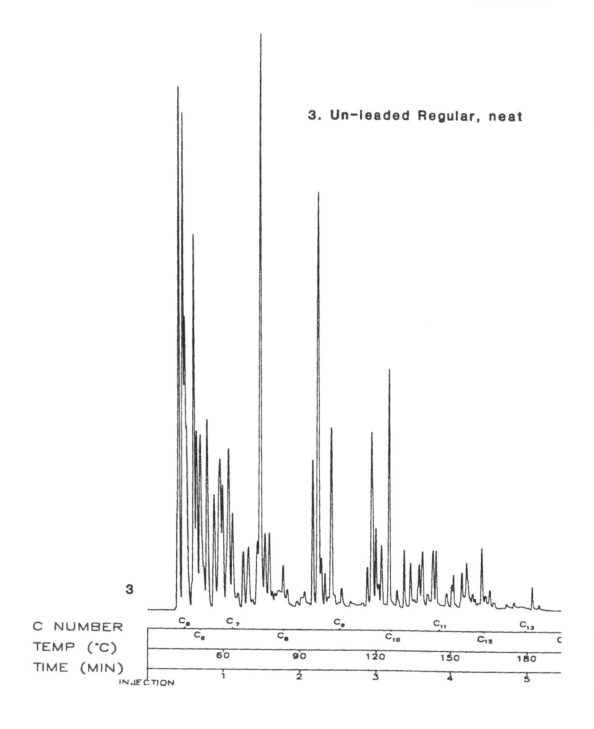

3. Un-leaded Regular, neat

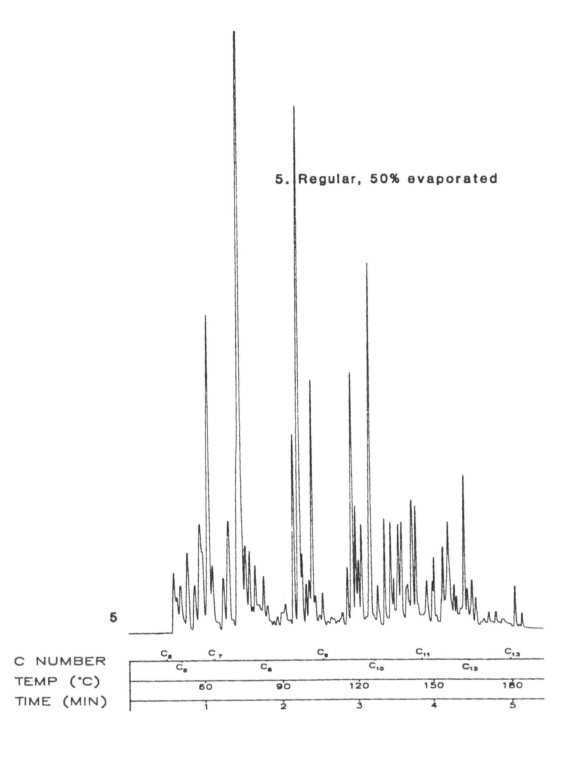

5. Regular, 50% evaporated

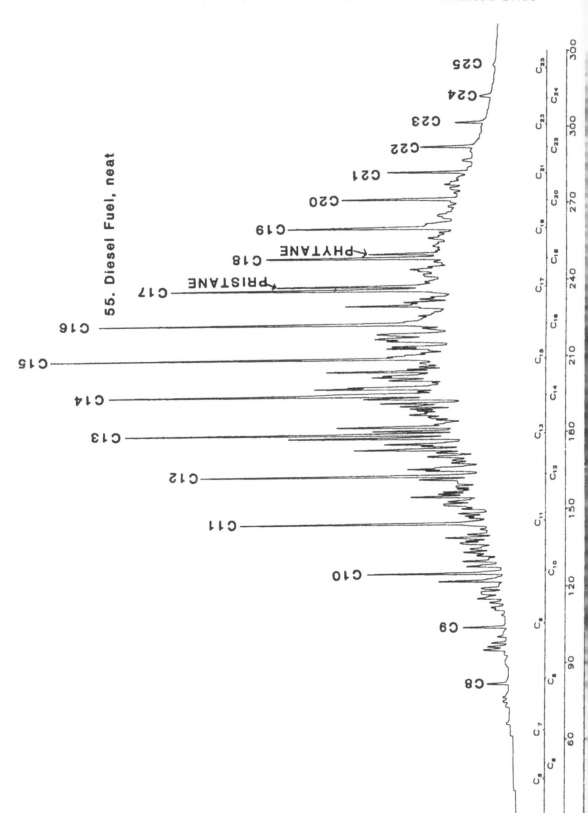

55. Diesel Fuel, neat

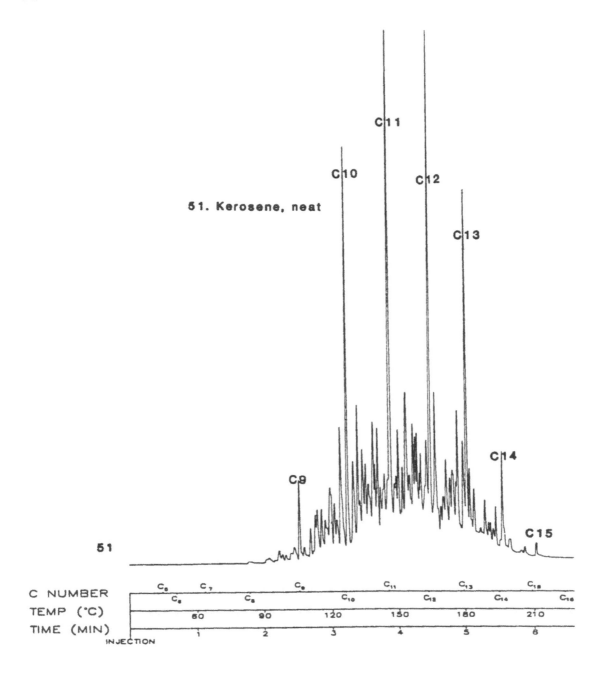

51. Kerosene, neat

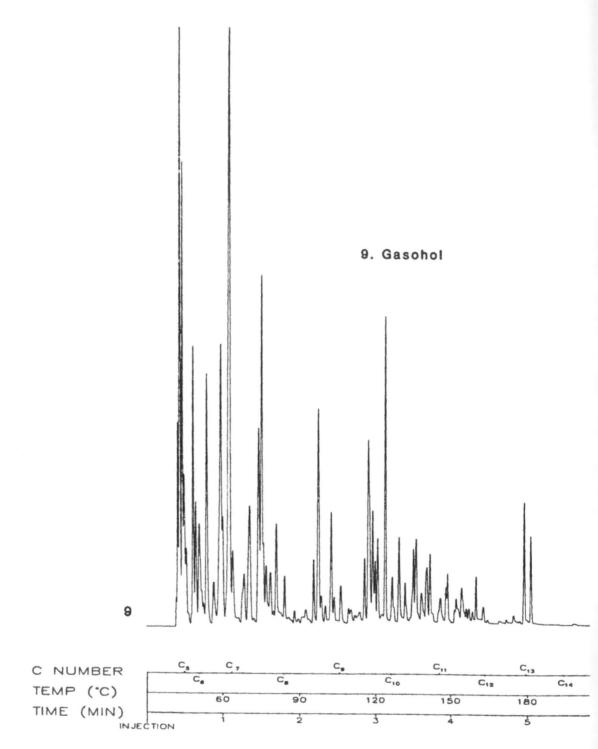

9. Gasohol

C NUMBER						
TEMP (°C)						
TIME (MIN)						

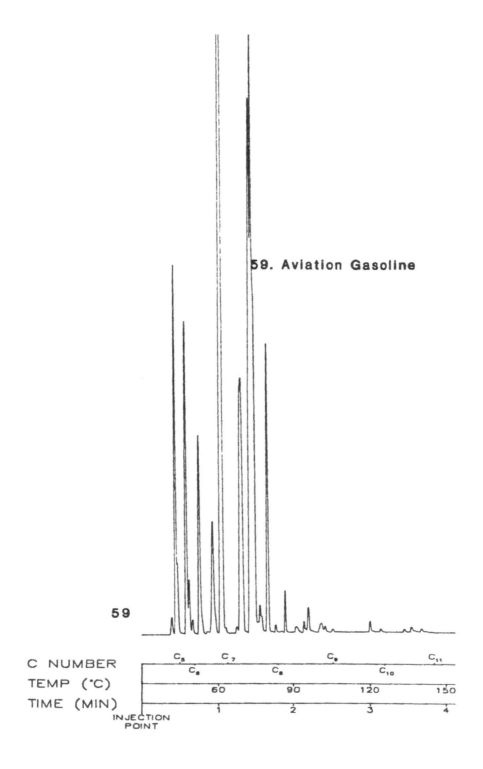

59. Aviation Gasoline

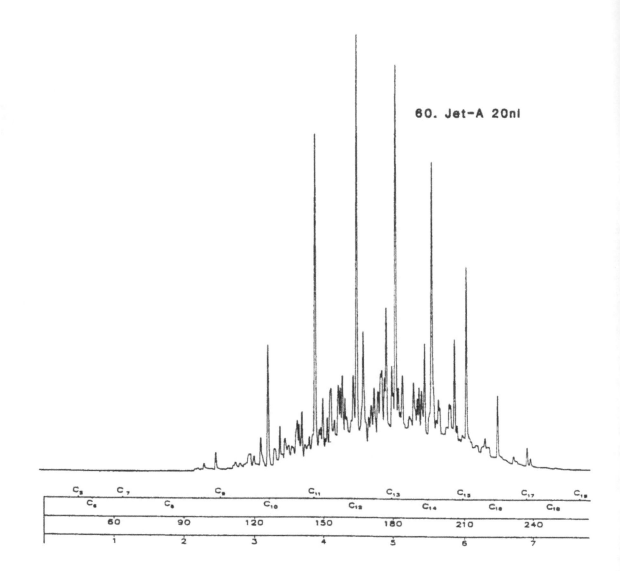

60. Jet-A 20nl

Appendix B: Summary of Federal Regulations

Petroleum Storage Sites

The authority to establish regulations for the operation of underground storage tank facilities is provided by Subtitle I of the 1984 Hazardous and Solid Waste Amendments to the Resource Conservation and Recovery Act. The regulations are published in the Code of Federal Regulations, Title 40, Section 280; 40 CFR §280. The first five subparts detail the regulations for operating a UST system. Subparts E, F, and G, quoted here relate to the requirements for dealing with a release, either a spill or a leak from an underground system.

The petroleum exemption applies to all petroleum based products, including those that contain a listed or characteristic hazardous substance. The exemption also applies to petroleum contaminated soils and debris from excavation of a petroleum contaminated site. However, if the soil or groundwater is contaminated with listed or characteristic hazardous wastes not derived directly from petroleum, then the soil must be treated as a hazardous material, not as a petroleum product.

Subpart E — Release Reporting, Investigation, and Confirmation

§280.50 Reporting of releases.

Owners and operators of UST systems must report to the implementing agency within 24 hours, or another reasonable time period specified by the implementing agency, and follow the procedures in §280.52 for any of the following conditions:

(a) The discovery by owners and operators or others of released regulated substances at the UST site or in the surrounding area (such as the presence of free product or vapors in soils, basements, sewer and utility lines, and nearby surface water).

(b) Unusual operating conditions observed by owners and operators (such as the erratic behavior of product dispensing equipment, the sudden loss of product from the UST system, or an unexplained presence of water in the tank), unless system equipment is found to be defective but not leaking, and is immediately repaired or replaced; and,

(c) Monitoring results from a release detection method required under §280.41 and §280.42 that indicate a release may have occurred unless:

   (1) The monitoring device is found to be defective, and is immediately repaired, recalibrated or replaced, and additional monitoring does not confirm the initial result; or

   (2) In the case of inventory control, a second month of data does not confirm the initial result.

§280.51 Investigation due to off–site impacts.

When required by the implementing agency, owners and operators of UST systems must follow the procedures in §280.52 to determine if the UST system is the source of off-site impacts. These impacts include the discovery of regulated substances (such as the presence of free product or vapors in soils, basements, sewer and utility lines, and nearby surface and drinking waters) that has been observed by the implementing agency or brought to its attention by another party.

§280.52 Release investigation and confirmation

Unless corrective action is initiated in accordance with Subpart F, owners and operators must immediately investigate and confirm all suspected releases of regulated substances requiring reporting under §280.50 within 7 days, or another reasonable time period specified by the implementing agency, using either the following steps or another procedure approved by the implementing agency:

(a) *System test.* Owners and operators must conduct tests (according to the requirements for tightness testing in §280.43(c) and §280.44(b)) that determine whether a leak exists in that portion of the tank that routinely contains product, or the attached delivery piping, or both.

   (1) Owners and operators must repair, replace or upgrade the UST system, and begin corrective action in accordance with Subpart F if the test results for the system, tank, or delivery piping indicate that a leak exists.

   (2) Further investigation is not required if the test results for the system, tank, and delivery piping do not indicate that a leak exists and if environmental contamination is not the basis for suspecting a release.

   (3) Owners and operators must conduct a site check as described in paragraph (b) of this section if the test results for the system, tank, and

delivery piping do not indicate that a leak exists but environmental contamination is the basis for suspecting a release.

(b) *Site check.* Owners and operators must measure for the presence of a release where contamination is most likely to be present at the UST site. In selecting sample types, sample locations, and measurement methods, owners and operators must consider the nature of the stored substance, the type of initial alarm or cause for suspicion, the type of backfill, the depth of ground water, and other factors appropriate for identifying the presence and source of the release.

    (1) If the test results for the excavation zone or the UST site indicate that a release has occurred, owners and operators must begin corrective action in accordance with Subpart F;

    (2) If the test results for the excavation zone or the UST site do not indicate that a release has occurred, further investigation is not required.

§280.53 Reporting and cleanup of spills and overfills.

(a) Owners and operators of UST systems must contain and immediately clean up a spill or overfill and report to the implementing agency within 24 hours, or another reasonable time period specified by the implementing agency, and begin corrective action in accordance with Subpart F in the following cases:

    (1) Spill or overfill of petroleum that results in a release to the environment that exceeds 25 gallons or another reasonable amount specified by the implementing agency, or that causes a sheen on nearby surface water; and

    (2) Spill or overfill of a hazardous substance that results in a release to the environment that equals or exceeds its reportable quantity under CERCLA (40 CFR Part 302).

(b) Owners and operators of UST systems must contain and immediately clean up a spill or overfill of petroleum that is less than 25 gallons or another reasonable amount specified by the implementing agency, and a spill or overfill of a hazardous substance that is less than the reportable quantity. If cleanup cannot be accomplished within 24 hours, or another reasonable time period established by the implementing agency, owners and operators must immediately notify the implementing agency.

[Note: Pursuant to §§302.6 and 355.40, a release of a hazardous substance equal to or in excess of its reportable quantity must also be reported immediately (rather than within 24 hours) to the National Response Center under sections 102 and 103 of the Comprehensive Environmental Response, Compensation, and Liability Act of 1980 and to appropriate state and local authorities under Title Ill of the Superfund Amendments and Reauthorization Act of 1986.1]

Subpart F — Release Response and Corrective Action for UST Systems Containing Petroleum or Hazardous Substances

§280.60 General.

Owners and operators of petroleum or hazardous substance UST systems must, in response to a confirmed release from the UST system, comply with the requirements of this subpart except for USTs excluded under §28010(b) and UST systems subject to RCRA Subtitle C corrective action requirements under section 3004(u) of the Resource Conservation and Recovery Act, as amended.

§280.61 Initial response

Upon confirmation of a release in accordance with §280.52 or after a release from the UST system is identified in any other manner, owners and operators must perform the following initial response actions within 24 hours of a release or within another reasonable period of time determined by the implementing agency:

(a) Report the release to the implementing agency (e.g., by telephone or electronic mail);

(b) Take immediate action to prevent any further release of the regulated substance into the environment; and

(c) Identify and mitigate fire, explosion, and vapor hazards.

§280.62 Initial abatement measures and site check.

(a) Unless directed to do otherwise by the implementing agency, owners and operators must perform the following abatement measures:

(1) Remove as much of the regulated substance from the UST system as is necessary to prevent further release to the environment;

(2) Visually inspect any aboveground releases or exposed belowground releases and prevent further migration of the released substance into surrounding soils and ground water;

(3) Continue to monitor and mitigate any additional fire and safety hazards posed by vapors or free product that have migrated from the UST excavation zone and entered into subsurface structures (such as sewers or basements);

(4) Remedy hazards posed by contaminated soils that are excavated or exposed as a result of release confirmation, site investigation, abatement, or corrective action activities. If these remedies include treatment or disposal of soils, the owner and operator must comply with applicable state and local requirements;

(5) Measure for the presence of a release where contamination is most likely to be present at the UST site, unless the presence and source of the release have been confirmed in accordance with the site check required by §280.52(b) or the closure site assessment of §280.72(a). In selecting sample types, sample locations, and measurement methods, the owner and operator must consider the nature of the stored substance, the type of backfill, depth to ground water and other factors as appropriate for identifying the presence and source of the release; and

(6) Investigate to determine the possible presence of free product, and begin free product removal as soon as practicable and in accordance with §280.64.

(b) Within 20 days after release confirmation, or within another reasonable period of time determined by the implementing agency, owners and operators must submit a report to the implementing agency summarizing the initial abatement steps taken under paragraph (a) of this section and any resulting information or data.

§280.63 Initial site characterization.

(a) Unless directed to do otherwise by the implementing agency, owners and operators must assemble information about the site and the nature of the release, including information gained while confirming the release or completing the initial abatement measures in §280.60 and §280.61. This information must include, but is not necessarily limited to the following:

(1) Data on the nature and estimated quantity of release;

(2) Data from available sources and/or site investigations concerning the following factors: surrounding populations, water quality, use and approximate locations of wells potentially affected by the release, subsurface soil conditions, locations of subsurface sewers, climatological conditions, and land use;

(3) Results of the site check required under §280.62(a)(5); and

(4) Results of the free product investigations required under §280.62(a)(6), to be used by owners and operators to determine whether free product must be recovered under §280.64.

(b) Within 45 days of release confirmation or another reasonable period of time determined by the implementing agency, owners and operators must submit the information collected in compliance with paragraph (a) of this section to the implementing agency in a manner that demonstrates its applicability and technical adequacy, or in a format and according to the schedule required by the implementing agency.

§280.64 Free product removal.

At sites where investigations under §280.62(a)(6) indicate the presence of free product, owners and operators must remove free product to the maximum extent practicable as determined by the implementing agency while continuing, as necessary, any actions initiated under §§280.61 through 280.63, or preparing for actions required under §§280.65 through 280.66. In meeting the requirements of this section, owners and operators must:

(a) Conduct free product removal in a manner that minimizes the spread of contamination into previously uncontaminated zones by using recovery and disposal techniques appropriate to the hydrogeologic conditions at the site, and that properly treats, discharges or disposes of recovery byproducts in compliance with applicable local, State and Federal regulations;

(b) Use abatement of free product migration as a minimum objective for the design of the free product removal system;

(c) Handle any flammable products in a safe and competent manner to prevent fires or explosions; and

(d) Unless directed to do otherwise by the implementing agency, prepare and submit to the implementing agency, within 45 days after confirming a release, a free product removal report that provides at least the following information:

(1) The name of the person(s) responsible for implementing the free product removal measures;

(2) The estimated quantity, type, and thickness of free product observed or measured in wells, boreholes, and excavations;

(3) The type of free product recovery system used;

(4) Whether any discharge will take place onsite or offsite during the recovery operation and where this discharge will be located;

(5) The type of treatment applied to, and the effluent quality expected from, any discharge;

(6) The steps that have been or are being taken to obtain necessary permits for any discharge; and

(7) The disposition of the recovered free product.

§280.65 Investigations for soil and groundwater cleanup.

(a) In order to determine the full extent and location of soils contaminated by the release and the presence and concentrations of dissolved product contamination in the ground water, owners and operators must conduct investigations of the release, the release site, and the surrounding area possibly affected by the release if any of the following conditions exist:

(1) There is evidence that groundwater wells have been affected by the release (e.g., as found during release confirmation or previous corrective action measures);

(2) Free product is found to need recovery in compliance with §280.64;

(3) There is evidence that contaminated soils may be in contact with ground water (e.g.; as found during conduct of the initial response measures or investigations required under §§280.60 through 280.64);

and

(4) The implementing agency requests an investigation, based on the potential effects of contaminated soil or ground water on nearby surface water and groundwater resources.

(b) Owners and operators must submit the information collected under paragraph (a) of this section as soon as practicable or in accordance with a schedule established by the implementing agency.

§280.66 Corrective action plan.

(a) At any point after reviewing the information submitted in compliance with §280.61 through §280.63, the implementing agency may require owners and operators to submit additional information or to develop and submit a corrective action plan for responding to contaminated soils and ground water. If a plan is required, owners and operators must submit the plan according to a schedule and format established by the implementing agency. Alternatively, owners and operators may, after fulfilling the requirements of §280.61 through §280.63, choose to submit a corrective action plan for responding to contaminated soil and ground water. In either case, owners and operators are responsible for submitting a plan that provides for adequate protection of human health and the environment as determined by the implementing agency, and must modify their plan as necessary to meet this standard.

(b) The implementing agency will approve the corrective action plan only after ensuring that implementation of the plan will adequately protect human health, safety, and the environment. In making this determination, the implementing agency should consider the following factors as appropriate:

(1) The physical and chemical characteristics of the regulated substance, including its toxicity, persistence, and potential for migration;

(2) The hydrogeologic characteristics of the facility and the surrounding area;

(3) The proximity, quality, and current and future uses of nearby surface water and ground water;

(4) The potential effects of residual contamination on nearby surface water and ground water;

(5) An exposure assessment; and

(6) Any information assembled in compliance with this subpart.

(c) Upon approval of the corrective action plan or as directed by the implementing agency, owners and operators must implement the plan, including modifications to the plan made by the implementing agency. They must monitor, evaluate, and report the results of implementing the plan in accordance with a schedule and in a format established by the implementing agency.

(d) Owners and operators may, in the interest of minimizing environmental contamination and promoting more effective cleanup, begin cleanup of soil and ground water before the corrective action plan is approved provided that they:

(1) Notify the implementing agency of their intention to begin cleanup;

(2) Comply with any conditions imposed by the implementing agency, including halting cleanup or mitigating adverse consequences from cleanup activities;

and

(3) Incorporate these self-initiated cleanup measures in the corrective action plan that is submitted to the implementing agency for approval.

§ 280.67 Public participation.

(a) For each confirmed release that requires a corrective action plan, the implementing agency must provide notice to the public by means designed to reach those members of the public directly affected by the release and the planned corrective action. This notice may include, but is not limited to, public notice in local newspapers, block advertisements, public service announcements, publication in a state register, letters to individual households, or personal contacts by field staff.

(b) The implementing agency must ensure that site release information and decisions concerning the corrective action plan are made available to the public for inspection upon request.

(c) Before approving a corrective action plan, the implementing agency may hold a public meeting to consider comments on the proposed corrective action plan if there is sufficient public interest, or for any other reason.

(d) The implementing agency must give public notice that complies with paragraph (a) of this section if implementation of an approved corrective action plan does not achieve the established cleanup levels in the plan and termination of that plan is under consideration by the implementing agency.

Subpart G — Out–of–Service UST Systems and Closure

§280.70 Temporary closure.

(a) When an UST system is temporarily closed, owners and operators must continue operation and maintenance of corrosion protection in accordance with §280.31, and any release detection in accordance with Subpart D. Subparts E and P must be complied with if a release is suspected or confirmed. However, release detection is not required as long as the UST system is empty. The UST system is empty when all materials have been removed using commonly employed practices so that no more than 2.5 centimeters (one inch) of residue, or 0.3 percent by weight of the total capacity of the UST system, remain in the system.

(b) When an UST system is temporarily closed for 3 months or more, owners and operators must also comply with the following requirements:

(1) Leave vent lines open and functioning;

and

(2) Cap and secure all other lines, pumps, manways, and ancillary equipment.

(c) When an UST system is temporarily closed for more than 12 months, owners and operators must permanently close the UST system if it does not meet either performance standards in §280.20 for new UST systems or the upgrading requirements in §280.21, except that the spill and overfill equipment requirements do not have to be met. Owners and operators must permanently close the substandard UST systems at the end of this 12-month period in accordance with §§280.71 and 280.74, unless the implementing agency provides an extension of the 12-month temporary closure period. Owners and operators must complete a site assessment in accordance with §280.72 before such an extension can be applied for.

§280.71 Permanent closure and changes-in-service.

(a) At least 30 days before beginning either permanent closure or a change-in-service under paragraphs (b) and (c) of this section, or within another reasonable time period determined by the implementing agency, owners and operators must notify the implementing agency of their intent to permanently close or make the change–in–service, unless such action is in response to corrective action. The required assessment of the excavation zone under §280.72 must be performed after notifying the implementing agency but before completion of the permanent closure or a change-in-service.

(b) To permanently close a tank, owners and operators must empty and clean it by removing all liquids and accumulated sludges. All tanks taken out of service permanently must also be either removed from the ground or filled with an inert solid material.

(c) Continued use of an UST system to store a non-regulated substance is considered a change–in–service. Before a change-in-service, owners and operators must empty and clean the tank by removing all liquid and accumulated sludge and conduct a site assessment in accordance with §280.72.

[Note: The following cleaning and closure procedures may be used to comply with this section:

(A) American Petroleum Institute Recommended Practice 1604, *Removal and Disposal of Used Underground Petroleum Storage Tanks;*

(B) American Petroleum Institute Publication 2015, *Cleaning Petroleum Storage Tanks;*

(C) American Petroleum Institute Recommended Practice 1631, *Interior Lining of Underground Storage Tanks,* may be used as guidance for compliance with this section; and

(D) The National Institute for Occupational Safety and Health *Criteria for a Recommended Standard for Working in Confined Space,* may be used as guidance for conducting safe closure procedures at some hazardous substance tanks.

§280.72 Assessing the site at closure or change–in–service.

(a) Before permanent closure or a change–in–service, owners and operators must measure for the presence of a release where contamination is most likely to be present at the UST site. In selecting sample types, sample locations, and measurement methods, owners and operators must consider the method of closure, the nature of the stored substance, the type of backfill, the depth to ground water, and other factors appropriate for identifying the presence of a release. The requirements of this section are satisfied if one of the external release detection methods allowed in §280.43(e) and (f) is operating in accordance with the requirements in §280.43 at the time of closure, and indicates no release has occurred.

(b) If contaminated soils, contaminated ground water, or free product as a liquid or vapor is discovered under paragraph (a) of this section, or by any other manner, owners and operators must begin corrective action in accordance with Subpart F.

§280.73 Applicability to previously closed UST systems.

When directed by the implementing agency, the owner and operator of an UST system permanently closed before December 22, 1988 must assess the excavation zone and close the UST system in accordance with this Subpart if releases from the UST may, in the judgment of the implementing agency, pose a current or potential threat to human health and the environment.

§280.74 Closure records

Owners and operators must maintain records in accordance with §280.34 that are capable of demonstrating compliance with closure requirements under this Subpart. The results of the excavation zone assessment required in §280.72 must be maintained for at least 3 years after completion of permanent closure or change–in–service in one of the following ways:

(a) By the owners and operators who took the UST system out of service;

(b) By the current owners and operators of the UST system site; or

(c) By mailing these records to the implementing agency if they cannot be maintained at the closed facility.

Appendix C: The Unified Soil Classification System

One of the more important pieces of information that goes into the CAP is the types and distribution of soils in the vadose zone. This information is essential to estimate contaminant migration rates, potential threats to public health and safety, off–site impacts, need for containment measures, and potential for liability.

The Unified Soil Classification System, used by soil geologists to classify soils, is presented in **Figure C–2**. For reference a generalized soil column is given in **Figure C–1**. This figure is the same as **Figure IV–1** from Chapter IV. Finally, **Figure C–3** gives the classification scheme for fine–grained soils.

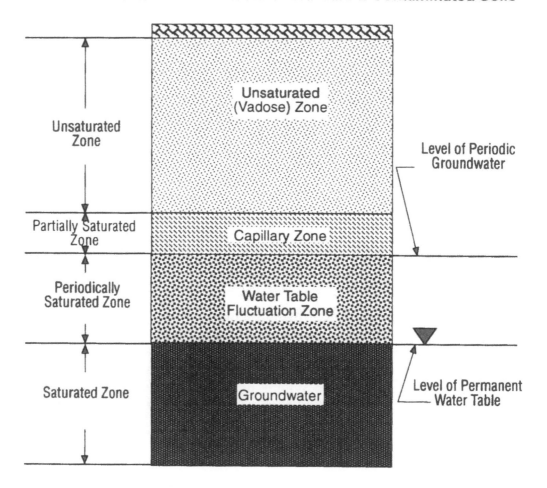

Figure C-1. Generalized Soil Column. The vadose zone is normally unsaturated, even in periods of heavy runoff or snow melt. The saturated zone is the highest level of permanent groundwater. Intermediate zones include the capillary fringe, which has bulk liquid water as the major phase; the fluctuation zone is saturated on a seasonal or occasional basis. Hydrocarbon mobility decreases as saturation increases.

Figure C–2. (Facing page) Unified Soil Classification System. An integral part of a site assessment, well logs will contain soil classification notation. Soils should be classified as completely as possible according to physical parameters and appearance. Example: Silty sand – gravelly; about 20% hard angular gravel particles ⅓ inch maximum size; rounded and subangular sand grains coarse to fine; about 15% nonplastic fines with low dry strength; well–compacted and moist; alluvial sand; (SW).

Unified Soil Classification System

Major Divisions			Symbol	Typical Names	Field Descriptions
Coarse-Grained Soils — More than half of material is larger than no. 200 sieve size.	Gravels — More than half of the coarse fraction is larger than no. 4 sieve size.	Clean Gravels (No fines)	GW	Well-graded gravels, gravel-sands mixtures, little or no fines	Name; Approx. % sand & gravels; Size; Angularity; Surface conditions; Hardness of coarse grains; Local or other name; Other descriptive information; Symbol; In undisturbed soils stratification, compactness, cementation, moisture content, drainage characteristics
			GP	Poorly-graded gravels, gravel-sand mixtures, little or no fines	
		Gravels with fines	GM	Silty gravels, poorly graded gravel-sand-silt mixtures	
			GC	Clayey gravels, poorly graded gravel-sand-clay mixtures	
	Sands — More than half of the coarse fraction is smaller than no. 4 sieve size	Clean Sands (No fines)	SW	Well-graded sands, gravelly sands, little or no fines	
			SP	Poorly graded sands, gravelly sands, little or no fines	
		Sands with fines	SM	Silty sands, sand-silt mixtures	
			SC	Clayey sands, sand-clay mixtures	
Fine-Grained Soils — More than half of the material is smaller than no. 200 sieve.	Silts and Clays	Low Liquid Limit (<50)	ML	Inorganic silts and very fine sands, rock flour, silty or clayey fine sands, or clayey silts, with slight plasticity	Name; Degree of plasticity; Size of coarse grains; Color of wet sample; Odor, if any; Local or other name; Other descriptive information; Symbol; In undisturbed soils structure, stratification, compactness, cementation, moisture content, drainage characteristics
			CL	Inorganic clays of low to medium plasticity, gravelly clays, sandy clays, silty clays, lean clays	
			OL	Organic silts and organic silty clays of low plasticity	
		High Liquid Limit (>50)	MH	Inorganic silts, micaceous or diatomaceous fine sandy or silty soils, elastic silts	
			CH	Inorganic clays of high plasticity, fat clays	
			OH	Organic clays of medium to high plasticity, organic silts	
			Pt	Peat and other highly organic silts	

Notes: (to accompany **Figure C–2**)

1. Boundary Classification: Soils possessing characteristics of two groups are designated by combinations of group symbols. For example, GW–GC, for well–graded gravel–sand mixtures with clay binder.

2. All sieve sizes on this chart are U.S. Standard.

3. The terms *silt* and *clay* are used, respectively, to distinguish materials exhibiting lower plasticity from those with higher plasticity. The minus no. 200 sieve material (0.074 mm grain size) is silt if the liquid limit and plasticity index plot below the "A" line on the plasticity chart and is clay if the liquid limit and plasticity index plot above the "A" line on the chart (see **Figure C–3**).

4. For a complete description of the Unified Soil Classification System, see *Technical Memorandum No. 3-357,* available from Office, Chief of Engineers, Waterways Equipment Station, Vicksburg, MS, March, 1953.

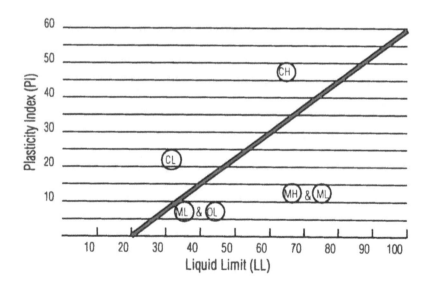

Figure C–3. Classification Chart for Fine–Grained Soils.

Appendix D: Documents for Environmental Sampling

One of the most important steps in collecting information about a site is proper documentation. Documentation is important for several reasons:

- Memory failure. In the time that elapses between collecting information and writing the report it is easy to forget what was done or why it was done. Documentation is an essential memory aid. A videotape is an excellent supplement to completed forms.

- Justification of billing. If the client questions a bill or parts of a bill, documentation is available for support and justification. Documentation does not guarantee clients will not complain, it just helps explain the reasons.

- Support for depositions and testimony. Sooner or later most sites end up in dispute; if not actual litigation, at least attorney–negotiated settlements. Documentation is absolutely essential to support arguments for the good guys and refute arguments for the bad guys.

Documentation is simplified by having the appropriate forms available for use. As information is gathered the appropriate form should be completed on the spot, not later back at the office. For example, sample forms should be completed as the individual samples are being collected. Chain–of–custody must be completed on the spot.

Advances in Geopositioning Satellite System (GPS) instruments make well location according to latitude and longitude accurate to within a meter. Wells can be located more accurately than the USGS topographical maps used for reference.

Forms should include as many variations and circumstances as possible. That is, documentations forms should be "overcomplete." Information filled out on the back of an envelope is usually unacceptable. One problem encountered by many analytical labs is that clients do not always know the correct analysis to request. State agencies are increasingly specific about the method and technique required for specific data. Method 418.1 is one that is often modified.

Therefore, sample request forms should have either very specific state or EPA Method numbers, or specific analyses, such as BTEX–GC, or BTEX–GC/MS. Alternatively, sample requests can be combined with Chain–of–Custody forms.

Several sample documents are presented in this Appendix. No claims are made regarding completeness, nor are the forms entirely original. Most have been kindly provided by environmental companies and have been used as is, or have been adapted.

Figure D–1. (Below and following pages) Sample Documents. The first document is a typical well log that should be completed as the bore is being drilled. The second is the crucial Chain–of Custody record that must be completed properly. Finally, there is a sample request form. Individual labs may have their own.

Environmental Sample Chain of Custody Record

ABC Environmental, Inc.
Environmental Sampling Division
1234 West 100th Ave.
West City, Colorado 88888
(303)555-1234

Page _____ of _____

Project Name:

Project Description:

Project Invoice Number:

Client _____

Client Contact: _____

Sample Destination: _____

Project Location:

Sampler #1 (Signature) _____ Date _____

Sampler #2 (Signature) _____ Date _____

Analysis columns:
- Volatile Organic Compounds
- Volatile Aromatic Hydrocarbons
- Acid Extractable Compounds
- Base/Neutral Extractable Cmpds
- Pesticides
- PCBs
- Metals
- Total Petroleum Hydrocarbons
- Phenols
- TCLP — Organics
- TCLP — Inorganics
- Ignitability
- Corrosivity
- Reactivity
- Other-Specify

Sample No.	Reference	Date	Time	Sample Description	(analysis columns)
			AM/PM		

Sample seals intact at destination? Yes No None

Initials

Remarks

Relinquished by (Signature): _____ Date _____ Time _____ AM/PM   Received by (Signature): _____ Date _____ Time _____ AM/PM

Relinquished by (Signature): _____ Date _____ Time _____ AM/PM   Received by (Signature): _____ Date _____ Time _____ AM/PM

ABC Environmental Co., Inc.
1234 Main Street
City, State 00000

SAMPLE ANALYSIS REQUEST FORM

Contract
Number _____

Sample Location Data

Facility Name	Ref. Grid	Complete address
Sampling Location		

Sample Collector

Sampler #1 Name	Signature	Date
Sampler #2 Name	Signature	Date

Sample Identification

Sample Number	Sampling Date	Longitude
Sample Point ID	Time Sampling Start	Latitude
Sample Code	Time Sampling Finish	Elevation FT METERS (Circle One)

Sampling Process

Collection Method	Sample Type
Sampler Type	Sample Matrix
Sample Container	Preservative(s)
Well Diameter	Water Elevation FT METERS (Circle One)

Field Conditions

Sample Temperature °C	Matrix Temperature °C
Air Temperature (Start) °C	Storage Container Temp (Start) °C	Coolant
Air Temperature (Finish) °C	Storage Container Temp (Finish) °C	Weather Conditions

Analyses Requested

Parameters	Container No.	Detection Limits	Emergency YES/NO
1.			
2.			
3.			
4.			
5.			
6.			
7.			
8.			
9.			
10.			
11.			

Authorized by	Title	Date

Laboratory

Lab Name	Person Accepting Sample	Time
Lab ID #	Holding Times OK OVER (Circle One)	Date

Chain–of–Custody Implemented	YES NO	Initials

Appendix E: CERCLA Case Law

The following cases represent recent clarifications of some of the questions raised in Fleet Factors, and in lender liability. The list is not necessarily complete, but will provide relevant information. The research was done by Mike Glade, Esq.

Fleet Factors Decision

United States of America, Plaintiff,

vs.

Fleet Factors Corporation, et al., Defendants.
[821 F. Supp. 707 (S.D. GA, 1993)]

The United States sued textile printing plant's secured creditor under CERCLA seeking to recover response and enforcement costs associated with hazardous substance removal. The District Court held that:

(1) creditor was liable as owner or operator at the time of hazardous waste disposal, and

(2) creditor was not liable for having arranged for disposal of hazardous substances.

The secured creditor forfeits protection afforded by secured creditor exemption from CERCLA owner or operator liability if such creditor either holds ownership indicia other than primarily to protect security interest or be participating in management. The failed printing plant's secured creditor was entitled to protection of secured creditor exemption from CERCLA owner or operator liability with respect to the period between plant's shutdown and creditor's employment of industrial liquidator to auction plant's equipment and machinery; although creditor was only entity making affirmative decisions at plant during that time, creditor engaged only in activities reasonably related to accomplishing shipment of plant's remaining inventory and, to extent practicable, creditor acted as outsider interested only in seeing inventory sold.

Textile printing plant's secured creditor's protection under secured creditor exemption from CERCLA owner or operator liability was voided by postforeclosure handling of plant's chemical drums by industrial liquidator that creditor hired to auction plant's equipment and machinery; environmental threat posed by drums was apparent and serious, and handling of those drums was not done in accordance with National Contingency Plan (NCP) of under supervision of the NCP on–scene coordinator.

Lender Liability

U.S. vs. McLamb
[1993 U.S. App. LEXIS 23967 (4th Cir. Sept. 17, 1993)]

In *U.S. vs. McLamb*, the court affirmed the dismissal of a bank from a claim based upon the security interest exemption under CERCLA. The bank had purchased the subject property at a foreclosure sale. It subsequently took prompt steps to resell the property. Interestingly, the court did not address the EPA's lender liability rule, stating that the bank's protection from liability was clear under the CERCLA security interest exemption from the definition of owner contained in Section 101(20)(A) of CERCLA.

Petroleum Byproducts and The Petroleum Exclusion

Cose vs. Getty Oil Company
[4 F. 3rd (9th Cir. August 11, 1993)]

In *Cose vs. Getty Oil Company*, the Ninth Circuit elaborated on the well–established dichotomy between unadulterated, uncontaminated petroleum products (including leaded gasoline), which are excluded from CERCLA, and petroleum derived materials contaminated with other substances such as heavy metals or elevated levels of petroleum constituents, which are not exempt from CERCLA. In *Cose* the 9th Circuit held that because crude oil tank bottoms, which contained the listed substance Chrysene (a PAH), were clearly "waste" material, and not "petroleum," these tank bottoms were subject to CERCLA jurisdiction. This decision considered the fact that petroleum is one of the several "fractions" of crude oil expressly exempt from coverage. The decision is important because it defines the barrier between petroleum products, which are entitled to the petroleum exclusion, and petroleum wastes, which are not.

A pipeline pumping facility owned by Getty Oil Co. was located in Tracy, California. The sump for the pumping station was located on nearby property called the "Gravel Pit." About once a week, the crude oil tank bottoms from the station's storage tanks were drained and dumped into the Gravel Pit. Getty Oil closed the facility in 1968. In May 1974, the Coses purchased the Gravel Pit from Getty Oil. The complaint alleges that in 1987 a layer of "subsurface asphalt or tar–like material" was discover beneath a surface layer of topsoil. The petroleum material was found to contain 28 ppm Chrysene, a listed hazardous waste and a known carcinogen.

The Coses filed suit under CERCLA to recover "response costs" needed to clean up the property. In response Getty Oil contended that Getty Oil could not have disposed of a "hazardous substance" on the property because CERCLA excludes from its definition of hazardous substances crude oil tank bottoms. The district court agreed and the case was appealed.

The appellate court noted the following established points:

- The petroleum exclusion applies to crude oil, petroleum feedstocks, and refined petroleum products, even if a specifically listed hazardous substance, such as Chrysene, is indigenous to such products.

- The petroleum exclusion does apply to unrefined and refined gasoline even though certain of its indigenous components and certain additives during the refining process have themselves been designated as hazardous substances within the meaning of CERCLA.

- EPA does *not* consider materials such as waste oil to which listed CERCLA hazardous substances have been added to be within the petroleum exclusion. (Final Rule, April 4, 1985).

- If a specifically listed hazardous substance is indigenous to petroleum and is present as a result of the release of petroleum, such substance will fall within the petroleum exclusion unless it is present at a concentration level that exceeds the concentration level that naturally occurs in the petroleum product.

The Court's interpretation of crude oil tank bottoms is the following:

- Crude oil tank bottoms are comprised of water and sedimentary solids that settle out of the crude oil and create a layer of waste at the bottom of storage tanks.

- Crude oil bottoms are never "subjected to various refining processes" and are not used "for producing useful products." The substance is simply discarded waste.

Accordingly the district court's decision was reversed and damages were awarded.

The following cases are relevant, but are not related directly to topics in Chapter 5.

Parent/Subsidiary Liability

Lansford-Coaldale Joint Water Authority vs. Tonolli Corp.
[4 F. 3d 1209 (3d Cir. Sept. 17, 1993)]

In *Lansford-Coaldale Joint Water Authority v. Tonolli Corp.*, the Third Circuit sided with a growing majority of courts holding that the appropriate test for "operator" liability of a parent or shareholder corporation is whether it exercised "pervasive control" over the subsidiary entity, not the more sweeping "authority to control" test adopted by a few courts. Adopting what appears to be a growing trend, the court also refused to designate the sister company as an "owner" under traditional corporate veil piercing standards, although the court remanded the case for further factual findings as to whether sister corporation was an operator and whether the parent corporation was an owner or operator.

Mixed Wastes

Louisiana Pacific Corp. v. ASARCO, Inc.
[812 F. Supp. 1528 (E.D. Cal. 1992)]

Negating the impact of *U.S. v. Iron Mountain Mines, Inc.*, , the Ninth Circuit in *Louisiana Pacific Corp. v. ASARCO, Inc.*, [37 E.R.C. 1345 (9th Cir. Sept. 23, 1993)] held that those "Bevill wastes" excluded under one category of the definition "hazardous substances," [see 42 U.S.C. 9601(14)(C)], are nevertheless hazardous if separately listed constituents, such as heavy metals, are found in this material (in this case, smelter slag). In the prior Iron Mountain decision, a California district court judge ruled that Congress did intend to exempt mine waste from regulation under CERCLA by virtue of the Bevill exclusion. The Louisiana Pacific court, reviewing a lower court decision which had summarily sided with the previous majority view reflected in *Eagle-Picher Indus. v. EPA*, [759 F. 2d 922 (D.C. Cir. 1985)], relied in part on the Supreme Court's agency deference standard articulated in Chevron in upholding EPA's interpretation of the statute. The Iron Mountain decision was not mentioned.

United States v. Alcan Aluminum Corp.
[964 F. 2d 252 (3rd Cir. 1992)]
United States v. Alcan Aluminum Corp.
[990 F. 2d 711 (2d Cir. 1993)]

The complex rulings in *United States vs. Alcan Aluminum Corp.* and *United States vs. Alcan Aluminum Corp.*, combine to reinject causation into the CERCLA analysis "through the back door" by allowing a defendant to prove, at the liability stage of a case, that its waste either did not or could not contribute to the release or, if such contribution did occur, that the environmental harm caused by its waste was divisible and therefore subject to apportionment. *Alcan II* is noteworthy for holding also that a defendant who contributed no more than "background levels" of hazardous substances and did not "concentrate" can also escape liability. Adding an additional spin on these decisions, the court in *In the Matter of Bell Petroleum Services, Inc.,* [3 F. 3d 1889 (5th Cir. 1993)] also indicated that the "divisibility" inquiry should be resolved if possible at the liability stage, but noted that the district court should have discretion as to the best stage for such a determination. The court also held that the principles of the Restatement (2nd) of Torts, allowing a defendant to avoid the imposition on joint and several liability by proving the amount of harm it caused, was the appropriate approach. As importantly, the court in *Bell Petroleum* concluded that existence of several competing theories of apportionment was insufficient reason for the district court to reject all such theories, and found that sufficient evidence existed for apportionment based on one of the theories espoused by the defendant.

"Passive" Migration - Definition of "Disposal".

Joslyn Manufacturing Co. vs. T.L. James and Company
[Civil Action No. 87-2054, 1993, U.S. Dist. LEXIS 12343 (W.D. La. July 29, 1993)]

At the district court level, another court has recently sided with what appears to be the growing trend by finding that the passive migration of contaminants in soils and groundwater does not constitute "disposal" necessary for past owner/operator liability to attach. In *Joslyn Manufacturing Co. v. T.L. James and Company,* the court refused to impose Superfund liability on a prior owner of a contaminated site on the basis of the mere act of precipitation causing hazardous materials to leach through the soil. The court distinguished *Nurad vs. William E. Hooper and Sons,* [966 F. 2d 837 (4th Cir. 1992)], the leading passive migration case to the opposite effect, based on the fact that *Nurad* involved an underground storage tank that was actively leaking. Thus, a "passive landowner" who nevertheless owns property while a release or leak occurs might still be liable under the *Joslyn* rationale despite having done nothing during the ownership period to cause the leak or release.

Perhaps most striking point about the *Bell* decision, *supra*, was the separate holding that EPA's decision to develop a $300,000 water pollution treatment facility was arbitrary and capricious, since there was no basis in the record for suggesting that such a treatment plant was necessary. This aspect of *Bell* is also notable for its rejection of EPA's position that the recoverability of its costs hinged solely on consistency with the NCP, not on the reasonableness or necessity of expending these funds, a view largely embraced in *U.S. v. Hardage*, [982 F. 2d 1436 (10th Cir. 1992)].

Accuracy: The difference between measured and referenced values. Accuracy measures how close a series of measurements comes to a "true" value. Precision measures the variance in a group of measurements.

ACGIH: American Council of Governmental Industrial Hygienists

Acute toxicity: Harmful effects of a chemical or product that occur within a relatively short time frame (hours to months).

Additive interferences: Interferences caused by sample constituents which generate a signal that adds to the analyte model.

Aerobic: Indicates that oxygen, O_2, is required or is present.

Aliphatic hydrocarbons: The class of hydrocarbons that contain no aromatic rings. The class includes alkanes, alkenes, alkynes, and cyclic hydrocarbons.

Alkanes: The family of hydrocarbons composed of molecules that contain no carbon–carbon multiple bonds. Alkanes are also known as saturated or paraffin hydrocarbons.

Alkenes: The family of hydrocarbons composed of molecules that contain one or more carbon–carbon double bond. Members of this family are also known as olefins. Alkenes comprise a significant portion of gasoline range hydrocarbons.

Alkylaromatic: Aromatic compounds containing alkyl substituents. The more important compounds are benzene–based with one or more methyl, ethyl, propyl, or i-propyl substituents.

Alkynes: The family of hydrocarbons composed of molecules that contain one or more carbon-carbon triple bonds. Also called acetylenic hydrocarbons after the simplest member, acetylene. Alkynes are not significant in remedial operations.

Anaerobic: Indicates that oxygen, O_2, is not required or is absent.

Aromatic hydrocarbons: The family of hydrocarbons composed of molecules whose structures are based on benzene; that is, 6–membered unsaturated rings, with conjugated double bonds. Nearly all aromatic hydrocarbons are toxic to some extent. Aromatic and alkylaromatic hydrocarbons form a significant fraction of gasoline and diesel range hydrocarbons.

AST: Aboveground Storage Tank. Storage tanks for petroleum products that are not 10% or more underground. At the time of writing ASTs are not regulated at the federal level as USTs are. Regulations are promised, but not yet in place. A LAST is a Leaking Aboveground Storage Tank.

ASTM: Formerly American Society for Testing and Materials. The organization sets international standards for everything from steel to Phase I assessments. Recently the name was changed to simply ASTM to reflect a broader mandate and an international scope.

Background samples: Also called background controls. Samples collected from the surrounding environment to establish background levels of contamination. Because of the ubiquitous nature of petroleum, background controls are appropriate for petroleum contaminated sites. Background samples must be collected according to the same protocols as normal samples.

BACT: Best Available Control Technology. The most appropriate technology available for control of a waste.

BAT: Best Available Technology. The most appropriate technology available for disposal or treatment of a waste.

BDAT: Best Demonstrated Available Technology. In 1986 EPA issued regulations requiring that hazardous wastes be treated to levels achievable by BDAT. Congress has specifically voided the consideration of land treatment as BDAT by defining it as land disposal, which is prohibited. BDAT determines the post–treatment MCLs for contaminants. Note land treatment for petroleum contaminated soils is permitted because of the petroleum exclusion.

Bentonite: An aluminosilicate clay that is formed by the weathering of feldspars and, thus, is found around granitic formations. Bentonites swell when wet and from a dense cement–like material. It is commonly used to fill around well casings and in containment structures.

Benzene: C_6H_6. The prototype for a large class of aromatic hydrocarbons. Benzene is a listed hazardous substance unless it occurs in gasoline.

Blank: A sample that contains the analyte of interest but in other respects has the same composition as the actual sample. Used to determine the extent of contamination or error.

BTEX: Benzene, Toluene, Ethylbenzene, Xylene. A special type of volatile organic analysis that uses gas chromatography to analyze for the principal aromatic components of gasoline.

BOD: Biological Oxygen Demand. A type of pollutant defined in the Clean Water Act. BOD is a rough measure of the organic waste load in a stream or surface water. The greater the organic pollutant burden, the greater the demand for oxygen to oxide the waste, and the greater the BOD.

CH_3OH: Methanol. The simplest alcohol and one of the most toxic. Methanol has been added to gasoline to reduce CO production in non–attainment areas.

C_6H_6: Benzene. The prototype for a large class of aromatic hydrocarbons. Benzene is a listed hazardous substance unless it occurs in gasoline.

$C_6H_5CH_3$: Toluene or Methylbenzene. An aromatic hydrocarbon found in gasoline blends. More toxic than aliphatic hydrocarbons, but less toxic than benzene.

$C_6H_5CH_2CH_3$: Ethylbenzene. An alkylbenzene found in gasoline. Ethylbenzene is a specifically listed hazardous substance unless it occurs in gasoline.

$C_6H_4(CH_3)_2$: Xylenes. Alkylbenzenes found in gasoline. Xylenes is plural since there are three isomers. The xylenes are listed hazardous substances unless they occur in gasoline.

C_6H_5OH: Phenol. An aromatic alcohol that occurs naturally in petroleum and petroleum products. Phenol is also the prototype for a class of aromatic alcohols having similar chemical structures.

CAA: Clean Air Act of 1990. The latest in a series of air quality laws. CAA sets air quality standards and maximum contaminant levels, and mandates state implementation programs to achieve compliance.

CAP: Corrective Action Plan. A document required by federal regulations in the event of a confirmed release. Must include remedial plans and other relevant information.

Capillary zone: Also called the capillary fringe. The subsurface zone between the water table and the vadose zone that is increasingly saturated with increasing depth due to capillary forces drawing water up from the saturated zone.

Capture zone: The volume influenced by the negative pressure at a wellhead. It is the space from which contaminated air or water is drawn into an extraction well. The more porous the soil, the larger capture zone.

CDPHE: Colorado Department of Public Health and Environment. The new title of the Colorado Department of Health.

CERCLA: Comprehensive Emergency Response, Compensation and Liability Act. Passed in 1980 in response to discovery of widespread contamination by hazardous chemicals. CERCLA sets liability standards for environmental impairment and authorizes identification and remediation of abandoned waste sites.

CERCLIS: Comprehensive Emergency Response, Compensation and Liability Information System.

CFR: Code of Federal Regulations. The government document in which all federal regulations are published. Each Title or Chapter is concerned with a different federal department or agency. Title 40 is "protection of the environment."

Chronic toxicity: Harmful effects of a chemical or product that take a relatively long (years) time frame to appear.

CLP: Contract Laboratory Program. A program developed by the U.S. EPA to make uniform the quality of outside analytical services for Superfund remediation. CLP is not a typo for TCLP, and it is definitely not a "certification" program for laboratories. CLP simply means that on a given day a particular laboratory performed a set of analyses within allowable limits.

COD: Chemical Oxygen Demand. An approximate measure of the toxic chemical burden in a surface water. COD estimates the amount of oxygen required to oxidize toxic chemicals.

Combustible gas indicator: Also called an explosimeter. Detects whether the ambient atmosphere is at an explosive level.

Comparability: The confidence with which one data set can be compared to another.

Composite samples: A non–discrete sample composed of more than one specific sample collected at various sampling points and/or at different points in time. The practice of combining two or more samples to make an "average" sample. The difficulty with composite samples is whether or not the composite is representative of the sampled site. May not be acceptable to regulatory agencies.

Cone of depression: A cone–shaped depression that is formed in a water table when groundwater is removed.

Confidence interval: A range of values that can be declared with a specific degree of confidence to contain the correct or true value for a population.

Confined aquifer: An aquifer whose upper and/or lower boundaries are confined by an impermeable geologic formation; e.g., clay or shale.

Contamination: A substance inadvertently added to the sample during the sampling and/or the analytical process.

Control: A type of sample, of the environment or of a population, against which the results of a procedure are judged. In this book controls and blanks, which are two aspects of the same process, are used synonymously.

Convenience sample: A type of field sample chosen on the basis of accessibility, expedience, cost, efficiency, or other reasons not connected with sampling protocols. Convenience samples must be justified.

CWA: Clean Water Act of 1987. One of the major regulatory acts to control the spread of pollutants. CWA requires NPDES permits to discharge pollutants from a point source.

Cycloalkanes: A class of alkanes containing saturated rings of carbons with no carbon–carbon multiple bonds. Cycloalkanes occur naturally in petroleum and are formed during catalytic reforming of petroleum hydrocarbons.

Data quality: The magnitude of error associated with a particular set of data. Also the legal defensibility of a data set.

De Minimus: A term defined in §122(g) of CERCLA as a minimal contribution of a contaminant or contaminants by a PRP to a Superfund site.

Dissolved product: The water soluble fuel constituents, specifically benzene, toluene, ethylbenzene, and the xylenes.

Downgradient: In a direction parallel to the natural direction of flow of groundwater. Petroleum normally flows with groundwater and spreads downgradient. (See upgradient).

DNAPLs: Dense Non–Aqueous Phase Liquids. Contaminants that have a liquid density greater than 1 g/mL, or a specific gravity greater than 1. DNAPLs are difficult to remedy since they sink when they encounter groundwater.

DQOs: Data Quality Objectives. Qualitative and quantitative statements developed by data users to specify the quality of data needed from a particular data collection activity

DRE: Destruction and Removal Efficiency. A parameter to evaluate the treatment potential of thermal treatment facilities.

Duplicates: Also called "dupes." The practice of collecting duplicate samples from the same specific sampling site. Duplicates are analyzed separately to ensure accuracy and to estimate possible contamination. A means of independently checking on a laboratory's accuracy.

E50.01 and *E50.02:* Standards for Transaction Screens and Phase I assessments, respectively, published by ASTM.

EDB: Ethylene dibromide. $C_2H_2Br_2$. An additive to leaded gasolines to scavenge lead oxides produced in the combustion process.

EDC: Ethylene dichloride. $C_2H_2Cl_2$. An additive to leaded gasolines to scavenge lead oxides produced in the combustion process.

EHS: Extremely Hazardous Substances. Defined in EPCRTKA §302(2) as a list published the U.S. EPA that is the same as Appendix A of "Chemical Emergency Preparedness Program Interim Guidance." The EPA publishes two lists: Hazardous Substances and Extremely Hazardous Substances. The threshold planning quantity for EHS is much lower than for hazardous substances.

EIS: Environmental Impact Statement. A document required by NEPA before any change in status for a federal property.

EP–toxicity: Extraction Procedure Toxicity. The former analytical procedure used to identify wastes with a potential to leach toxic contaminants into groundwater. The procedure, also called "EP–Tox," was replaced by TCLP.

EPA: The United States Environmental Protection Agency. The chief federal agency charged with setting regulations to protect the environment. The U.S. EPA has been proposed as a federal department.

EPCRTKA: Emergency Planning and Community Right To Know Act. Passed as part of SARA. Among other things, this Act requires notification of the section of the public that may be affected by a contaminated site, storage of hazardous substances, or a remedial action.

Equipment blanks: A special type of field blank used primarily as a qualitative check for contamination rather than as a quantitative measure.

ERNS: The Emergency Response Notification List. A list available from the U.S. EPA or state implementing agency of emergency responses to incidents that affect the environment. Used in a Phase I or initial site assessment.

Ethylbenzene: $C_6H_5C_2H_5$. An alkylbenzene that occurs in gasoline. One of four listed hazardous aromatic compounds that are analyzed for in a BTEX analysis.

Facility: Defined in CERCLA ¶101(1) a facility is (A) any building, structure, installation, equipment, pipe or pipeline, well, pit, pond, lagoon, impoundment, ditch, landfill, storage container, motor vehicle, rolling stock, or aircraft, or (B) any site or area where a hazardous substance has been deposited, stored, disposed of, or placed, or otherwise come to be located; but does not include any consumer product in consumer use or any vessel.

Federal Water Pollution Control Act: See FWPCA.

FID: Flame Ionization Detection. Also called an Organic Vapor Analyzer, OVA. A field instrument which uses a hydrogen flame to detect hydrocarbons. The detection range extends up to 23 ev ionization energy.

Field blank: A blank used to provide information about contaminants that may be introduced during sample collection, storage, and transport. (Compare Transport blank).

Fluctuation zone: A region of the subsurface immediately above the lowest permanent water table that is subject to periodic saturation as the water table rises and falls. Fluctuation may be seasonal, as with snow melt, or periodic, as with tides or rainfall.

FOPs: Field Operating Procedures. A set of written guidelines for collecting, documenting, storing, and transporting samples in the field.

FR: Federal Register. The document in which all proposed or final federal regulations are published.

FWPCA: Federal Water Pollution Control Act. The Act establishes procedures to prevent or contain the discharge of oil into "navigable waters or adjoining shorelines." This is one of the laws that currently affect AST operations.

Ganglia: Liquid phase hydrocarbon residuals that have broken up into globules due to fluctuating groundwater levels. Ganglia are persistent, long–term sources of hydrocarbon contamination, both in soils and in groundwater.

GC: Gas Chromatograph or Gas Chromatography. An instrument or instrumental method, respectively, for analyzing volatile and semi–volatile organic compounds. The compound(s) to be analyzed are entrained in a carrier gas and passed through a column containing a stationary phase which acts as a filter. The compounds separate and are detected by flame ionization or photoionization. The chromatogram is interpreted to identify individual compounds.

GC/MS: Gas Chromatography/Mass Spectroscopy. Gas chromatography is described above. In GC/MS the separated compounds are sent into a mass spectrometer, which acts as a detector. The compounds are ionized and passed through an electromagnet which separates the ions according to molecular weight. The ions are detected and identified, usually by consulting reference (library) spectra.

Geostatistical sampling: A type of sampling procedure that is guided by the assumed properties of the random field and prior, or early, estimates of the covariance or variogram functions. A means of determining where and when to sample.

Grab sample: A discrete aliquot that is representative of one specific sample site at a specific point in time. Compare with composite sample.

H_2SO_4: Sulfuric acid. A common preservative for field samples. Sulfuric acid is an aqueous solution of SO_3 and can be unstable.

HASWA: Hazardous and Solid Waste Amendments. An act passed in 1984 as amendments to RCRA. Incorporated requirements for regulating operations of UST systems into RCRA and exempted petroleum products from the hazardous substances definitions.

HCl: Hydrochloric acid. A common preservative for field samples. Concentrated HCl is 36 M (molar). The normal field 1:1 dilution is 18 M or 648 g HCl per liter of water. Since pure HCl is a gas, the normal procedure is to dilute concentrated HCl.

HCS: Hazardous Communication Standard. An OSHA standard (29CFR §1910) requiring communication of hazardous material risks to workers in regulated facilities.

Headspace: The air space at the top of a water or soil sample. The amount of head space for a volatile analyte should be zero. (See ZHE). Headspace is something to avoid when sampling for VOAs.

HNu: A particular brand name for a PID field instrument.

HTTT: High Temperature Thermal Treatment. A variety of non–in situ methods for treating petroleum contaminated soils by incinerating the hydrocarbons and/or the soils.

Interferences: Compounds whose presence obscures the measurement of the analyte of interest by the introduction of an unrelated signal where the analyte is measured.

IR: Infrared. The spectral region in which is found the C—H vibrational, stretching, and bending frequencies. Petroleum hydrocarbons are detected by the presence of C—H bonds.

ISC: Initial Site Characterization. An investigation required by federal UST regulations (40 CFR §280) after confirmation of a spill or release of a regulated substance.

Kg: Kilogram. 1,000 grams = 2.2 pounds. The standard unit for soil samples.

Kinematic viscosity: A measure of the ability of a fluid to move through interstitial spaces in soils. Viscosity is a measure of flow rate.

LAST: Leaking aboveground storage tank. A large volume tank storing petroleum products that is not 10% or more underground.

LEL: Lower Explosive Limit. The lower limit of flammability for an oxygen/ hydrocarbon mixture. Below this value a mixture is too rich to burn.

LNAPL: Light Non–Aqueous Phase Liquids. Contaminants that are not soluble in water and are less dense than water. LNAPLs float or perch on groundwater.

LTTS: Low Temperature Thermal Stripping. A variety of non–in situ methods to treat petroleum contaminated soils. Soils are heated to volatilize hydrocarbons which are subsequently burned or collected.

LUFT: Leaking Underground Fuel Tank. Guidelines for site assessment, cleanup, and closure for underground storage tanks published by the State of California.

LUST: Leaking Underground Storage Tank. Used generically to refer to any leaking underground storage tank. Specifically the LUST Trust Fund is the federal trust fund used to reimburse states for expenses incurred in investigating or remediating orphaned or abandoned contaminated sites.

MACT: Maximum Achievable Control Technology. Section 112(b) of CAA requires EPA to promulgate emission standards which require installation of MACT to control sources of 189 chemicals considered to be harmful to the environment.

Matched–matrix field blank: The most common type of field blank. It is used to estimate incidental or accidental contamination of a sample during the collection procedure. See field blank.

Matrix: The solvent in which a sample is collected.

Matrix blank: Used to determine the presence of the analyte in the matrix when the solvent is not deionized water.

Matrix spike: A precisely known amount of an analyte is added to the matrix in use in collection of samples. This is a control or blank to estimate the error in analytical procedures.

MCL: Maximum Contaminant Level. The Safe Drinking Water Act requires EPA to set MCLs in water delivered to users of public water systems. Also, the maximum amount of a contaminant detectable by normal analytical methods.

Methanol: CH_3OH. An additive to gasoline and other consumer products. Methanol is added to gasoline to increase the octane number. Methanol is highly soluble in water and corrosive due to its acidic nature.

μg: Microgram. Micro = one millionth. $1/1,000,000$ or 10^{-6} gram.

μg/Kg: Microgram per kilogram. 10^{-6} grams per kilogram. Used in solid samples. 1 mg/Kg = 1 ppb.

μg/L: Microgram per liter. 10^{-6} grams per liter. Used in water samples. 1 mg/L = 1 ppb.

mg: Milligram. Milli = one thousandth = $1/1,000$ = 10^{-3}.

mg/Kg: Milligram per kilogram. 10^{-3} grams per kilogram. Used as a unit of measurement in solid samples. 1 mg/Kg = 1 ppm.

mg/L: Milligram per liter. 10^{-3} grams per liter. Used in water samples. 1 mg/L = 1 ppm.

Middle distillates: The kerosene fraction derived from petroleum refining. A range of hydrocarbons from C_{12} to about C_{24}. Middle distillates mainly produce diesel and jet fuels.

Mobile sources: Non–stationary pollution sources. Defined in Clean Air Act, mobile sources means automobiles and other vehicles.

MSDS: Material Data Safety Sheets. Information required by OSHA §651 and EPCRTKA §311(a)(1). The owner or operator of any facility that is required to provide MSDSs must also submit them to local, state, and federal emergency response agencies. These sheets contain relevant health and safety information about a chemical or product. Information relates to toxicity, treatment, flammability, reactivity, etc.

MTBE: Methyl Tertiary-Butyl Ether. A common gasoline additive used to increase octane numbers. MTBE is an ether and is apparently nontoxic to humans or the environment. It is both water–soluble and very volatile.

NEPA: National Environmental Protection Act. Passed in 1970 NEPA established the Council on Environmental Quality and required an environmental impact statement prior to a change in status on any federal property.

ng: Nanogram. Nano = one billionth or $1/1,000,000,000$; nanogram = one billionth of a gram or 10^{-9} gram.

ng/Kg: nanogram per kilogram. 10^{-9} grams per kilogram, or one billionth of a gram per kilogram, or $1/1,000,000,000$ gram per kilogram. 1 ng/Kg = 1 ppb.

ng/L: nanogram per liter. 10^{-9} grams per liter, or one billionth of a gram per liter, or $1/1,000,000,000$ gram per liter. 1 ng/L = 1 ppb.

NIOSH: National Institutes of Occupational Safety and Health. A government funded organization that sets permissible levels of exposure to hazardous substances. A unit of the National Institutes of Health.

NPDES: National Pollutant Discharge Elimination System. A permit system defined under CWA to restrict or control discharges into surface waters.

NPL: National Priorities List. The "Superfund" list defined under CERCLA. Making this list is bad news for companies.

O_2: Oxygen. A gas humans breathe; also used by bacteria in respiration. A nutrient required in microbial biodegradation processes. (See aerobic and anaerobic).

Octane Scale: A scale to quantify the combustion rate for gasolines. Zero is defined as the combustion rate for n–heptane, 100 is the combustion rate for iso–octane. Octane is measured by either the motor or research method.

Olefins: Hydrocarbons having one or more carbon–carbon double bonds. The IUPAC term for olefin is alkene. Olefins are abundant in gasolines.

OPA: Oil Pollution Act. Passed in 1990 in the aftermath of the *Exxon Valdez* oil spill, OPA requires spill prevention, control, and countermeasure (SPCC) plans to protect surface waters.

OSHA: Occupational Safety and Health Administration. Established by the Occupational Safety and Health Act.

OUST: Office of Underground Storage Tanks. Fondly referred to out in the Regions as "Headquarters."

OVA: Organic Vapor Analyzer. A field instrument to detect hydrocarbons. An OVA is a spectrophotometer modified for field work.

Oxygenated fuels: A term applied to gasoline blends which contain an alcohol or ether additive to decrease the production of carbon monoxide. Current additives are MTBE, ethanol, and methanol. MTBE and ethanol are current choices.

PAH: Polycyclic Aromatic Hydrocarbons. Also called polyaromatic hydrocarbons. Petroleum hydrocarbons in the C_{12} to C_{25} range containing multiple, fused benzene rings. PAHs are found naturally in middle distillate fuels.

Paraffin hydrocarbons: The family of hydrocarbons composed of molecules that contain no carbon–carbon multiple bonds. Paraffin hydrocarbons are also known as saturated hydrocarbons or alkanes.

PCBs: Polychlorobiphenyls. Ubiquitous compounds found throughout the environment in electrical transformers, and manufacturing processes. Resistant to normal degradation processes.

PCE: Polychloroethylene. The common name for tetrachloroethylene.

PCP: Pentachlorophenol. An aromatic alcohol used chimerically as a wood preservative. A hazardous substance.

Pellicular: Hydrocarbons adsorbed onto soil particles. Pellicular hydrocarbons are very slow to be removed by any in situ method.

Petroleum: A natural product composed of up to several hundred hydrocarbons. Crude petroleum is refined (distilled) to produce gasoline, diesel, and jet fuel among other products.

Petroleum exclusion: In RCRA and CERCLA petroleum products are explicitly excluded from the definition of hazardous substances. Uncontaminated petroleum products are regulated substances, not hazardous.

Phenol: C_6H_5OH. An aromatic alcohol that occurs naturally in petroleum and petroleum products. Phenol is also the prototype for a class of aromatic alcohols having similar chemical structures.

pH: A measure of the hydrogen ion concentration, and hence, the acidity of a solution. pH is defined as $-\log[H^+]$. Since it is a logarithmic scale, each unit change corresponds to a tenfold change in acidity. The pH scale is not defined for solids. Soil pH is actually the pH of soil water.

PID: Photoionization Detector. A field instrument to detect hydrocarbons. PIDs are modified spectrophotometers. Vapors are drawn into an ionization chamber where they are ionized by photons and detected. PIDs have an ionization potential limit of about 11 ev.

ppb: Parts per billion. μg/L or μg/Kg. A means of expressing very small fractions in terms of more conveniently–sized numbers. 0.001% = 10.000 ppb.

ppm: Parts per million. mg/L or mg/Kg. A means of expressing very small fractions in terms of more conveniently–sized numbers. 1% = 10,000 ppm.

ppt: Parts per trillion. ng/L or ng/Kg. A means of expressing very small fraction in terms of more conveniently–sized numbers. .000001% = 10,000 ppt. PPT is close to the limit of even research instrument resolution and hence is rarely encountered.

POTW: Publicly Owned Treatment Works. A waste water treatment facility funded through the Clean Water Act. Discharging contaminated ground-water through the sanitary sewer system requires a discharge permit from a POTW.

Precision: measure of the variability of measurements of the same quantity by the same method. Accuracy measures how close a series of measurements come to a "true" value. Precision measures the variance in a group of measurements.

Protocol: Thorough written description of the detailed steps and procedures involved in the collection of samples.

PRP: Potentially Responsible Party. The conditions for PRPs are defined in CERCLA. A PRP has an interest in a contaminated site as an owner, operator, transporter, or contributor of contaminated wastes.

PVC: Polyvinylchloride. A polymer made from vinyl chloride used to manu-facture "plastic" pipe. PVC pipe is commonly used in the construction of wells.

Quality assurance: Part of the process of ensuring the validity of environmental data. QA procedures include the use of appropriate protocols, statistics, redundancy checks, and documentation.

QAMS: Quality Assurance Management Staff.

QC: Quality control. Part of the process of ensuring the validity of environmen-tal data. QC procedures include the use of blanks and controls.

Radius of influence: The area of pressure gradient induced by a negative pressure at a wellhead from which contaminants can be withdrawn. Also called capture zone or zone of influence.

Random sample: A sample selected such that any portion of the sampled population has an equal probability of being chosen. To guarantee representativeness randomness, must be selected mathematically.

RCRA: Resource Conservation and Recovery Act. Originally passed in 1976 and amended in 1986, RCRA is the principal act under which the U.S. EPA regulates hazardous wastes, landfills, and USTs.

RCRACAL: Resource Conservation and Recovery Act Corrective Action List. A list, available through the U.S. EPA, of all RCRA sites for which corrective action has been undertaken or ordered. Used in a Phase I document search.

Reagent blank: Also called a method blank. A blank or control that contains any reagent(s) used in the sample preparation, extraction, and/or analysis procedure.

Recovery spike: A spike used to estimate the variance in a laboratory's analytical procedures. A precise amount of an analyte is added to the appropriate matrix and analyzed.

Regulated substance: Petroleum or a product derived from petroleum are exempt from the (Subtitle C) hazardous substances definition in CERCLA and RCRA; instead they are defined as regulated substances. Likewise, soils contaminated with petroleum are exempt from hazardous waste disposal requirements.

Remedy or *Remedial action:* Defined in CERCLA §101(24) these terms mean those actions consistent with permanent remedy taken instead of or in addition to removal actions in the event of a release or threatened release of a hazardous substance into the environment.

Replicate sample: Multiple samples taken under comparable conditions for the purpose of comparison; also called duplicates, but different from splits.

Reportable quantities: Defined in CERCLA §102(a) and (b) [SARA], RQs are minimum amounts of hazardous substances that, if released into the environment, must be reported to appropriate authorities. Unless otherwise specified, an RQ is one pound. (See RQ).

Representative sample: A sample that can be expected to reflect the properties of the parent population. A representative sample may be a stratified sample or a random sample, depending on the objective of the sampling plan.

RI/FS: Remedial Investigation/Feasibility Studies. Defined in CERCLA as preliminary investigations of a hazardous waste site.

RQ: Defined in CERCLA §102(a) and (b) [SARA], RQs are minimum amounts of hazardous substances that, if released into the environment, must be reported to appropriate authorities. Unless otherwise specified, an RQ is one pound.

Safe Drinking Water Act: See SDWA.

Sample: A portion of material selected to represent a larger body of material. Analytical data obtained from a sample merely estimate the quantity or concentration of a constituent or property of the parent material.

Sampling: An attempt to choose and extract a representative portion of a physical system from its surroundings. (See also geostatistical sampling).

Sampling blank: Also called a trip or travel blank. A blank consisting of the sampling media used for collection of field samples.

Sampling error: The part of the total error (the estimate from a sample minus the population value) associated with using only a fraction of the population and extrapolating to the whole, as distinct from analytical or test error, it arises from a lack of homogeneity in the parent population.

Sampling plan: A predetermined procedure for the selection, withdrawal, preservation, transportation, and preparation of the portions to be removed from a population and used as samples.

Sampling preparation blank: A blank run before and/or after a series of sample processing procedures, such as mixing, stirring, extraction, etc. It is a control on errors in laboratory procedures.

SARA: Superfund Amendments and Reauthorization Act. Passed in 1986 SARA includes amendments to CERCLA and EPCRTKA. SARA contains numerous reporting requirements and is often referred to as a standalone entity

Selective sample: A sample that is deliberately chosen using a sampling selection plan that screens out materials with certain specified characteristics, or selects samples with certain characteristics.

SemiVOA: Semi Volatile Analysis. A compound that is less volatile than a VOA; usually must be analyzed by liquid chromatography rather than gas chromatography.

SIP: State Implementation Plans. Required under CAA. States must develop plans to implement the provisions of CAA by November 12, 1993.

SDWA: Safe Drinking Water Act. Amended in 1986 guarantees safe drinking water to all Americans. Requires EPA to set MCLs for water delivered to users of public drinking water systems. Under this Act the EPA has defined analytical procedures for many analytes (see Chapter 7 and References).

SOPs: Standard Operating Procedures. Detailed, written plans for collection of data and information at a site. Includes FOPs, sampling protocols, laboratory analyses, and monitoring plans.

SPCC: Spill Prevention, Control, and Countermeasure. Plans specified by the Oil Pollution Control Act. SPCCs are emergency preparedness plans in the event of a spill or release.

Spike: A control sample in which a precisely known amount of an analyte is added to the appropriate matrix. Used to estimate error in laboratory methods. Common spikes are matrix spikes and recovery spikes.

Splits: The practice of dividing a sample into aliquots for duplicate analysis at different laboratories. A quality control procedure to ensure accuracy. (See Replicates).

STEL: TLV–STEL. Threshold Limit Value–Short Term Exposure Limit. Defined by NIOSH as the concentration to which workers can be exposed continuously for a short period of time without suffering from 1) irritation, 2) chronic or irreversible tissue damage, or 3) narcosis of sufficient degree to increase the likelihood of accidental injury, impair self-rescue or materially reduce work efficiency, and provided that the daily TLV–TWA is not exceeded. It is not a separate independent exposure limit; rather, it supplements the TLV–TWA where recognized acute effects from a substance whose toxic effects are primarily of a chronic nature. STELs are recommended where the toxic effects are primarily of an acute nature. A STEL is defined as a 15–minute, time–weighted exposure which should not be exceeded during a workday. (See also *TLV*).

Stratified sample: A sample consisting of portions obtained from identical subparts (strata) of the parent population. Within each stratum samples are taken randomly.

Subtitle C: The portion of RCRA that deals with hazardous substances and wastes. A Subtitle C waste is a hazardous waste and must be disposed of in a permitted facility.

Subtitle D: The portion of RCRA that deals with solid waste disposal facilities (sanitary landfills).

Subtitle I: The portion of RCRA that deals with underground storage tanks. A Subtitle I waste is a regulated waste.

Superfund: Established by CERCLA, the "Superfund" provides funding to clean up the most severely contaminated sites across the U.S. Sites qualify by being placed on the National Priorities List.

Superfund Law: A common appellation for CERCLA. See CERCLA and Superfund.

Surrogate: Also called a recovery spike. A control sample containing a known amount of an analyte. The sample is analyzed as a QA/QC measure.

SW–846: Test Methods for Evaluating Solid Waste, Physical/Chemical Methods. A 4–volume protocol of approved test methods, sampling, and monitoring guidance for use in solid waste analyses.

SWDA: Solid Waste Disposal Act. SWDA was the original waste disposal act passed in 1965 to regulate solid and hazardous waste disposal at landfills and the problem of resource recycling, especially used motor oil. SWDA was amended in 1970, 1973, 1976, 1984, and 1986. On each occasion the law was expanded and made more inclusive.

System blank: Also called an instrument blank. A measure of the instrument background response in the absence of a sample. Insures that the instrument is not responding to residual contaminants.

TC rule: Toxicity Characteristics rule. A set of criteria for defining hazardous waste. (See TCLP).

TCE: Trichloroethylene. A common solvent used in industrial processing. A persistent pollutant.

TCLP: Toxicity Characteristic Leaching Procedure. An analytical extraction and test to determine the leaching potential in landfilled hazardous contaminants in liquid and solid wastes. TCLP tests for both inorganic and organic contaminants. Officially adopted Sept. 25, 1990.

Technical Standards: 53 FR 370082, Sept. 23, 1988; amended 55 FR 17753, April 27, 1990 and published in 40 CFR Part 280. The technical requirements for owners and operators of UST systems.

TEL: Tetraethyllead. An additive to leaded gasolines to raise octane numbers. TEL and TML use was restricted by the EPA after 1973.

TEPH: Total Extractable Petroleum Hydrocarbons. Also called TEH. Refers to EPA Method 625 for drinking water or Method 8250 for soils.

TLV: Threshold Limit Values. There are three exposure levels defined by NIOSH for a toxic substance. *a) TLV–TWA.* Threshold Limit Value–Time–Weighted Average. Defined as the time–weighted average concentration for a normal 8–hour workday and a 40–hour workweek, to which nearly all workers may be repeatedly exposed, day after day, without adverse effect. *b) TLV–STEL.* Threshold Limit Value–Short Term Exposure Limit. Defined as the concentration to which workers can be exposed continuously for a short period of time without suffering from 1) irritation, 2) chronic or irreversible tissue damage, or 3) narcosis of sufficient degree to increase the likelihood of accidental injury, impair self–rescue or materially reduce work efficiency, and provided that the daily TLV–TWA is not exceeded. It is not a separate independent exposure limit; rather, it supplements the TLV–TWA where recognized acute effects from a substance whose toxic effects are primarily of a chronic nature. STELs are recommended where the toxic effects are primarily of an acute nature. TWA values are recommended where the toxic effects are primarily of a chronic nature. *c) TLV–C.* Threshold Limit Value–Ceiling. Defined as the concentration that should not be exceeded during any part of the working exposure.

TML: Tetramethyllead. An additive to leaded gasolines to raise octane numbers. TEL and TML use was restricted by the EPA after 1973.

Toluene: $C_6H_5CH_3$. Methylbenzene. An aromatic hydrocarbon found in gasoline blends. More toxic than aliphatic hydrocarbons, but less toxic than benzene.

TPH: Total Petroleum Hydrocarbons. Refers to EPA Method 418.1 or EPA Method 8015 which describe the procedures for determining and quantifying the petroleum hydrocarbon content of a sample. Because of interferences, 8015 is preferred.

TPQ: Threshold Planning Quantities. Amounts of hazardous substances defined in EPCRTKA §302(3)(A)(ii) that may be kept on premises without notification under EPCRTKA. Under the interim requirement the TPQ for all EHS is 2 pounds.

Transport blank: A control sample used to estimate sample contamination from the container and preservative during transportation and storage. (Compare field blank).

TRPH: Total Residual Petroleum Hydrocarbons. A modified EPA method 418 that measures heavier petroleum hydrocarbons. It is a substitute for Oil and Grease.

Trust Fund: May be either the federal LUST Trust Fund or a state reimbursement fund. These funds are funded through taxes on petroleum products and are to reimburse states (federal fund) or owners/operators (state funds) for costs incurred during remediation.

TSCA: Toxic Substances Control Act. A federal law to control the use and manufacture of pesticides and manufactured chemicals.

TSD: Treatment Storage and Disposal. A permitted facility under RCRA to accept, treat, and dispose of or store hazardous wastes.

TVPH: A GC/MS method to quantify all gasoline components (not just BTEX) in the C_5 to C_{10} range. As such it is much more time–consuming and expensive than BTEX.

TWA: TLV–TWA. Threshold Limit Value–Time–Weighted Average. Defined by NIOSH as the time–weighted average concentration for a normal 8–hour workday and a 40–hour workweek, to which nearly all workers may be repeatedly exposed, day after day, without adverse effect. TWA values are recommended where the toxic effects are primarily of a chronic nature.

UEL: Upper Explosive Limit. The highest percentage of O_2 relative to a given amount of hydrocarbon that will burn. Above the UEL the mixture is too lean to burn. (Compare LEL).

Umpire sample: A sample taken, prepared, and stored in an agreed–upon manner for the purpose of settling a dispute.

Upgradient: In a direction 180° to the natural flow of groundwater. Petroleum normally flows with the groundwater flow, and rarely spreads upgradient. (See downgradient).

USTs: Underground Storage Tanks or Underground Storage Tank Systems. Any system including tank and associated piping and equipment that routinely contains product that is 10% or more underground.

VOA: Volatile Organic Analysis. A method of identifying volatile compounds by gas chromatography or GC/MS. VOCs are entrained in a stream of helium and separated.

VOC: Volatile Organic Compounds. Volatile compounds that are detected by gas chromatography. There is no exact boundary between VOA and semiVOA. In hydrocarbons gasoline components are VOAs and diesel fuels are semiVOAs.

Volatility: A qualitative term for vapor pressure which is the tendency of a liquid to enter the vapor phase. For dissolved liquids the vapor pressure is proportional to the concentration of the dissolved component. In general the volatility of hydrocarbons is a function of molecular weight. Lighter hydrocarbons are more volatile.

WPCA: Water Pollution Control Act. Passed in 1965 and amended in 1972 this act addressed industrial water pollution by providing for area wide waste treatment plants (POTWs) and management plans.

WQA: Water Quality Act of 1987. Also called the Clean Water Act of 1987. This is the most recent "Clean Water Act". See CWA.

Wetlands: Areas that are inundated for a specified part of each year. Under CWA, projects that will change wetlands must have a §404 permit.

Xylenes: $C_6H_4(CH_3)_2$. Dimethylbenzene. A group of three aromatic isomers found in gasoline blends. The xylenes are listed hazardous substances unless they are found in gasoline.

ZH: Zero Headspace. A sampling container that is filled by injecting liquid through a Teflon® septum until the vessel is entirely full — no air bubbles. Also a sample container that is filled from a bailer or other sampler until the liquid meniscus is above the top of the vessel. The Teflon–lined top is screwed on so there is no residual air space.

ZHE: Zero Headspace Extractor. A container that allows for liquid/solid separation with the elimination of headspace, which prevents loss of volatile constituents.

Index

A

Abandoned milling operations
    in document review   105
Acceptable error levels   121
Accuracy and precision   140
    figure   140
    glossary   313
Acenaphthalenes
    in diesel fuels   74
Acetylenic hydrocarbons   48
ACGIH
    glossary   313
Achromobacter   210
Acid characteristic
    of phenol   59
Acinetobacter   210
Active and passive bioremediation   9
Acute toxic effects   70
    of alkanes   71
    of alkylbenzenes   70
    of benzene   70
    of ethylbenzene   71
    of fuel oil   74
    of n–hexane   71

    of middle distillates   74
    of octane   73
    of iso–pentane   73
    of toluene   71
    of xylenes   71
Acute toxicity   68–70
    glossary   313
Additive interferences
    glossary   305
Additives
    gasoline   73
Adjacent property
    in emergency response   154
Adsorbed hydrocarbons   74
    in volatilization   202
Adsorbed phases
    hydrocarbons   80
Adsorption
    rates of   80
Aerial photos
    Phase I   109
Aerobic
    conditions   210, 241, 244
    glossary   313
Aerobic microbial oxidation   207–212

Bioreactors   211
Bioremediation   206–213
   comments   206
   considerations   173
   equipment
    figure   208
   evaluation parameters   175
   fact sheet   174
   health and safety   207
   nutrient system   208
   of diesel   212
   landfarming   9, 240
    evaluation parameters   167
    fact sheet   166
    recommendations   243
   rates of   80
   recommendations   173
   requirements   207
Biorestoration   208
Bioventing   199
Biphenyl   51
Black box   100
Black hole
   Fleet Factors   98–101,
Blank   133–135
   and controls   134
   glossary   306
Blow counts
   well development   235
BOD pollutants   26
   glossary   315
Boiling point distribution   61
   of petroleum products   65
Bore hole   235
   figure   232
   well casing diameters   235
Bottlenecks
   in regulatory agencies   157
Branched hydrocarbons   45
   figure   45–46
   octane number   46
BTEX
   contaminant fluctuations   82
   free gasoline   199
   glossary   315
   in diesel   8

   laboratory methods   144
   risk assessment   162
Building permits
   Phase I   109
Bulk hydrocarbon phases   80
Bulk liquid phase
   hydrocarbons   80
   phase diagram   86
Butane   43
   empirical formula of   45
   family of alkanes   44
   flow rates   87
   structure of   44
   volatilization   201

C

CH_3OH 54, 58
   glossary   315
$(CH_3)_4Pb$ 73
$(CH_3CH_2)_4Pb$ 73
C_2H_6 42
$C_2H_4Br_2$ 55, 73
$C_2H_4Cl_2$ 55, 73
C_6H_6 50–53
$C_6H_5CH_3$ 51
   glossary   315
$C_6H_5CH_2CH_3$ 52
   glossary   315
C_6H_{12} 49
$C_6H_4(CH_3)_2$ 52
C_6H_5OH 58, 59
   glossary   315
C_7H_{16} 42
C_8H_{18} 46
C–H bonds   38
CAA   26
   glossary   315
CAP   3–4, 35, 155–171
   overview   155
   planning   158
    figure   163
   federal regulations   35, 291–295
   preparation   155
   problems with   155–156

site assessment 105
Capillary zone
    defined 79
    glossary 315
Capture zone
    figures 77–79
    glossary 307
    recovery wells 230
    volatilization 199
Carbon dioxide 40–41
Carbon monoxide 58
Case Law
    CERCLA 307–310
    conflicting 99
    liability 97
Catalytic converters
    volatilization 202
CDPHE
    glossary 316
Central nervous system 69
    see also CNS
CERCLA 22, 58
    and RCRA 92
    concept of liability 96–97
    defenses 24, 94
    definitions 18, 93
    glossary 316
    lender liability amendment 101
    liability 23
    overview 94
    petroleum exclusion 19, 93
    Phase I assessments 91
    philosophy 93
CERCLIS
    glossary 316
    Phase I 109
CFR
    29 CFR 15, 29
    40 CFR 15, 32, 285–293
    Code of Federal Regulations 15
    glossary 316
Chain–of–custody 131–132
    checklist 132
    record 132
    sample document 304
Chain of legislation

RCRA
    figure 19
Checklist
    chain–of–custody 132
    field logbook 129
    site assessment 117
    well documentation 235
Chemical (detergent) extraction 9, 216–217
    comments 216
    considerations and recommendations 216
    evaluation parameters 179
    fact Sheet 178
    requirements 216
    technical basis 216
Chemical oxygen demand
    COD 26
    glossary 316
Chemical structure
    of alkanes 41–44
    of alkenes 47
    of alkylbenzenes 49
    of benzene 50
    of cycloalkanes 49
    of ethylbenzene 49
    of phenol 56
    of octane
    of toluene 51
    of xylenes 52
Chronic effects
    of alkanes 71
    of alkylbenzenes 70
    of benzene 70–71
    of ethylbenzene 71
    of fuel oil 74
    of n–hexane 71
    of middle distillates 74
    of octane 73
    of iso–pentane 73
    of toluene 71
    of xylenes 71
Chronic exposure
    to benzene 70–71
    to hexane 71
    to toluene 71

STEL 319
TLV 320
Chronic toxicity 68–72
  glossary 316
City directory listings
  Phase I 109
Class A flammable liquid 1, 61, 67
Clean Air Act 26, 54–58
  "big three" 92
  of 1977 26
  of 1987 26
  of 1990 26, 58
Cleanup guidelines 162–166
  criteria 160–162
  in the planning phase 159, 162
  figure 163
  guidelines 166
Climatic conditions 130
Closure 157
  figure 163
  planning for 111
  problems with 166
CLP
  glossary 316
C:N:P ratio
  bioremediation 210
CNS depression 70
  from octane 67
  from toluene 71
  from xylenes 71
Coastal Zone Management Act of 1972
  30
Code of Federal Regulations 15
Coefficient of permeability 84
Collection
  date and time 130
Collection frequency 126
Collector
  chain–of–custody 132
Colorado
  fuel release 20
  generic regulations 10
Combustible gas indicator
  glossary 316
Combustion 45
  of hydrocarbons 39–41

Comments
  bioremediation 206
  chemical extraction 216
  excavation and land filling 238
  HTTT 250
  landfarming 242
  land treatment
    groundwater 244
  linear interception 219
  LTTS 248
  passive remediation 215
  soil washing 214
  vitrification 216
  volatilization 199
Commercial bacteria 210
Commercial sources
  historical review
    Phase I 110
Community right to know
  EPCRTKA 25
Comparability
  glossary 316
Complete degradation
  bioremediation 208
Components of gasoline. 56
  see also Appendix A 270
Composite samples 131
  glossary 316
Composting 244
Comprehensive Environmental Response,
  Compensation, and Liability Act 23–
  25, 93–102
  amendments 24
  case law 97–99, 307–310
  concept of liability 96–97
  definitions in 93–95
  and environmental assessments 102,
    105
  environmental legislation 1, 5, 7, 93–
    94
  EPA rule 101
  financial 100–101
  information system 109
  overview 92–93
  secured creditor exemption 95–96
Concentration levels 121

Conditions
    for excavation   238
Cone of depression   230
    figure   231
    glossary   316
Confidence interval
    glossary   316
Confined aquifer
    glossary   316
Considerations   9
Considerations, tables of
    for asphalt incorporation   253
    for excavation and landfill   240
    for groundwater extraction and treatment   226
    for high temperature thermal treatment   251
    for in situ bioremediation   211
    for in situ vitrification   217
    for isolation and containment   221
    for landfarming   240
    for land treatment   243
    for leaching and extraction   215
    for linear interception   219
    for low temperature thermal stripping   249
    for passsive remediation   213
    for solidification and stabilization   222
    for volatilization   206
Consultant's day
    OUST program   158
Containers
    number of
        chain–of–custody   132
Containment   217–223
    isolation and   220–223
    emergency response   156–157
    gasoline   7
    planning   163
    temporary   170
    solidification and stabilization   222–224
    vitrification   217
Contaminant characteristics
    estimating   167

figure   163, 165
    in the planning phase   157
    pathways   83, 87
    phases   80–84
    risk assessment   166
Contaminant migration   88
    figure   89
Contaminant movement   76
Contaminant phases   80–84
    distribution   86–88
        figure   84, 204
        table   85
    equilibria   202–204
    mobility   84, 165
    table   167
Contaminated groundwater
    extraction and treatment   224–226
    land treatment   244
Contamination   132–135
    gasoline   7
    glossary   317
    in sampling   130–131, 132–135
    migration   89
    off–site sources   126
    phases   84–87
    risk   160
    sources
        in emergency response   156
Contamination information
    in initial site characterization   114
Contamination plumes
    CO_2 mapping   210
Corrective Action Plan   155–168
    see also CAP
    federal regulations   33–35, 285–294
    Phase I   103
    planning   127–128
Corynebacterium   210
Creosote   60
Crude petroleum   39, 53
    deposits   39
    refining   53–57
Cumene   53
    in middle distillates   74
CWA
    glossary   317

Cyclic hydrocarbons 48
 cyclobutane 48
 cyclopentane 49
 cyclopropane 48
 family of 48
Cycloalkanes 48–49
Cycloparaffins 49, 73

D

Data
 flawed
  sampling design   121
 scientific quality vs legal quality   123
 usable 127
Data acquisition 166–167
 goals 164
Data documentation   126–130
Data flow
 in sampling and analyses
  figure   128
Data quality
 glossary   317
Data quality assessments   126
Data quality objectives
 see DQOs   126
Data quality requirements   126
Data reporting   126
Data validation   126
Date of sampling
 in Field Log Books   129
Deadlines
 UST compliance 1–2
DCE
 second level assessment   118
De Minimus
 glossary   317
Decision tree
 contaminant determination
  figure   120
Decontamination
 procedures   130–133
Defense
 liability 24, 96–97
 in CERCLA 96
 innocent purchaser 97

Definitions in CERCLA 93–97
Degradation
 biological 207–214
 of heavy hydrocarbons   212
Degree of saturation 76
Density of use
 Phase I 107
Detailed site characterization 115
  see also second level characterization
1,2-Dichloroethane
 see also DCE
 second level assessment 118
Detection limits   121, 125
Diesel fuels 60
 boiling point 65
 bioremediation of 212
 decision tree 120
 grades 60
 health effects 73–74
 kinematic viscosity 66
 lab methods 145–148
 PAHs 53
 solubility 62
 subsurface contaminant   75, 212
 viscosity 63
 volatilization   199, 205
Dimethylamine
 second level assessment   118
Dissolved phase
 hydrocarbons 80
Distribution
 of contaminant phases 86
  figure 79
 subsurface contaminants 84
Diversion ditches   212
DNAPLs 9, 38, 87
Document review   92
 in initial site characterization   113
 Phase I   102
Documentation
 chain–of–custody 132
 sampling   125–126
 site assessment 117–118
 well construction 235–236
Dose/response assessment 162
Downgradient

glossary   317
DQOs   123, 126
    glossary   317
Drinking water wells   8
    emergency response   156
    in initial site characterization   107,
        113
    in planning
    in risk assessments   160
Dry cleaners
    and environmental risk   104
Due diligence
    defined   102
    in Phase I assessments   106
Dumps and landfills
    and environmental risk   104
Duplicates   130
    glossary   317

E

E50.01   8, 105
    glossary 317
E50.02   8, 105
    glossary 317
EDB   73
    glossary   287
    health effects   72
EDC   70
    glossary   287
    health effects   72
EHS   24
    glossary   318
EIS   16
    glossary   318
Elevation
    of groundwater   235
    of surface
        well development   234
    to top of monitoring well casing
        well documentation   235
Emergency Planning, Community Right
    to Know Act   25
    EPCRTKA
        glossary   318
Emergency response   156–157

federal regulations   288
figure   163
state regulations   112
Emergency Response Notification System
    see ERNS   108
Endangered Species Act of 1973   31
Emulsions
    soil leaching   214–215
Environmental Assessments   105–110
    bottlenecks   167
    initial site characterization   111, 161
    federal regulations   291
    ISC report   113–114
    Phase I   105, 108
    Phase II   110
    Phase III   110
    overview   102
    second level   115–119
Environmental damage   67
Environmental effects   40
Environmental impact statement   16
    see also EIS
Environmental media   121
Environmental policy   5
Environmental Protection Agency   17
    see also EPA
    see also U.S. EPA
EP–toxicity
    glossary   318
EPA   17, 123
    glossary   318
    regulations   285–294
EPA Rule
    lender liability   101
EPCRTKA   28
    glossary   318
Equilibrium shifts
    in remediation   203
Equipment
    sample withdrawal   130
Equipment blanks
    glossary   318
Era of the environment   4
ERNS list   109
    glossary   318
    Phase I   109

Ethanol 59
  see also ethyl alcohol
  volatilization   199
Ethyl alcohol 58
  ethanol (*syn.*) 59
Ethylbenzene 52, 71
  chemical structure 52
  health effects 71
  flammability 62
  in middle distillates 74
  solubility 63
  volatility 200
Ethylene dibromide 73, 118
  see also EDB
Ethylene dichloride 73, 118
  see also EDC
Ethylene glycol
  laboratory analysis 147
  second level assessment   118
Evaluation parameters
  for asphalt incorporation 195
  for excavation 187
  for high temperature thermal treatment 193
  for in situ bioremediation 175
  for in situ passive remediation 177
  for in situ vitrification 181
  for isolation and containment 183
  for landfarming 189
  for low temperature thermal treatment 191
  for leaching and chemical extraction 179
  for solidification and stabilization 185
  for volatilization (soil venting) 173
Excavation
  and land filling   238–240
  exposure routes   239
  fact sheet 187
  health and safety   239
  in situ vs non–in situ   237
  summary and considerations table   240
Exemption
  secured creditor 93

Exposure
  in excavation   239
  table 72
  to benzene 70–71
  to EDB and EDC 73
  to ethylbenzene 71
  to gasoline 67
  to hexane 71
  to jet fuel 74
  to octane 73
  to *i*–pentane 73
  to toluene 71
  to xylenes 71
Exemption
  secured creditor   95, 95–96
Exposure assessment 164
Exposure pathways 164
  in excavation   239
Extraction wells
  bioremediation   208
  volatilization   199
Extremely hazardous substances 24
  glossary 318

F

Facility
  definition in CERCLA 18, 93
  glossary   318
Facility history
  in initial site characterization   113
Fact Sheet
  for asphalt incorporation 194
  for excavation 186
  for high temperature thermal treatment 192
  for in situ bioremediation 174
  for in situ passive remediation 176
  for in situ vitrification 180
  for isolation and containment 182
  for landfarming 188
  for low temperature thermal treatment 190
  for soil washing and chemical extraction 178
  for solidification and stabilization 184

for volatilization (soil venting) 172
Factors
    in bioremediation 209
Fatty tissues
    hydrocarbon solubility in 70
Federal Aid Highway Act of 1968 32
Federal documents
    records review
        Phase I 108
Federal Environmental Pesticide Control
    Act of 30
Federal Insecticide, Fungicide and Roden-
    ticide 30
Federal Land Policy and Management Act
    of 1976 32
Federal Pesticide Act of 1978 30
Federal Register 15
Federal Water Pollution Control Act 30
    amendments of 30
    FWPCA
        glossary 311
    of 1972 26
FEPCA 28
Fiberglass
    well casing 231
FID 117
    glossary 318
Field analysis data 130
Field analysis method 130
Field blanks 134
    glossary 318
    in QA/QC 151
Field logbook checklist
    figure 129
Field Operating Procedures
    see FOPs 125
Field sample containers 130
    internal temperatures 130
Field sampling 130–131
    data validity 150
    planning 126
    operating procedures 127–130
    uncertainty in 130
Financial outcomes
    and lender liability 100

Fire department hazardous materials
    responses
        Phase I 109
Fire department MSDS files
    Phase I 109
Fire insurance directories
    Phase I 109
Fire or explosive hazards 67, 200
Fish and Wildlife Coordination Act of
    1958 31
Flame ionization detector 117
    see also FID
Flammability data
    table of 62
Flash point 61, 62
    table of 62
Flavobacterium 211
Fleet Factors 98
    impact on financial community 100
Flow rates
    for groundwater 88
    for hydrocarbons 88
    groundwater 113
Flow velocities
    hydrocarbons 87
Fluctuation zone
    defined 80
    figure 77, 78, 79
    glossary 319
FOPs 125
    glossary 319
Fossil fuel combustion
    Clean Air Act 58
Fossil fuels 39
FR 15
    glossary 319
Fractured strata 7
Frequency of collection 130
Fuel oil
    no. 1 73
    no. 2 74
    no. 3 72
Fuels 40
FWPCA 30
    glossary 319

G

Ganglia 82
    glossary 319
Gas chromatography 141
Gasoline 37–39, 57–58, 67
    components of 269
    hazardous components of 56
    refining 55
    regular blends 57
    subsurface contaminant 75
    table of components 56, 269
    vapor pressure 67
Gasoline vapors 61
Gasoline contaminated soils
    analysis of 139, 145
    and RCRA 22
    health effects 200
GC 141
    and MS 145, 274
    GC/MS 141
    glossary 319
    low resolution 142
Generalized soil column
    figure 77
    microview
        figure 78
Generators
    of hazardous wastes 18
Geostatistical sampling
    glossary 319
Goals
    planning 155
    of remediation 2, 163
Golden Eagle Protection Act of 1962 31
Grab sample 131
    glossary 319
Grain alcohol 58
    see also ethanol
Grain surface area 76
Granulated activated charcoal filters
    volatilization 201
Grid identification
    table 129
Grid index
    in Field Log Book 129

Grid reference
    chain–of–custody 132
Grid size 127
Groundwater
    depth
        volatilization 198
    land treatment for 244
Groundwater extraction and treatment
    222–226
    pump and treat 223
    comments 224
    considerations 226
    contaminant fluctuation 82
    extraction well 233
    health and safety 225
    requirements 226
    technical basis 223
    well design 229–234
Groundwater flow direction
    well development 231
Groundwater levels
    well documentation 234
Guidelines 125, 166

H

H_2O_2
    in bioremediation 211
H_2SO_4
    glossary 319
    preservative 137
Half–life
    of petroleum hydrocarbons 212
Hand–driven probes 199
HASWA 18
    glossary 319
Hazard identification 160, 166
Hazardous and Solid Waste Amendments
    18
    see also HASWA
Hazardous materials 5
Hazardous substances
    definition in CERCLA 18, 22, 93
    early identification 106
Hazardous waste 18
Hazardous waste site list

Phase I   109
HCl
    glossary   320
HCS
    glossary   320
Head space
    glossary   320
Health and safety
    asphalt incorporation   254
    bioremediation   200
    excavation and land filling   239
    HTTT 251
    landfarming   242
    land treatment 245
    linear interception   220
    LTTS   249
    passive remediation   212
    soil leaching   214
    soil vacuuming   200
    solidification and stabilization 222
    vitrification   219
    volatilization 216
Health and safety information
    in initial site characterization   114
Health effects 66–74
    in volatilization   200
    of hydrocarbons   69
    table 72
Heating oil
    health effects 74
    volatility   199
    volatilization   201
Heavy naphtha 57
Henry's law   199–207
Hexane 71, 73
    flow rates 87
    health effects 71–72
    volatilization   200, 201
High probability
    for environmental risk   104
High level exposure 65, 67
High temperature thermal treatment 193,
    246–248, 249–251
    see also HTTT
    comments 250
    considerations 251

evaluation parameters 194
fact sheet 193
health and safety 251
requirements 251
technical basis 250
Historic Sites Act of 1935 32
Historical maps and plats
    Phase I 109
Historical Review
    checklist
        table 108
    Phase I 102
Historical site use 126
HNu
    glossary   320
Hollow–stem augers   233
HTTT 193, 246–248, 249–251
    see also High Temperature Thermal
        Treatment
    comments 251
    considerations 250
    evaluation parameters   194
    fact sheet 193
    health and safety 251
    requirements 251
    technical basis 250
Hydraulic conductivity 84
Hydrocarbon
    adsorbed
        in volatilization   202
    anerobic oxidation   241
    biodegradation 174–175, 207–212
    combustion of   39
    contaminants
        gasoline 7
        potential routes, figure 88
    exposure 68
    kinematic viscosities 63, 66
    microbial oxidation   241
    migration 88–89
        figure 89
    phases 80–84
    physical properties   60
    representations of   44
    soils 79
    toxicity 67

table 72
solubility 61
solubility in disolved phase   200
    Henry's Law   200
table of solubilities 68
vapors
    fire hazards 200
viscosity 66
volatility 61
Hydrocarbons
    alkanes 43, 69
        butane family of 41
        pentane family of 42
    aromatic compounds 46–47, 67
        alkylbenzenes 50–52
        benzene 50
        ethylbenzene 52, 53
        toluene 51
        xylenes 51, 52
    cyclic hydrocarbons 45
        family of 45
        cycloalkanes 45–46
        cycloalkenes 45
        cyclobutane 45
        cycloparaffins 70
        cyclopentane 45
        cyclopropane 45
    isomers of 44–45
    polycyclic aromatic 53
    the simplest   41
    structures of 41–52
    the simplest 41
Hydrocracking 57
Hydrogen peroxide   211
    use in bioremediation   211
Hydrogeology
    in initial site characterization
        site   115

I

Identification
    of contaminants   139
Immobile phase
    partitioning   87
In situ bioremediation
    see bioremediation
In situ chemical extraction
    see chemical extraction
In situ leaching
    see leaching and chemical extraction
In situ methods 171–185, 197–236
In situ passive bioremediation 212–214
    comments 212
    considerations and recommendations
        213
    evaluation parameters 195
    fact sheet 194
    health and safety 212
    requirements 213
    technical basis 212
In situ vitrification 9, 217–219
    comments 217
    considerations and recommendations
        218
    evaluation parameters 181
    fact sheet 180
    health and safety 216
    requirements 217
    technical basis 216
In situ vs. non–in situ 197–198, 237–238
Incineration 250
    see also HTTT
Infiltration galleries
    in bioremediation   210
Information
    data flow 128
    generating 123
Inhalation exposure
    in middle distillates 74
    of hydrocarbon vapors   69
    to toluene 72
Initial site characterization 111–115
    checklist 113–114
    flow chart 165
    Phase I assessments   91
    report
        table   113
    summary   115
Innocent purchaser defense 97, 105
    and environmental assessments 102
    in CERCLA 97

Installation
    well 117
Interferences
    glossary 320
Internal temperatures 130
Interstitial space 76
Interviews
    Phase I 109
Interviews
    Phase I 109
Intrusive sampling
    Phase I 103
IR
    glossary 320
    methods 144–147
ISC
    glossary 320
Iso isomer 45
Isooctane 46
Iso–pentane
    health effects 73
Iso–propyl alcohol
    volatilization 199
Isolation and Containment 182, 220–222
    comments 220
    considerations and recommendations
        221
    evaluation parameters 182
    fact sheet 183
Isolation and containment
    and bioremediation 212
Isomeric hydrocarbons 44
isomers 45
Isolation and containment 9
Isomeric hydrocarbons 42
Isomers 42
    of the butane family 37
    of the octane family 43

J

Jet A 74
Jet fuels 73–74
    health effects 73–74
    Jet A 74
    JP–4 74

    solubility 62
    viscosity 63
JP–8 74
    kinematic viscosity 66
    subsurface contaminant 75
    volatilization 199
Joint and several liability 23
    defined in CERCLA 96

K

Kerosene 57, 73–74
Kerosene–based fuels 74
Ketones
    volatilization 199
Key Site Manager
    Phase I 109
Kg
    glossary 320
Kinematic Viscosity 61, 66
    glossary 320
    table of 66
    volatilization 198

L

Lab analysis requirements
    in the planning phase 159, 165
Laboratory results
    second level assessment 118
Landfarming 208, 240–244
    comments 242
    considerations and recommendations
        243
    evaluation parameters 189
    fact sheet 188
    health and safety 242
    requirements 242
    technical basis 241
Landfarming
    of diesel 212
Land filling
    and excavation 238–240
Land treatment 244–246

comments   244
considerations   245
health and safety   244
requirements   244
technical basis   244
Landfill or solid waste disposal sites list
    Phase I   109
Landfill storage   238
Landfilling
    summary and considerations   240
LAST
    glossary   320
Leaching and chemical extraction   214–
    215
    comments   214
    considerations and recommendations
        215
    evaluation parameters   179
    fact sheet   178
    health and safety   214
    requirements   215
    technical basis
Lead
    in gasoline   73
Leak detection   1, 33
Leaking ASTs
    emergency response   156
Leaking underground storage tank   25,
        33–34, 157, 285–287
    see also LUST
Legal environment   89
Legal environment   91
Legal outcomes
    and lender liability   99
Legislation
    proposed for lender liability   101
LEL
    glossary   320
Lender Liability
    EPA Rule   101
    Fleet Factors   98
Letting nature take its course
    bioremediation   208
Liability
    avoidance   8
    CERCLA   23, 91, 94

CERCLA defenses   24
CERCLA definition   96
changing concept   96–97
defense
    new standard   92
Light Non–Aqueous Phase Liquids   80
    see LNAPLs
Linear interception   218–219
    comments   218
    considerations and recommendatins
        219
    health and safety   218
    requirements   218
    technical basis   218
Liquid chromatography   142
Liquid phase hydrocarbon
    ganglia   82
Lithologic conditions
    documentation   234
Liver damage
    from benzene   70
LNAPLs   9, 38, 87
    glossary   320
    in soils   80
Local (County) documents
    historical review
        Phase I   109
Local agencies
    historical review
        Phase I   109
Location of well   229
Log books   125–131
    checklist   129
    field log book   129
    legal documents   129
    master log book   128
Low probability
    of environmental risk   103
Low Temperature Thermal Strip-
    ping   247–249
    see also LTTS
    comments   247
    considerations and recommendations
        248
    evaluation parameters   191
    fact sheet   190

health and safety 249
   requirements 249
   technical basis 247
Lower explosive limit 57, 61
LTTS 246
   comparison with HTTT 246–247
   glossary 312
LUFT
   glossary 320
LUST 25
   glossary 321
LUST lists
   Phase I 109

M

Master Log Book 128
Matched–matrix field blank
   glossary 321
Material Data Safety Sheets 28
Matrix
   glossary 321
Matrix blank
   glossary 321
Matrix spike
   glossary 321
Maximum contaminant levels 166
   defined 28
   groundwater extraction 236
MCL
   glossary 321
Medical supply facilities
   and environmental risk 104
mesitylene 53
   solubility 53
Methanol 59–60, 80
   see also methyl alcohol (*syn.*)
   chemical structure 58
   glossary 321
   second level assessment 118
   volatilization 199
methyl alcohol 58
Methyl *tert*–butyl ether 59
   chemical structure 58
   second level assessment

see MTBE 118
mg 143–144
   glossary 321
mg/Kg 143–144
   glossary 313
mg/L 143–144
   glossary 321
Microview of petroleum contaminated
   soils
   figure 81
Microbial oxidation
   of hydrocarbons 241
Microbial process
   and the vadose zone 76
   and soil particle size 74
   particle size 76
Micrococcus 212
Micronutrient control
   in bioremediation 210
Micronutrients
   in bioremediation 210
Middle distillates 55, 73
   glossary 321
   health effects of 74
Migration
   migration routes for hydrocar-
      bons 88
   of dissolved phase hydrocarbons 80
   potential for
      initial site characterization 113
Migration routes
   in Phase I 107
   phase, figure 87
Minimum search distance 108
Mining operations
   and environmental risk 105
Mixed waste sites
   volatilization 199
Mobile phase
   partitioning 87
Mobile sources 27
   glossary 321
Moisture control
   in bioremediation 212
Molecules
   foreign 70

shapes 44
Monitoring network design   126, 127
Monitoring
    parameters   126
    performance 168–171
Monitoring wells 211–225, 228–236
    design 228–231
    dimensions 230
    figure 230
    in the planning phase 159
    Phase I 103
    placement 231
MSDS   28
    glossary   322
MTBE   59
    glossary   322
    recovery in volatilization   204
    second level assessment   118
Muck and truck   238
Mycobacterium   212

N

Nanogram
    glossary   314
Naphthalene   74
    chemical structure 54
    health effects 74
        table 72
    in diesel   74
    naphthalenic hydrocarbons
        flow rates   88
National Environmental Protection
    Act   15
    see also NEPA
National Fire Protection Association 1
National Historic Preservation Act of 1966
    32
National Pollutant Discharge Elimination
    System 25
    see also NPDES
National Priorities List 2, 22
    see also NPL
    in Phase I assessment 108
National Trails System Act of 1968 31

National Wildlife Refuge System Adminis-
    tration 31
NEPA   15
    glossary   322
ng 143–144
    glossary   322
ng/Kg 143–144
    glossary   322
ng/L 143–144
    glossary   322
NIOSH
    glossary   322
No. 1 diesel   60
No. 2 diesel   60
No. 4 diesel   60
Nocardia   211
Non-In Situ Soil Treatment Technolo-
    gies   237
Normal heptane 42
Normal isomer 45
Notification 33, 288–289
    implementing agency 33, 288–289
    in emergency response 156
    glossary   322
NPDES 26
    permits 176, 178
NPL   22
    glossary   322
NPL site list
    Phase I   108
Nutrient System
    bioremediation   209

O

O_2 209–210, 241
    bioremediation 209–210
    glossary   322
Occupational Safety and Health Act 28
    Occupational Safety and Health
        Administration
        see OSHA 15
Octane 73
    health effects 73
    isomers of 46
    in volatilization   201

Octane Scale 57
    glossary 322
Off-site sampling
    second level assessment 119
Office of Solid Waste and Emergency
        Response 21
Office of Underground Storage Tanks 21
Oil Pollution Act of 1990 28
Olefins 47, 49
    glossary 322
Operating procedures 127–130
Operator
    as PRP 112
    defined in CERCLA 20, 94
Organic chemistry 42
Organic horizon 76
Organic solvents
    second level assessment 119
OSHA 28
    glossary 322
OTS 28
OUST 21
    glossary 322
OVA
    glossary 322
Overview
    environmental assessment 102
Owner
    initial site characterization
        as PRP 112
Owner or operator
    definition in CERCLA 20, 94
Oxidative degradation 208
Oxygen control
    in bioremediation 211
Oxygen delivery
    in bioremediation 211
Oxygenated fuel additives 59
Oxygenated Fuels 58
    glossary 322
Ozonolysis
    volatilization 202

P

PAH 53, 54, 74
    chemical structures 54
    glossary 322
    flow rates 88
    in diesel 74
Paint operations
    and environmental risk 104
paper processing plants
    and environmental risk 104
Paraffin hydrocarbons 46–49, 73
    glossary 322
Parameters
    requested for analysis 130
Particle size
    of soils 746
Partition
    hydrocarbon migration 86–89
        table 86, 87
    in a soil column 84
Partitioning coefficient 199
Passive bioremediation 176–177, 212–
        214
    comments 212
    considerations and recommenda-
        tions 213
    evaluation parameters 177
    fact sheet 176
    health and safety 212
    requirements 213
    technical basis 212
Pathways 75–80
    adsorption 79
    multiple 82–83
    migration
        figure 89
    volatilization 83
PCBs 51
    glossary 323
    second level assessment 118
PCE
    glossary 323
PCP 60
    glossary 323
Pellicular

glossary   323
hydrocarbons   82
Pentachlorophenol   60
   see also PCP
Pentane 43, 45–46
   chemical structure
      figure 45
   flow rates   87
   volatilization   201
Performance evaluation 168–171
   flow chart
      figure 169
   in planning
      figure 163
Permeability   76, 88
   coefficient of   84
   figure 88
Permitted well records 8, 109
   Phase I   109
Permitting
   land fills   239
Permitting requirements
   for land fills   238
Pesticides
   second level assessment   119
Petroleum   38
   glossary   323
   insolubility   80
Petroleum contamination
   court decisions   91
   emphasis on gasoline 6, 7
Petroleum exclusion   18–22, 91–93
   glossary   323
   in CERCLA   18, 22, 93
Petroleum products   39, 60
   boiling points
      table of   65
   health effects
      table of   72
   included in ISC   105
   refining
      figure 55
   solubilities
      table of   63
pH
   glossary   323

pH limits
   in bioremediation   211
   relative to soil water   76
   soil
      in landfarming   242
pH control
   in bioremediation   211
Phase
   bulk hydrocarbon   82
   liquid hydrocarbon   82
   equilibria
      in volatilization   202–204
Phase I environmental assessments   8, 91,
   105–106
   basic requirements   106
   environmental assessment
      defined   102
   records review   108
Phase I site inspection   107–108
   checklist 107
Phase II environmental assessment   8, 92,
   103, 109–111
   environmental assessment
      defined   103
   initial site characterization 111
   terminology 92
Phase III assessments   103, 110
   defined   103
Phase migration routes 88–89
   figure 88
   table   87
Phases 81–87, 202–205
   distribution in soils
      figure   79
   equilibria
      figure   204
Phenanthrenes
   chemical structure 54
   in diesel fuels   74
Phenol
   chemical structure   59, 60
   glossary   323
Phenyl   51
Photo processing operations
   and environmental risk   104
Photosynthesis   39

Physical inspection
    Phase I   102
Physical properties
    of hydrocarbons   61
Physical Properties of Hydrocarbons   60
PID   117
    glossary   323
Planning 121, 158–168
    and environmental assessments 106–
        108
    data acquisition   166–167
    environmental conditions
        ASTM Standard Practice   105
    evaluation flow chart
        figure   160
    integrated
        figure   163
    overview   155–156
    planning group
        flow chart   164
    project priorities   158–159
    sampling   126
    steps   159
Plating operations
    and environmental risk   104
Plume
    establishing extent   113
Polyaromatic hydrocarbons   53
    see also PAHs
Polychlorobiphenyl   51
    see also PCBs
Polycyclic aromatic hydrocarbons   53
    characteristic   53
    chemical structures   54
    in diesel fuels   74
    see also PAHs
Polynuclear aromatic hydrocarbons   50
    see also PAHs
    see also polycyclic aromatic hydrocar-
        bons (syn.)
    see also polynuclear aromatic hydro-
        carbons
Porosity 76
    of soils
        volatilization   198
Potentially responsible parties   23

    see PRPs   93
POTW   25
    glossary   323
ppb   143
    glossary   324
ppm   143
    glossary   324
ppt   143
    glossary   324
Precision
    and accuracy
        figure   140
    glossary   324
Preservatives   130
Principle
    Heisenberg uncertainty   130
Printing operations
    and environmental risk   104
Priorities
    in the planning phase   158
Probes
    hand driven   199
Problems With Sampling   138–139
Procedures
    decontamination   130
Product Recovery
    with groundwater
        figure   231
Product release
    in initial site characterization   112
    regulations 32–35, 285–288
Production of CO_2 40
Program auditing   126
Project description   126, 127
Project goals   127
Project organization   126
Project planning 121, 158–168
    and environmental assessments 106–
        108
    data acquisition   166–167
    environmental conditions
        ASTM Standard Practice   105
    evaluation flow chart
        figure   160
    integrated
        figure   163

overview 155–156
planning group
　flow chart 164
project priorities 158–159
sampling 126
steps 159
Project strategy
　in the planning phase 158
Property transactions
　CERCLA liability 91
Protection of national resources 5
Protocol
　glossary 324
PRPs 23, 91
　CERCLA 91
　definition 93–94
　glossary 324
Pseudomonas 210
Public health 41
Public safety exposure
　in ASTM Standard practice 106
Pumping rate 130
Purge volume 130
Putt's First Law 121
PVC
　glossary 324
Published sources
　document reviews 108–109
　historical reviews 108–109
　Phase I 102–103, 107
　record review 108–109

Q

QA/QC 121, 150–152
　components 150–152
　deliverables 121
　glossary 324
　in initial site characterization 116
　in the planning phase 157
　　figure 121
　problems 139
　protocols 150–152
　sampling 135
　techniques
　　checklist 152–153

flow chart 151
QAMS
　glossary 324
QAPjPs 126
Quality assurance project plans
　see QAPjPs 126

R

Radionuclides
　early identification 106
Radius of influence 84
　glossary 324
　volatilization 198
Random sample
　glossary 324
Rates
　in volatilization 207
RCRA 17, 57
　and CERCLA 92
　"big three" 92
　definitions 18
　glossary 324
　petroleum exclusion 19, 93
　UST regulations 20
RCRA generators list
　Phase I 108
RCRA, Subtitle C regulations 57
RCRACAL
　glossary 325
Reagent blank 130
　glossary 325
Recognized environmental conditions 105
Recommendations (considerations and)
　asphalt incorporation 253
　excavation and landfilling 240
　groundwater extraction and treatment 226
　high temperature thermal treatment 250
　in situ bioremediation 206
　in situ vitrification 218
　isolation and containment 220
　landfarming 242

land treatment for contaminated
   groundwater 245
leaching and chemical extraction 214
linear interception 219
low temperature thermal stripping
   247
passive bioremediation 212
solidification and stabilization 222
volatilization 205
Record.
   chain–of–custody 132
Records review 90, 102
   Phase I 108
   resources
      table 108–109
Recovery Spike
   glossary 325
Recovery system
   vapor and groundwater 235
Reduction
   chemical 38
Reference grid 127
Refinery
   diagram 55
   petroleum 53
Refining Petroleum 53
Regional geology
   in initial site characterization 114
Regulated 19–23, 37
   glossary 325
   petroleum exclusion 22
   substances
      in RCRA 19, 22–23
      in CERCLA 23
Regulatory agencies 157–158
   figure 163
Regulatory guidelines 125, 162–166
   figure 161
Regulatory jurisdictions 125
Release 33–35, 286–288
   indicators 111
Relevance 151
   of sampling to site needs 125
Remedial action 35, 288–291
   glossary 325
Remedy

glossary 325
Replicate sample 151
   glossary 325
Reportable quantities 25
   glossary 325
Reportable quantities 25
Representative petroleum products 62
Representative sample 126
   glossary 325
Requirements 246
   asphalt incorporation 253
   chemical extraction 216
   excavation 239
   groundwater extraction and treatment
      226
   HTTT 251
   in situ bioremediation 209
   in situ vitrification 217
   isolation and containment 220
   landfarming 242
   land treatment for groundwater 244
   linear interception 220
   LTTS 249
   passive remediation 214
   soil leaching 216
   solidification and stabilization 222
   thermal treatments 246
   volatilization 201
Residual contamination 76
Resolution Trust Corporation 7
   see also RTC
   glossary 325
Resource Conservation and Recovery Act
   1, 18–23
   see RCRA
Resources
   document reviews 108–109
   historical reviews 108–109
   Phase I 102–103, 107
   record review 108–109
Retroactive liability 24
   defined in CERCLA 96
RI/FS
   glossary 325
Risk
   financial 95

Risk analysis 103–105
   financial 93
   property 103
Risk assessment 160–162, 166
   flow chart 161
   management 166
   steps in 162
   vs. risk assurance 158
Risk characterization 166
Risk management 165
Routes
   potential migration routes 88
RQ 25
   glossary 325
RTC 7
   glossary 325
Rules of the road
   for log books 129

S

Safe Drinking Water Act
   glossary 325
Sample
   blank 130
   containers 130
   distribution 130
   glossary 317
   location 129
   number
      chain–of–custody 132
   reference number
      in Field Log Book 129
   reference system 128
   storage 131
   transporter 130
   uncertainty 131
   withdrawal equipment 130
   withdrawal procedure 130
Sample collection location
   in Field Log Book 129
Sampling 119–121
   analyses
      data flow, figure 128
   area
      table 129

bias 131
blank 130
   glossary 325
design 119
error
   glossary 326
frequency 122
glossary 326
plan
   glossary 326
   in initial site characterization 115
preparation blank
   glossary 326
procedures 125–126
protocols 119–121, 126–127
   figure 124
   second level assessment 118
site location
   table 129
wells 238–231
Sampling wells 238–231
   Phase I 103
Sampling wheel
   figure 124
Sanitation department records
   Phase I 109
SARA 18, 24–25
   definitions 24
   glossary 326
SARA Title III 25
SARA Title III Reports
   Phase I 109
Scoop and run
   see excavation
Screened interval 234
SDWA
   glossary 326
Search Radius 106–107
Seasonally Saturated Soils
   figure 82
Second level assessment 92, 111
Seasonally saturated soils
   figure 80
Secured creditor exemption 95–96
   definition in CERCLA 95
   EPA rule 101–102

Selective sample
glossary 326
Semi-volatile compounds
separation of 142–143
Semi-volatiles
separation of 142
SemiVOA
glossary 326
Separation
laboratory methods 141–142
of semi-volatiles 142
of volatile compounds 141
Sequence
well sampling 130
Service stations
and environmental risk 104
recovery system 235
Shipping containers
internal temperatures 130
SIP 24, 26
glossary 326
Site address
in Field Log Books 129
Site assessment
goals 106
Site assessment checklist
table 117
Site Characterization
initial 111–115
Site characterization
well development 231
Site description 127
in initial site characterization 114
Site geology
second level assessment 117
Site grid
well documentation 234
Site hydrogeology
second level assessment 117
Site reconnaissance 107
checklist
figure 107
Site–specific characteristics
in the planning phase 157
Smearing 80
Soil 75–90

characteristics 76
second level assessment 117
classification 90, 298–302
contaminant migration 90–91
gas surveys 126
gases 199
leaching 215–216
permeability 84–87
pH 76
porosity
volatilization 198
types 75
zones
figure 77, 78
Soil classification system 90, 299–302
Soil gas surveys 126
Soil gases 199
Soil leaching 215–216
Soil moisture content
volatilization 198
Soil permeability 84
Soil pH 76
Soil temperature
bioremediation 210–211
volatilization 198
Soil vapor surveys
well development 231
Soil–water interface 76
Soil washing
see leaching and chemical extraction
Soils
microview
figure 81
Solid Waste Disposal Act 17
see also SWDA
Solubility 61, 63
of selected hydrocarbons
table 68
Henry's law
figure 200
Solidification and stabilization
comments 220
considerations and recommendations
221
evaluation parameters 185
fact sheet 184

health and safety 220
    requirements 220
    technical basis 221
Solubility and viscosity data
    representative petroleum prod-
        ucts 63
solvents 69
SOP 125
    glossary 326
    flow chart 163
    well documentation 234
SPCC 26
    glossary 326
Species control
    in bioremediation 211
Specific gravity 61
Sphere of influence 233
Spike
    glossary 326
Spill Prevention, Control, and Counter-
        measures 26
    see also SPCC
Spills 32, 285
Splits
    glossary 326
Stabilization 9, 185, 222
Stainless steel
    casing 233
    well casing 233
Standard Operating Procedures
    see SOP 125
State agencies
    site assessments 91
State documents
    historical review
        Phase I 109
State groundwater survey maps
    Phase I 109
State implementation programs 25, 27
State NPL equivalent site list
    Phase I 109
State subsurface geology maps
    Phase I 109
Static water level depth
    table 129
Statistical validity 125, 126

STEL
    TLV–STEL
        glossary 327
Storm drainage
    emergency response 156
Straight–chain alkanes 43, 44
    degradation of 212
Stratifed sample
    glossary 327
Streamlining
    OUST program 156
Strict liability 23
    defined in CERCLA 96
Stripping coefficient
    in Henry's law 199
Subsurface characteristics 75–90
Subsurface information
    in initial site characterization 114
Subsurface Phase Equilibria
    figure 205
Subsurface profiles
    figure 116
Subtitle C 18
    glossary 327
    problems with aromatic com-
        pounds 94
Subtitle D 18
    glossary 327
Subtitle I 20
    glossary 327
Summary tables
    for asphalt incorporation 254
    for excavation and landfill 240
    for high temperature thermal treat-
        ment 250
    for in situ bioremediation 212
    for in situ vitrification 218
    for isolation and containment 221
    for landfarming 240
    for land treatment 245
    for leaching and extraction 216
    for linear interception 219
    for low temperature thermal stripping
        248
    for passsive remediation 213

for solidification and stabilization
222
for volatilization 206
Superfund 2, 22
glossary 327
Superfund Amendments and Reauthorization Act 18–19
see SARA
Superfund law
see CERCLA
glossary 327
Surface area 76
Surface tension 79
Surrogate
glossary 327
Surveys
Phase I 109
SW–846 123
glossary 327
SWDA 17
glossary 327
System blank
glossary 327

T

Tailings piles
and environmental risk 105
Tax records
Phase I 109
TC rule
glossary 328
TCE
glossary 328
TCLP 131, 149, 319
glossary 328
Technical Basis
asphalt incorporation 254
chemical extraction 215
excavation and landfilling 238
groundwater treatment 223
HTTT 251
in situ bioremediation 207
in situ vitrification 217
isolation and containment 220
landfarming 241

land treatment for contaminated
groundwater 245
leaching 214
linear interception 218
LTTS 249
passive remediation 213
solidification and stabilization 221
thermal treatments 247
volatilization 200
Technical Standards
glossary 328
Technology 167–168
decision tree
figure 168
flow chart 161
screen
figure 170
selection 153
TEL 73
glossary 328
health effects 73
Temperature control
in bioremediation 211
Temperature of soils 211
Temperatures
in LTTS 246
TEPH 131, 320
Tetraethyllead 73
see also TEL
second level assessment 118
Tetramethyllead 73
see also TML
glossary 328
health effects 73
Thermal treatments 246–247
requirements 246
technical basis 246
Third party contamination
CERCLA 91
Threshold exposure limit 61
see also TLV
benzene 67
Threshold planning quantities 24
TLV
glossary 328
TLV–C

Threshold Limit Value–Ceiling
  glossary 328
TML 73
  glossary 328
Toluene 71
  chemical structures 52
  chronic effects 71
  flashpoint 62
  flow rates 87
  glossary 328
  health effects 71
    table 72
  in middle distillates 74
  potential problems 94
  solubility 63, 68
  time–weighted exposure 51
  viscosity 63
  volatility 62
  volatilization 201, 205
Toxic pollutants 26
Toxic Substances Control Act 28
Toxicity 67–74
  acute 68
  chronic 69
Toxicological properties
  of petroleum hydrocarbons 67
TPH 136, 145–147
  decision tree 120
  figure 146, 147
  glossary 329
TPQ 24
  glossary 329
Trace element
  requirements in bioremediation 210
Transaction screen 105
Transport blank 134
  glossary 329
Trip blank 129, 134
TRPH
  glossary 329
Trust Fund 25
  glossary 329
TSCA 28
  glossary 329
TSD
  glossary 329

TSD facilities 238
  Phase I 108
TSD facility 238
TVPH
  glossary 329
TWA 71
  time–weighted average
    glossary 329

U

U.S. Environmental Protection Agency 1
U.S. EPA 2
Underground petroleum storage tanks 1–2
UEL
  glossary 329
Umpire sample
  glossary 329
Unsaturated Hydrocarbons 47
Upgradient
  glossary 329
  sampling 230
Upgradient wells 230
  DNAPLs 87
Usable data
  obtaining
    figure 127
Used oil 149–150, 245
USGS
  groundwater survey maps
    Phase I 109
  subsurface geology maps
    Phase I 109
  topographical maps
    Phase I 109
UST 1, 20
  glossary 329
  regulation 41, 288
UST lists
  Phase I 106
UST regulations 39, 291–297
Utility corridor 8, 89

V

Vadose zone
  figure 77, 78
  the unsaturated zone 76
Vapor extraction 201
Vapor extraction
  contaminant fluctuation 82
Vapor migration 91
Vapor phase 202, 204
  flashpoints 61
  hydrocarbons 80, 199
Vapor pressure 61, 66, 67
  Henry's law 199
Vapor pressure 57, 63, 65
  and flash points 57
Variance 130
Vertical movement
  of water 82
Viscosity 84
Viscosity Data 63
Vitrification 218–219
VOA 145–147
  glossary 330
VOC 145–147
  glossary 330
Volatile compounds 60
  separation of 141
Volatility 61
  glossary 330
  table of 62
Volatility data 62
  for hydrocarbons 59
Volatility of Alkanes 64
Volatilization
  see soil vacuuming
Volatilization
  in landfarming 208
  migration pathway
    figure 83
  rates of 80
  rates 207

W

Water level depth
  measurement technique 129
Water Pollution Control Act 25
Water Quality Act of 1987 26
Water quality agency records
  Phase I 109
Water Bank Act of 1970 30
Water table
  fluctuations 82
Well
  casings
    size of 233
  depth
    table 129
  development 233
    checklist 234
  documentation 234
  recovery
    figure 231
  second level assessment 118
Well evacuation
  equipment 130
  procedure 130
Well sampling sequence 130
Wetlands 26
  emergency response 156
  glossary 330
  SPCC 26, 28
Wild and Scenic Rivers Act of 1968 30
Wild Horse and Burro Protection Act of
  1971 29
Wilderness Act of 1964 30
Wood alcohol 58
  see methanol (syn.)
Worker exposure
  health and safety
    asphalt incorporation 254
    bioremediation 200
    excavation and land filling 239
    HTTT 251
    landfarming 242
    land treatment 245
    linear interception 220
    LTTS 249

passive remediation 212
soil leaching 214
soil vacuuming 200
solidification and stabilization 222
vitrification 219
volatilization 216
in ASTM Standard Practice 103, 106
risk analysis 160–161
TLV 326–327
Workplace exposures 326–327
to benzene 70
WPCA
glossary 330
WQA
glossary 330

X

Xylene 53, 71–74
chemical structures 52
chronic effects 71
flashpoint 62
glossary 330
health effects 71
table 72
solubility 63, 68
viscosity 63
volatility 62
volatilization 201, 205

Z

Zero headspace
container 136
glossary 330
Zone of influence
volatilization 198

Milton Keynes UK
Ingram Content Group UK Ltd.
UKHW051945071024
449327UK00026B/2177

9 780367 449544